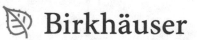

Compact Textbooks in Mathematics

This textbook series presents concise introductions to current topics in mathematics and mainly addresses advanced undergraduates and master students. The concept is to offer small books covering subject matter equivalent to 2- or 3-hour lectures or seminars which are also suitable for self-study. The books provide students and teachers with new perspectives and novel approaches. They may feature examples and exercises to illustrate key concepts and applications of the theoretical contents. The series also includes textbooks specifically speaking to the needs of students from other disciplines such as physics, computer science, engineering, life sciences, finance.

- **compact**: small books presenting the relevant knowledge
- **learning made easy**: examples and exercises illustrate the application of the contents
- **useful for lecturers**: each title can serve as basis and guideline for a semester course/lecture/seminar of 2–3 hours per week.

More information about this series at http://www.springer.com/series/11225

Markus Land

Introduction to Infinity-Categories

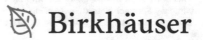

Markus Land
Department of Mathematical Sciences
University of Copenhagen
Copenhagen, Denmark

ISSN 2296-4568 ISSN 2296-455X (electronic)
Compact Textbooks in Mathematics
ISBN 978-3-030-61523-9 ISBN 978-3-030-61524-6 (eBook)
https://doi.org/10.1007/978-3-030-61524-6

Mathematics Subject Classification: 18N60, 18N50, 18N55

This textbook has been reviewed and accepted by the Editorial Board of Mathematik Kompakt, the German language version of this series.

This book is published under the imprint Birkhäuser, www.birkhauser-science.com, by the registered company Springer Nature Switzerland AG.
The registered company address is: Gewerbestrasse 11, 6330 Cham, Switzerland

Preface

This book grew out of lecture notes for the courses "Introduction to ∞-categories" and "∞-categories" that I taught at the University of Regensburg in the winter term 2018/2019 and the summer term 2019, respectively. Most of the material is not original; on many occasions throughout the first chapters of the book, I follow the arguments of Rezk [Rez20]. Further sources of inspiration are, of course, Lurie's book "Higher Topos Theory", [Lur09], Cisinski's book "Higher Categories and Homotopical Algebra", [Cis19], Joyal's paper about quasi-categories, [Joy08], and Haugseng's lecture notes on ∞-categories, [Hau17]. Additional survey papers are Rahn's (or Groth's, at the time of writing) "A short course on ∞-categories", as well as Gepner's "An introduction to higher categorical algebra", both of which appear as a chapter in [Mil19]. In particular, Gepner's paper is a very nice read before attacking Lurie's seminal work "Higher Algebra", [Lur17], on the topic.

When reading this book, it is beneficial, but not strictly necessary, to have had some exposure to ordinary category theory, although not much prior knowledge is needed and the most important concepts are recalled in the beginning of the book. For instance, it will be useful to have seen examples of categories, functors, adjunctions, colimits and limits. Some exposure to algebraic topology in the form of basic homotopy theory and the relation between topological spaces and simplicial sets is also helpful. In particular, knowing that many categories which appear in algebraic topology admit enrichments in topological spaces will help appreciate the construction of the coherent nerve, which produces from a category enriched in topological spaces an associated ∞-category and although a reader who is familiar with model categories will certainly recognise some arguments throughout the first two chapters of the book, no knowledge of model categories is necessary to follow the reasoning in this book.

Summing up, the material presented in this book is well suited for anyone with some background in homotopy theory and category theory and an interest in the basic theory of ∞-categories. The book can be the basis for a lecture course on the topic, covering roughly two terms of 4 h of lecture per week; the set of exercises can be used for parallel exercise sessions. Likewise, the book can be used for reading courses or, of course, as an individual read.

Copenhagen, Denmark   Markus Land

Acknowledgments

I want to thank Hoang Kim Nguyen and Denis-Charles Cisinski for many helpful discussions about the basics of higher category theory, and Harry Gindi for many valuable questions and comments during the lectures. Thanks also to Sebastian Wolf, Paul Bärnreuther, and Jonas Kleinöder for spotting errors in earlier versions of this manuscript, and to Søren Galatius, Thomas Nikolaus, and Irakli Patchkoria for comments on the notes. Many thanks go to Fabian Hebestreit and Christoph Winges for very helpful comments and corrections while they were teaching a similar class a year later, and to Sune Precht Reeh for many helpful comments while I was teaching this class in Copenhagen.

I wish to thank Moritz Kerz for suggesting to write this book and the editorial board of Mathematik Kompakt for supporting the proposal to do so.

Many, many thanks also go to Tobias Schwaibold for his thorough reading of the manuscript and for providing numerous suggestions which greatly enhanced the presentation of the material in this book.

Last but not least, many thanks to our TEX-pert for developing this class file!

Contents

Categories, Simplicial Sets, and Infinity-Categories

Motivation and Overview

In this book, we will discuss some of the foundations of $(\infty, 1)$-categories, henceforth simply called ∞-categories. Examples of ∞-categories are given by ordinary categories, and by topologically (or simplicially) enriched categories. Furthermore, topological spaces give rise to ∞-groupoids. In fact, the associated ∞-groupoid depends only on the (weak) homotopy type of the topological space, and any ∞-groupoid arises in this fashion—this is the content of the homotopy hypotheses which we will see in Chap. 2. The collection of all ∞-groupoids itself is a nice ∞-category which takes in some sense the role of the category of sets in ordinary category theory.

∞-categories arise naturally at various points in homotopy theory and derived (algebraic) geometry, and they have some very nice features:

(1) What were previously only constructions like homotopy (co)limits in model categories become objects with universal properties in ∞-categories.

(2) If we can show that certain objects have universal properties, then we can construct many interesting maps whose existence cannot be anticipated by merely considering the construction of the associated objects (or which rely on having a particular construction at hand), e.g., calculating natural maps from topological K-theory (viewed as a functor on C^*-algebras) or the Tate diagonal (which is of utmost importance in the new description of TC).

(3) Therefore, the ∞-categorical machinery often allows for good definitions of objects: For instance, the algebraic K-theory of a discrete ring can now be defined just as Quillen and Segal have imagined it. No $S^{-1}S$, Q, or $S_\bullet$ construction is needed; much rather, it is simply a group completion, just as we

© The Author(s), under exclusive license to Springer Nature Switzerland AG 2021

M. Land, *Introduction to Infinity-Categories*, Compact Textbooks in Mathematics,

https://doi.org/10.1007/978-3-030-61524-6_1

know it from K_0. This does not mean that these constructions are not useful—it is good to know that the group completion can be described in these terms.

(4) One very big advantage of ∞-categories is that certain collections of ∞-categories assemble into a nice new ∞-category, and that one can thus perform many categorical arguments at that level. Examples include the Goodwillie tower of a functor, the Goerss–Hopkins–Miller obstruction tower, the universal property of algebraic K-theory as a noncommutative localizing invariant, its variant for hermitian K-theory, the smash product of spectra, and many more.

(5) With the help of ∞-categories, it becomes intuitively clear that that the assignment sending a topological space or a scheme to certain categories (of sheaves or complexes of sheaves) satisfies descent properties, as, e.g., the faithfully flat descent theorem of Grothendieck for derived categories. Among other things, this fact leads to the modern treatment and understanding of descent theorems for K-theory. Analogously, it allows to define a motivic J-homomorphism, $K(S) \to \mathrm{Pic}(\mathrm{SH}(S))$, which is used, e.g., in motivic Atiyah duality.

(6) There is a nice treatment of generalized Poincaré duality for finite CW-complexes, based on studying locally constant sheaves of spectra on such spaces, which is in spirit very similar to what you might know from local systems on X (in fact, these are simply locally constant sheaves of abelian groups on X). This leads to very a nice picture of Poincaré duality for closed manifolds or, more generally, Poincaré duality complexes; and this approach makes most clear for what kind of cohomology theories there is Poincaré duality for such spaces, which is very useful, e.g., in surgery theory.

The above list may convince you that the theory of ∞-categories is useful in practice. However, it also has some drawbacks. In my opinion, the biggest one is the following (exaggerated on purpose):

⚠ It is practically impossible to construct anything by hand.

The main problem consists of constructing functors between ∞-categories. We will define ∞-categories as certain kinds of simplicial sets, and a functor will be a map of simplicial sets. With this definition, a functor cannot be defined by specifying associations on objects and morphisms, and then checking that units and composition are respected. We have to provide much more data in the first place, which can sometimes, but not always be done by hand. Therefore, we always need some machinery that allows us to construct such functors.

Typical examples of problems which we will face are the following:

(1) Let $\Delta^1 \to J$ be the inclusion of one morphism in the free standing isomorphism J. Given a morphism f in an ∞-category, does its classifying map factor through $\Delta^1 \to J$ if and only if f is an equivalence?

(2) Given a natural transformation between functors which is pointwise an equivalence, can we find an inverse of the transformation? Notice that forming inverses is not unique, so that the usual proof in ordinary categories does not work. Also note that being a natural transformation means more than a collection of maps which have a property (that certain squares commute).

(3) Given a collection of composable morphisms $x_0 \to x_1 \to x_2 \to \ldots$ in an ∞-category $\mathcal{C}$, can we find a functor $\mathbb{N} \to \mathcal{C}$ refining these data?

(4) Given a fully faithful and essentially surjective functor, can we find an inverse?

Notice that these are all questions whose answer in ordinary category theory is yes. One would expect them to hold in ∞-categories as well, and it is the objective of this book to prove precisely such kinds of results. In principle one could say that any reasonable fact from ordinary category theory should have an analog in ∞-categories, but we have to be very careful with what it means to be a reasonable fact. Here are some reasonable (depending on your background) statements:

(1) The category of commutative abelian groups/rings is a full subcategory of the category of groups/rings.

(2) The forgetful functor $\mathcal{C}_{x/} \to \mathcal{C}$ preserves connected colimits.

(3) The colimit over a constant functor $I \to * \xrightarrow{x} \mathcal{C}$ is given by x if I is connected.

All these statements are not true in the setting of ∞-categories. This has to do with the fact that commutativity in ∞-categories is not a property, but a datum. In particular, maps might or might not preserve such data. The other fact is more intuitive to understand: The analog will hold true if we replace "connected" by "contractible".

Our first goals in this book will be to address some of the above problems. At the same time, we will get to know a type of argument that is used in the (basics of the) theory: combinatorial arguments in simplicial sets. In the last part of the book, we will discuss another very important tool to construct functors: the straightening-unstraightening equivalence. We will not prove it, but we will discuss some aspects of it. Afterwards, we will use it to study adjunctions between ∞-categories and finally prove adjoint functor theorems.

1.1 Categories and Simplicial Sets

The purpose of this section is to get acquainted with the basic objects which we will be studying throughout the book: simplicial sets. These simplicial sets are most naturally considered as a category of presheaves on the simplicial indexing category Δ which we will define in a moment. We will then study basic notions in (ordinary) category theory and discuss the relation between simplicial sets and ordinary categories implemented by the nerve construction, which says that the category of categories fully faithfully embeds into simplicial sets, and that one can explicitly characterize which simplicial sets arise as nerves of categories.

In order to avoid talking about classes, we will work in a set theoretic setting which we will introduce shortly. In addition to the usual ZFC axioms (Zermelo-Fraenkel set theory plus the axiom of choice), we will assume another axiom, called a large cardinal axiom:

Axiom

For every cardinal κ there exists an inaccessible cardinal κ' with $\kappa' > \kappa$.

A cardinal κ is called inaccessible if the collection of sets $\mathcal{V}_{<\kappa}$ of hereditary cardinality less than κ (i.e., the set and all its members have cardinality less than κ) itself satisfies the ZFC axioms. The collection $\mathcal{V}_{<\kappa}$ is called a universe. It turns out that this large cardinal axiom cannot be proven from ZFC and, in fact, is logically independent. In particular, an inaccessible cardinal κ is larger than $\aleph_k$ for any k. From the axiom, we may fix a sequence

$$\kappa_0 < \kappa_1 < \kappa_2 < \ldots$$

of inaccessible cardinals and consider their associated universes $\mathcal{V}_{<\kappa}$.

Definition 1.1.1

A set is called

(1) small, if it is contained in $\mathcal{V}_{<\kappa_0}$;
(2) large, if it is contained in $\mathcal{V}_{<\kappa_1}$;
(3) very large, if it is contained in $\mathcal{V}_{\kappa_2}$;
(4) very very large, if it is contained in $\mathcal{V}_{<\kappa_3}$, and so on.

In this book, we will not encounter any sets other than small, large and very large sets.

Example The set of small sets is large. The set of large sets is very large, and so on.

Definition 1.1.2

A category $\mathcal{C}$ consists of a (possibly large) set of objects $\mathrm{ob}(\mathcal{C})$, and, for any two objects x and y, a (also possibly large) set $\mathrm{Hom}_{\mathcal{C}}(x, y)$ of morphisms, equipped with composition maps

$$\mathrm{Hom}_{\mathcal{C}}(x, y) \times \mathrm{Hom}_{\mathcal{C}}(y, z) \to \mathrm{Hom}_{\mathcal{C}}(x, z)$$

and identities $* \to \mathrm{Hom}_{\mathcal{C}}(x, x)$ for all objects x, satisfying associativity and unitality.

A category is called locally small if all hom-sets $\mathrm{Hom}_{\mathcal{C}}(x, y)$ are small, it is called small if it is locally small and the set of objects is also small. It is called essentially small if it is locally small and the set of isomorphism classes of objects is small.

Remark 1.1.3

In category theory, a category would usually be defined as to have a (possibly proper) class of objects and, for any two objects, a set of morphisms. In our language, this is what we call a locally small category. In general, however, we will not assume that a category is locally small.

Definition 1.1.4

A partially ordered set is a set P equipped with a reflexive, antisymmetric and transitive relation $\leq$. That is, $a \leq a$, if $a \leq b$ and $b \leq a$ then $a = b$, and if $a \leq b$ and $b \leq c$, then $a \leq c$. A map of partially ordered sets is a map of sets $f : P \to Q$ such that $x \leq y$ implies $f(x) \leq f(y)$. This defines a category PoSet whose objects are posets and whose morphisms are maps of posets.

Example *Finite linearly ordered sets:* The set $\{0, 1, \ldots, n\}$ is linearly ordered, $0 \leq 1 \leq \cdots \leq n$, and written as $[n]$. A general finite poset S is called linearly ordered if it is isomorphic to one of the $[n]$'s. Morphisms of linearly ordered sets are just morphisms of the underlying poset. We obtain a category LinOrdSet.

Example *The subset poset:* Let S be a set. Consider its set $\mathcal{P}(S)$ of subsets: $\mathcal{P}(S) = \{I \subseteq S\}$. This set is partially ordered by inclusion:

$$I \leq J \Leftrightarrow I \subseteq J$$

Definition 1.1.5

The category Δ is the full subcategory of the category PoSet of posets consisting of the linearly ordered set $[n]$ for all $n \geq 0$. Notice that a morphism from $[n]$ to $[m]$ is thus simply a weakly monotonic map.

Example There are two special maps in Δ, namely the face and the degeneracy maps. For every $n \geq 0$ and $0 \leq i \leq n$, the face maps are the maps

$$d_i : [n - 1] \to [n]$$

which are uniquely determined by the property that $i \notin \operatorname{Im}(d_i)$ and that d_i is injective. Furthermore, for $n \geq 1$ and $0 \leq i \leq n - 1$, the degeneracy maps are the maps

$$s_i : [n] \to [n - 1]$$

which are uniquely determined by the property $|s_i^{-1}(i)| = 2$ and that s_i is surjective. In other words, we have $s_i(i) = s_i(i + 1) = i$.

Definition 1.1.6

Let $\mathcal{C}$ be a category. We denote the category of functors $\mathcal{C}^{\mathrm{op}} \to \mathrm{Set}$ by $\mathcal{P}(\mathcal{C})$ and call it the category of presheaves on $\mathcal{C}$. An object $x \in \mathcal{C}$ determines a *representable* presheaf, namely the presheaf $\operatorname{Hom}_{\mathcal{C}}(-, x)$ which sends $y \in \mathcal{C}$ to the set of morphisms from y to x. This determines a functor $\mathcal{C} \to \mathcal{P}(\mathcal{C})$ which is called the *Yoneda embedding*.

Definition 1.1.7

A simplicial set is a presheaf on Δ, i.e., a functor $\Delta^{\mathrm{op}} \to \mathrm{Set}$. Given a simplicial set X, its set of n-simplices is given by $X([n])$ and will be written as X_n. An n-simplex x is called degenerate if there exists a surjection $\alpha : [n] \to [m]$ with $m \neq n$, and an n-simplex y such that $x = \alpha^*(y)$. Equivalently, x is degenerate if $x = s_i^*(y)$ for some $y \in X_{n-1}$ and some $0 \leq i \leq n - 1$. An n-simplex is called non-degenerate if it is not degenerate. We denote the category $\mathcal{P}(\Delta)$ of simplicial sets by sSet.

Definition 1.1.8

We let Δ^n be the simplicial set represented by $[n] \in \Delta$. Concretely, we have $(\Delta^n)_m = \operatorname{Hom}_{\Delta}([m], [n])$.

Definition 1.1.9

Let X be a simplicial set. We consider the equivalence relation $\sim$ on the set of 0-simplices X_0 which is generated by relating x and y if there exists a 1-simplex $f \in X_1$ such that $d_0(f) = x$ and $d_1(f) = y$. (This relation is reflexive but in general neither transitive nor symmetric.) Then we define the set $\pi_0^{\Delta}(X)$ as follows:

$$\pi_0^{\Delta}(X) = X_0/\!\sim$$

Lemma 1.1.10

The Yoneda lemma: Let $F : \mathcal{C}^{\mathrm{op}} \to \mathrm{Set}$ be a functor and $x \in \mathcal{C}$ an object. Then the map

$$\operatorname{Hom}_{\mathcal{P}(\mathcal{C})}(\operatorname{Hom}_{\mathcal{C}}(-, x), F) \to F(x)$$

given by sending η to $\eta(\mathrm{id}_x)$ is a bijection.

Proof The inverse is given by sending an element $s \in F(x)$ to the function $\mathrm{Hom}_{\mathcal{C}}(y, x) \to F(y)$ sending f to $f^*(s)$. It can be explicitly checked that this is a natural transformation and an inverse of the above-described map. $\qquad \square$

Lemma 1.1.11

The Yoneda embedding $\mathcal{C} \to \mathcal{P}(\mathcal{C})$ is fully faithful.

Proof This follows immediately from the Yoneda Lemma: The effect of the Yoneda embedding on morphisms is the map

$$\mathrm{Hom}_{\mathcal{C}}(x, y) \to \mathrm{Hom}_{\mathcal{P}(\mathcal{C})}(\mathrm{Hom}_{\mathcal{C}}(-, x), \mathrm{Hom}_{\mathcal{C}}(-, y))$$

given by sending f to

$$\mathrm{Hom}_{\mathcal{C}}(z, x) \xrightarrow{f_*} \mathrm{Hom}_{\mathcal{C}}(z, y).$$

It is readily seen that this map is inverse to the map as described in the Yoneda lemma, which is given by sending a map $f \in \mathrm{Hom}_{\mathcal{C}}(x, y)$ to the function $\mathrm{Hom}_{\mathcal{C}}(z, x) \to \mathrm{Hom}_{\mathcal{C}}(z, y)$ given by sending φ to $\varphi^*(f) = f_*\varphi$. $\qquad \square$

Corollary 1.1.12

For a simplicial set X, there is a canonical bijection

$$\mathrm{Hom}_{\mathrm{sSet}}(\Delta^n, X) \cong X_n.$$

Definition 1.1.13

A (co)limit of a functor $F : I \to \mathcal{C}$ is an object of $\mathcal{C}$, abbreviated as $\mathrm{colim}_I F$, equipped with maps $F(i) \to \mathrm{colim}_I F$, for every $i \in I$, which are compatible in the sense that for every morphism $i \to j$ in I, the diagram

$$\begin{array}{ccc} F(i) & \longrightarrow & \mathrm{colim}_I F \\ \downarrow & \nearrow & \\ F(j) & & \end{array}$$

commutes. This datum is required to satisfy the following universal property: Whenever a given further object $X \in \mathcal{C}$ is also equipped with maps $F(i) \to X$ which are compatible in the above-mentioned way, then there exists a unique morphism $\mathrm{colim}_I F \to X$ making the diagrams

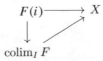

commute.

Dually, a limit of F is an object $\lim_I F$, equipped with maps $\lim_I F \to F(i)$ which are again compatible, satisfying the dual universal property: Whenever we are given an object X equipped with compatible morphisms $X \to F(i)$ for all $i \in I$, there exists a unique morphism $X \to \lim_I F$ which makes the obvious diagram commute.

> **Remark 1.1.14**
> Notice that such a universal property specifies an object up to a unique isomorphism. Notice also that the universal property refers to more than just the object $\mathrm{colim}_I F$. The reference maps are part of the data, and this is what makes the object unique up to a unique isomorphism.

Example A colimit of the empty diagram $\emptyset \to \mathcal{C}$ is an initial object, i.e., an object which admits a unique morphism to any other object. Dually, a limit of the empty diagram $\emptyset \to \mathcal{C}$ is a terminal object, i.e., an object which admits a unique morphism from any other object.

Example A colimit of a functor defined on the diagram $\bullet \leftarrow \bullet \to \bullet$ is called a *pushout*. Dually, a limit of a functor defined on the diagram $\bullet \to \bullet \leftarrow \bullet$ is called a *pullback*.

Example The quotient vector space V/U is a pushout of the diagram

$$
\begin{array}{ccc}
U & \longrightarrow & V \\
\downarrow & & \downarrow \\
0 & \longrightarrow & V/U
\end{array}
$$

> **Observation 1.1.15**
> One can phrase general (co)limits via initial and terminal objects. This point of view will be used later when we discuss limits and colimits in ∞-categories. Given a functor

(continued)

1.1.15 (continued)

$F : I \to \mathcal{C}$ we can consider the category of (co)cones of this functor. Given a category I we consider two new categories $I^\triangleleft$ and $I^\triangleright$, which are constructed from I by adding an initial respectively a terminal object. There is an obvious functor $I \to I^\triangleleft$ and $I \to I^\triangleright$. We can thus consider the functor categories $\mathrm{Fun}_F(I^\triangleleft, \mathcal{C})$ and $\mathrm{Fun}_F(I^\triangleright, \mathcal{C})$ of functors which restrict to F along the above-mentioned inclusion. They are called the categories of cones and cocones over F, respectively. One can then show that a colimit as described above is an initial object in the category of cones, and that a limit is a terminal object in the category of cocones.

The following lemma follows immediately from the definition of (co)limits and the fact established in Exercise 6 that the category Set is bicomplete (otherwise the statement does not make sense).

Lemma 1.1.16

Let $\mathcal{C}$ be a category and let $F : I \to \mathcal{C}$ be an I-shaped diagram in $\mathcal{C}$. Then, for every object $x \in \mathcal{C}$, there are canonical bijections

(1) $\mathrm{Hom}_{\mathcal{C}}(\mathrm{colim}_I F, x) \cong \lim_I \mathrm{Hom}_{\mathcal{C}}(F(i), x)$, and
(2) $\mathrm{Hom}_{\mathcal{C}}(x, \lim_I F) \cong \lim_I \mathrm{Hom}_{\mathcal{C}}(x, F(i))$.

Moreover, these bijections characterize (co)limits uniquely.

Definition 1.1.17

An *adjunction* consists of a pair of functors $(F : \mathcal{C} \to \mathcal{D}, G : \mathcal{D} \to \mathcal{C})$ together with a natural isomorphism between the two functors $\mathcal{C}^{\mathrm{op}} \times \mathcal{D} \to \mathrm{Set}$ given by

$$\mathrm{Hom}_{\mathcal{D}}(F(-), -) \text{ and } \mathrm{Hom}_{\mathcal{C}}(-, G(-)).$$

Given a functor $F : \mathcal{C} \to \mathcal{D}$ and a diagram $X : I \to \mathcal{C}$, there are canonical comparison maps

$$\mathrm{colim}_I F(X) \longrightarrow F(\mathrm{colim}_I X) \quad \text{respectively} \quad F(\lim_I X) \longrightarrow \lim_I F(X).$$

One says that F preserves colimits respectively limits, if the corresponding comparison map is an isomorphism.

Lemma 1.1.18

Left adjoints preserve colimits, and right adjoints preserve limits.

Proof Let $F: \mathcal{C} \to \mathcal{D}$ be a functor which admits a right adjoint, say G. Let $X: I \to \mathcal{C}$ be a diagram which has a colimit $\mathrm{colim}_I\, X \in \mathcal{C}$. We claim that F sends that colimit to a colimit of the diagram $I \to \mathcal{C} \to \mathcal{D}$. In formulas, we claim that the canonical map $\mathrm{colim}_I\, F(X(i)) \to F(\mathrm{colim}_I\, X(i))$ induced from the compatible maps $F(X(i)) \to F(\mathrm{colim}_I\, X(i))$, which are part of the datum of the colimit (and then applying F), is an isomorphism. In order to see this, it suffices to show that this canonical map induces a bijection on hom-sets for all other objects $y \in \mathcal{D}$:

$$\mathrm{Hom}_{\mathcal{D}}(F(\mathrm{colim}_I X(i)), y) \cong \mathrm{Hom}_{\mathcal{C}}(\mathrm{colim}_I X(i), Gy)$$

$$\cong \lim_I \mathrm{Hom}_{\mathcal{C}}(X(i), Gy)$$

$$\cong \lim_I \mathrm{Hom}_{\mathcal{D}}(F(X(i)), y)$$

$$\cong \mathrm{Hom}_{\mathcal{D}}(\mathrm{colim}_I F(X(i)), y)$$

By the Yoneda lemma, this completes the proof. The argument for the claim that right adjoints preserve limits is similar. $\square$

Lemma 1.1.19

Let $F: \mathcal{C} \to \mathcal{D}$ be a functor which admits right adjoints G and G'. Then there is a specified natural isomorphism between G and G' which is uniquely characterised by being compatible with the adjunctions. Hence, adjoints, if they exist, are unique up to a unique isomorphism.

Proof Consider the following two natural bijections:

$$\mathrm{Hom}_{\mathcal{C}}(Gx, G'x) \cong \mathrm{Hom}_{\mathcal{D}}(FGx, x) \cong \mathrm{Hom}_{\mathcal{C}}(Gx, Gx)$$

Then the identity of Gx corresponds to a natural transformation $G \to G'$. Applying the same trick to $\mathrm{Hom}_{\mathcal{C}}(G'x, Gx)$ shows that this must be a natural isomorphism. $\square$

Definition 1.1.20

A category is called (co)complete, if it admits (co)limits indexed over arbitrary small categories (statt (co)limits). It is called bicomplete if it is both complete and cocomplete.

Lemma 1.1.21

If $\mathcal{C}$ is bicomplete, then the (co)limit is left/right adjoint to the constant diagram functor. In particular, forming (co)limits determines a functor

$$\mathrm{Fun}(I, \mathcal{C}) \to \mathcal{C}.$$

Proof Let us spell out the colimit case. Consider the constant functor const: $\mathcal{C} \to \mathrm{Fun}(I, \mathcal{C})$. Now we specify for each functor $F \colon I \to \mathcal{C}$ an object, namely $\mathrm{colim}_I F$. Part of the datum of a colimit are compatible maps $\{F(i) \to \mathrm{colim}_I F\}_{\{i \in I\}}$ which are easily seen to assemble into a natural transformation

$$F \to \mathrm{const}(\underset{I}{\mathrm{colim}}\, F).$$

Next, we consider the composite

$$\mathrm{Hom}_{\mathcal{C}}(\underset{I}{\mathrm{colim}}\, F, X) \to \mathrm{Hom}_{\mathrm{Fun}(I,\mathcal{C})}(\mathrm{const}(\underset{I}{\mathrm{colim}}\, F), \mathrm{const}\, X) \to \mathrm{Hom}_{\mathrm{Fun}(I,\mathcal{C})}(F, \mathrm{const}\, X)$$

which is a bijection by the universal property of a colimit. The proof of the lemma thus follows from Exercise 10. The case of limits is completely analogous.   □

Lemma 1.1.22

Given an adjunction with $F \colon \mathcal{C} \to \mathcal{D}$ being left-adjoint to $G \colon \mathcal{D} \to \mathcal{C}$, and given a further auxiliary small category I, then the functors

$$F_* \colon \mathrm{Fun}(I, \mathcal{C}) \rightleftarrows \mathrm{Fun}(I, \mathcal{D}) \colon G_*$$

again form an adjoint pair. Here, F_* is given by post-composition with F, and likewise G_* is given by post-composition with G.

Proof The adjunction is determined by a counit map $\varepsilon \colon FG \to \mathrm{id}_{\mathcal{D}}$ and a unit map $\eta \colon \mathrm{id}_{\mathcal{C}} \to GF$ that satisfy the triangle identities of Exercise 9. We now use these maps to construct counit maps and unit maps for the pair of functors (F_*, G_*) as follows: Let $\varphi \in \mathrm{Fun}(I, \mathcal{D})$. We need to specify a natural map $\varepsilon_* \colon F_*(G_*(\varphi)) \to \varphi$ of functors $I \to \mathcal{D}$, so let $x \in \mathcal{E}$. We define the new counit ε_* to be the map

$$F(G(\varphi(x))) \xrightarrow{\varepsilon_{\varphi(x)}} \varphi(x).$$

It is easy to see that this is natural in φ, since ε itself is a natural transformation. Similarly, we define a natural transformation $\eta_* \colon \psi \to G_* F_*(\psi)$ given by

$$\psi(y) \xrightarrow{\eta_{\psi(y)}} G(F(\psi(y))).$$

Then, it is easy to see that the triangle identities are satisfied, because (ε, η) satisfy the triangle identities.   □

Proposition 1.1.23

If $\mathcal{C}$ is a bicomplete category, then $\mathrm{Fun}(I, \mathcal{C})$ is bicomplete as well. A (co)limit of the diagram $X : J \to \mathrm{Fun}(I, \mathcal{C})$ is given by the functor sending $i \in I$ to $\mathrm{colim}_J X(j)(i)$.

Proof Let us prove that $\mathrm{Fun}(I, \mathcal{C})$ is cocomplete. The completeness argument is similar (or can be formally deduced from this case by applying the functor $(-)^{\mathrm{op}}$ correctly). We claim that the composite

$$\mathrm{Fun}(J, \mathrm{Fun}(I, \mathcal{C})) \cong \mathrm{Fun}(I, \mathrm{Fun}(J, \mathcal{C})) \xrightarrow{\mathrm{colim}_J} \mathrm{Fun}(I, \mathcal{C})$$

is a colimit functor whose existence we wish to show. By Lemma 1.1.22 this functor has a right adjoint given by

$$\mathrm{const}_* : \mathrm{Fun}(I, \mathcal{C}) \to \mathrm{Fun}(I, \mathrm{Fun}(J, \mathcal{C})) \cong \mathrm{Fun}(J, \mathrm{Fun}(I, \mathcal{C})).$$

The proposition is proven once we have convinced ourselves that this is itself the constant functor (which follows immediately from the definition), since we can then allude to Lemma 1.1.21. $\square$

Corollary 1.1.24

The category of simplicial sets sSet is bicomplete.

Definition 1.1.25

Let $\mathcal{D}$ be a category and x an object of $\mathcal{D}$. The slice category $\mathcal{D}_{/x}$ has objects pairs $(d \in \mathcal{D}, \alpha d \to x)$, and morphisms from (d, α) to (d', α') consist of those maps $\beta : d \to d'$ in $\mathcal{D}$ such that the obvious triangle involving α, α', and β commutes. There is a canonical functor $\mathcal{D}_{/x} \to \mathcal{D}$ which sends a pair (d, α) to d.

If $F : \mathcal{C} \to \mathcal{D}$ is a functor and let $x \in \mathcal{D}$ be an object. We define the slice category associated to F and x by the pullback

$$
\begin{array}{ccc}
F_{/x} & \longrightarrow & \mathcal{D}_{/x} \\
\downarrow & & \downarrow \\
\mathcal{C} & \xrightarrow{\ F\ } & \mathcal{D}
\end{array}
$$

Thus, $F_{/x}$ has as objects pairs $(c \in \mathcal{C}, \alpha : F(c) \to x)$ and morphisms consist of those maps $\beta : c \to c'$ such that the obvious triangle involving α, α', and $F(\beta)$ commutes.

Sometimes, in particular when F is the inclusion of a subcategory, we simply write $\mathcal{C}_{/x}$ instead of $F_{/x}$, leaving the functor F understood. There is also an obvious dual notion of slice categories built from $D_{x/}$ whose objects are pairs $(d, \alpha : x \rightarrow d)$. In the following, we use the slice category associated to the Yoneda embedding $\mathcal{C} \rightarrow \mathcal{P}(\mathcal{C})$ of Definition 1.1.6.

Lemma 1.1.26

Every presheaf is a colimit of representables. More precisely, for every presheaf $F : \mathcal{C}^{\mathrm{op}} \rightarrow \mathrm{Set}$, *the tautological map*

$$\operatorname*{colim}_{X \in \mathcal{C}_{/F}} \operatorname{Hom}_{\mathcal{C}}(-, X) \rightarrow F$$

is an isomorphism.

Proof The proof relies once again on the Yoneda lemma, i.e., we show that this map induces a bijection on maps to an auxiliary presheaf G. For this purpose, we calculate

$$\operatorname{Hom}_{\mathcal{P}(\mathcal{C})}\Big(\operatorname*{colim}_{X \in \mathcal{C}_{/F}} \operatorname{Hom}_{\mathcal{C}}(-, X), G\Big) \cong \lim_{X \in \mathcal{C}_{/F}} \operatorname{Hom}_{\mathcal{P}(\mathcal{C})}(\operatorname{Hom}_{\mathcal{C}}(-, X), G)$$

$$\cong \lim_{X \in \mathcal{C}_{/F}} G(X).$$

It is not hard to see that the latter is in fact the set of natural transformations from F to G. $\square$

Let $i : \mathcal{C}_0 \subseteq \mathcal{C}$ be a small subcategory of a category, and let $\mathcal{D}$ be a bicomplete category.

Fact 1.1.27 The restriction functor

$$i^* : \mathrm{Fun}(\mathcal{C}, \mathcal{D}) \rightarrow \mathrm{Fun}(\mathcal{C}_0, \mathcal{D})$$

has a left adjoint denoted by $i_!$ and a right adjoint denoted by i_*. They are given as

$$i_!(F)(x) = \operatorname*{colim}_{c \in (\mathcal{C}_0)_{/x}} F(c)$$

and

$$i_*(F)(y) = \lim_{c \in (\mathcal{C}_0)_{x/}} F(x).$$

Notice that the slices are small by assumption, so that the colimits and limits exist. It is a nice exercise to convince yourself that this formula produces adjoints for the restriction functor i^*.

Definition 1.1.28

In the above situation, we call $i_!(F)$ the *left Kan extension* of F along i and $i_*(F)$ the *right Kan extension* of F along i.

Observation 1.1.29

The statement that the tautological map is an isomorphism shows that the identity of $\mathcal{P}(\mathcal{C})$ is a left Kan extension of the Yoneda embedding (along the Yoneda embedding).

Corollary 1.1.30

If $\mathcal{D}$ is a cocomplete category and $\mathcal{C}$ is a small category, then the canonical functor

$$\mathrm{Fun}^{\mathrm{colim}}(\mathcal{P}(\mathcal{C}), \mathcal{D}) \to \mathrm{Fun}(\mathcal{C}, \mathcal{D})$$

obtained by restriction along the Yoneda embedding is an equivalence. Here, $\mathrm{Fun}^{\mathrm{colim}}$ denotes the full subcategory of the functor category on those functors which preserve colimits.

Proof Given a functor $f \colon \mathcal{C} \to \mathcal{D}$, we want to construct a colimit-preserving functor $\hat{f} \colon \mathcal{P}(\mathcal{C}) \to \mathcal{D}$, such that $\hat{f}(X) = f(X)$ for $X \in \mathcal{C}$. By Lemma 1.1.26, given an object $F \in \mathcal{P}(\mathcal{C})$, we are forced to define

$$\hat{f}(F) = \operatorname*{colim}_{X \in \mathcal{C}_{/F}} f(X).$$

It is easy to check that this is in fact a functor: For $F \to G$ a morphism in $\mathcal{P}(\mathcal{C})$, there is an induced functor from the category $\mathcal{C}_{/F}$ to the category $\mathcal{C}_{/G}$ given by post-composition with the given morphism. Then it is not hard to see that taking the colimit produces a map

$$\operatorname*{colim}_{X \in \mathcal{C}_{/F}} f(X) \to \operatorname*{colim}_{X \in \mathcal{C}_{/G}} f(X).$$

Also, it can be readily seen that this is in fact a functor.

In order to see that this functor preserves colimits, we observe that $\hat{f}$ admits a right adjoint G given by

$$G(d)(X) = \mathrm{Hom}_{\mathcal{D}}(f(X), d).$$

In the proof of Lemma 1.1.26, we saw that the set of natural transformations between F and F' is given by

$$\text{Hom}_{\mathcal{P}(\mathcal{C})}(F, F') = \lim_{X \in \mathcal{C}/F} F'(X).$$

Therefore, we see that

$$\text{Hom}_{\mathcal{D}}(\hat{f}(F), d) \cong \lim_{X \in \mathcal{C}/F} \text{Hom}_{\mathcal{D}}(f(X), d) \cong \lim_{X \in \mathcal{C}/F} G(d)(X).$$

Hence $\hat{f}$ is left-adjoint to G and thus preserves colimits. $\square$

Corollary 1.1.31

Let X be a simplicial set. Then

$$X \cong \underset{[n] \in \Delta/X}{\text{colim}} \Delta^n.$$

Lemma 1.1.32

Let X be a fixed simplicial set. Then the functor sSet $\to$ sSet *sending Y to $X \times Y$ admits a right adjoint $\underline{\text{Hom}}(X, -)$ determined by the formula*

$$\text{Hom}_{\text{sSet}}(\Delta^n, \underline{\text{Hom}}(X, Z)) = \text{Hom}_{\text{sSet}}(\Delta^n \times X, Z).$$

Sometimes we will also write Z^X for $\underline{\text{Hom}}(X, Z)$.

Proof Mapping $[n]$ to the set on the right-hand side clearly determines a simplicial set which we call $\underline{\text{Hom}}(X, Z)$. Since it satisfies the adjunction property on representable simplicial sets, we can extend the adjunction to all simplicial sets. (Keep in mind that every simplicial set is a colimit of representables.) Notice that we use the fact that the functor $X \times - :$ sSet $\to$ sSet preserves colimits. This is certainly true in Set because the Hom-set provides a right adjoint.
$\square$

Definition 1.1.33

We let Top be the category whose objects are topological spaces and whose morphisms are continuous maps. The following are important objects for us: Let $n \geq 0$ be a natural number. The topological n-simplex Δ_{top}^n is the subspace of $\mathbb{R}_{\geq 0}^{n+1}$ consisting of those points whose coordinates add up to 1. The topological simplices form a cosimplicial space $[n] \mapsto \Delta_{\text{top}}^n$, where the induced maps are the unique affine linear maps that do what they should on vertices. More precisely, given $\alpha\colon [n] \to [m]$, the induced map $\alpha_*\colon \Delta_{\text{top}}^n \to \Delta_{\text{top}}^m$ is given by

$$\alpha_*(t_0, \ldots, t_n) = (v_0, \ldots, v_m),$$

where $v_i = \sum\limits_{j \mapsto i} t_j$.

Definition 1.1.34

The *singular simplicial set* of a topological space X is the simplicial set

$$\mathcal{S}(X) = \big([n] \mapsto \operatorname{Hom}_{\text{Top}}(\Delta_{\text{top}}^n, X)\big).$$

Definition 1.1.35

The geometric realization $|-|\colon \text{sSet} \to \text{Top}$ is the unique colimit-preserving functor which sends Δ^n to Δ_{top}^n. Concretely, the geometric realization of a simplicial set X is the topological space

$$|X| = \operatorname*{colim}_{\Delta^n \in \Delta/X} \Delta_{\text{top}}^n.$$

An even more concrete formula is given by

$$|X| = \Big(\coprod_{n \in \Delta} X_n \times \Delta_{\text{top}}^n\Big)/((f^*(x), t) \sim (x, f_*(t))$$

for $x \in X_n$, $t \in \Delta_{\text{top}}^m$ and $f\colon [m] \to [n]$ a morphism in Δ.

Proposition 1.1.36

The singular complex is right-adjoint to geometric realization.

Proof By definition of adjunctions, we need to specify a natural isomorphism of functors $\text{sSet}^{\text{op}} \times \text{Top} \to \text{Set}$ between

$$\operatorname{Hom}_{\text{Top}}(|X|, Y) \cong \operatorname{Hom}_{\text{sSet}}(X, \mathcal{S}(Y)).$$

But by the previous work, we know that these functors are equivalent to

$$\lim_{\Delta/X} \operatorname{Hom}_{\text{Top}}(\Delta_{\text{top}}^n, Y) \quad \text{and} \quad \lim_{\Delta/X} \operatorname{Hom}_{\text{sSet}}(\Delta^n, \mathcal{S}(Y)),$$

and the latter two are already isomorphic (by definition of $\mathcal{S}(Y)$) before forming the limit. $\square$

Definition 1.1.37

(1) The boundary $\partial\Delta^n$ is the subsimplicial set of Δ^n whose k-simplices consist of the non-surjective maps $[k] \to [n]$.

(2) For any subset $S \subseteq [n]$, the S-horn $\Lambda_S^n \subseteq \Delta^n$ consists of those k-simplices $f: [k] \to [n]$ where there exists an $i \in [n] \setminus S$ such that i is not in the image of f. A horn $\Lambda_j^n = \Lambda_{\{j\}}^n$ is called *inner horn* if $0 < j < n$ and it is called (left or right) *outer horn* if $j = 0$ or $j = n$, respectively.

(3) The spine $I^n \subseteq \Delta^n$ is given by those k-simplices $f: [k] \to [n]$ whose image is either of the form $\{j\}$ or of the form $\{j, j+1\}$.

Definition 1.1.38

The n-skeleton $\mathrm{sk}_n(X)$ of a simplicial set X is given by the simplicial set $i_! i^*(X)$, where $i: \Delta_{\leq n} \subseteq \Delta$ is the inclusion of the full subcategory on objects of cardinality $\leq n + 1$. Dually, the n-coskeleton $\mathrm{cosk}_n(X)$ of a simplicial set is given by the simplicial set $i_* i^*(X)$. This implies that the k-simplices of $\mathrm{cosk}_n(X)$ are given by

$$\mathrm{cosk}_n(X)_k = \mathrm{Hom}_{\mathrm{sSet}}(\mathrm{sk}_n(\Delta^k), X).$$

Lemma 1.1.39

(1) The skeleton $\mathrm{sk}_n(X)$ is isomorphic to the smallest sub-simplicial set of X whose set of k-simplices coincides with the ones of X for $k \leq n$.

(2) The functors sk_n and cosk_n are left- and right-adjoint to each other.

(3) There is the formula $\mathrm{cosk}_n(X) = \mathrm{Hom}_{\mathrm{sSet}}(\mathrm{sk}_n(\Delta^\bullet), X)$.

Proof

(1) It is easy to see that the map $\mathrm{sk}_n(X) \to X$ is levelwise injective, and that $\mathrm{sk}_n(X)_k = X_k$ for $k \leq n$. Given any other sub-simplicial set Z with this property, we have $\mathrm{sk}_n(X) = \mathrm{sk}_n(Z) \subseteq Z$, which yields the claim.

(2) Obvious, since adjoints compose.

(3) Obvious, by (2).

$\square$

Lemma 1.1.40

The geometric realization of the horn is a horn, and the geometric realization of the spine is a spine.

Proof This follows from the fact that $|\Delta^n| = \Delta^n_{\text{top}}$ and the following observations:

(1) $I^n = I^{n-1} \amalg_{\Delta^0} \Delta^1$;

(2) there is a coequalizer

$$\coprod_{0 \le i \le j \le n} \Delta^{[n]\backslash i} \times_{\Delta^n} \Delta^{[n]\backslash j} \rightrightarrows \coprod_{0 \le i \le n, i \ne k} \Delta^{[n]\backslash i} \longrightarrow \Lambda^n_i.$$

$\square$

Corollary 1.1.41
The geometric realization of a simplicial set is a CW-complex.

Proof Given a simplicial set X, define a filtration on $|X|$ through $|sk_n(X)| \subseteq |X|$. Since geometric realization commutes with colimits, we see that this is in fact a filtration of $|X|$ and the pushouts from above provide pushouts of geometric realizations. Next, we use that $|\partial \Delta^n| \cong S^{n-1}$ and $|\Delta^n| \cong D^n$. $\square$

Observation 1.1.42
A poset determines a category in the following way: Objects are the elements of the posets P, and for each pair of elements $x, y \in P$, we have

$$\text{Hom}(x, y) = \begin{cases} * & \text{if } x \le y \\ \emptyset & \text{else.} \end{cases}$$

Furthermore, a functor between categories associated to posets is the same thing as a map of posets, i.e., a map of sets respecting the partial ordering. This determines a fully faithful functor PoSet $\to$ Cat. It follows that we can view $[n]$ as a category. Sending $[n]$ to this category yields a cosimplicial small category.

Definition 1.1.43
The nerve of a category $\mathcal{C}$ is the simplicial set given by

$$[n] \mapsto \text{Fun}([n], \mathcal{C}),$$

i.e., it is given by taking functors out of the previous cosimplicial category to the given category.

Lemma 1.1.44

Δ^n *is isomorphic to the nerve of the category* $[n]$.

Proof Unravelling the definitions, we find that

$$(\Delta^n)_m = \mathrm{Hom}_\Delta([m], [n]),$$

whereas

$$\mathrm{N}([n])_m = \mathrm{Fun}([m], [n]).$$

Therefore, it suffices to recall that the functor $\mathrm{Posets} \to \mathrm{Cat}$ is fully faithful. $\square$

Definition 1.1.45

The classifying space BG of a group G is the geometric realization of the nerve of the group considered as groupoid with one object.

Definition 1.1.46

A *Kan complex* is a simplicial set X which has the extension property for horn inclusions $\Lambda^n_j \to \Delta^n$ for $0 \le j \le n$, i.e., where any lifting problem

admits a solution.

Lemma 1.1.47

The singular complex of a topological space is a Kan complex.

Proof By adjunction, there is an equivalence of the following two lifting problems:

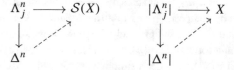

Next, we recall that the topological horn inclusion has a retract, so that the right lifting problem can be solved. $\square$

Fact 1.1.48 A Kan complex satisfies the extension property for any monomorphism of simplicial sets $K \to L$ which induces a weak equivalence on geometric realizations (these are called *anodyne maps*), see for instance [GJ09, Theorem 11.3].

Lemma 1.1.49

If the nerve of a category is a Kan complex, then the category is a groupoid.

Proof Since we can lift outer 2-horns, one can easily show that every morphism in $\mathcal{C}$ has a right and a left inverse and is therefore invertible itself.   □

Definition 1.1.50

Let $f, g \colon X \to Y$ be two maps of simplicial sets. We say that f and g are *homotopic* if there exists $H \colon X \times \Delta^1 \to Y$ such that H restricts to f and g. Given a pointed Kan complex (X, x) we define its simplicial homotopy groups as follows:

$$\pi_n^\Delta(X, x) = [(\Delta^n, \partial\Delta^n), (X, x)]_*$$

Fact 1.1.51 The homotopy relation is in fact an equivalence relation if Y is a Kan complex. The simplicial homotopy groups of a Kan complex agree with the ordinary homotopy groups of the geometric realization. In particular, they are groups for $n \geq 1$ and abelian groups for $n \geq 2$. For more details about simplicial homotopies and simplicial homotopy groups, see the book of Goerss and Jardine, [GJ09, I.6 and I.7].

Theorem 1.1.52

For a simplicial set X, the following three conditions are equivalent:

(1) X has unique extensions for $\Lambda_j^n \to \Delta^n$ if $0 < j < n$.
(2) X has unique extensions for $I^n \to \Delta^n$ for $n \geq 2$.
(3) X is isomorphic to the nerve of a category.

Proof We will show that $(1) \Leftrightarrow (2) \Leftrightarrow (3)$.

In order to show $(3) \Rightarrow (2)$, we consider a category $\mathcal{C}$ and its nerve $N(\mathcal{C})$. Recall that its n-simplices are given by $\mathrm{Fun}([n], \mathcal{C})$, and thus by chains of composable morphisms. Face and degeneracies are given by composition and inserting identities. In particular, the restriction along the spine inclusion I^n picks out precisely the morphisms, so that restriction along the spine induces an bijection between $\mathrm{Fun}([n], \mathcal{C})$ and $\mathrm{Hom}_{\mathrm{sSet}}(I^n, N(\mathcal{C}))$.

In order to show $(2) \Rightarrow (3)$, consider a simplicial set X which has unique liftings against spines. We define a category $\mathcal{C}$ as follows: The objects are given by X_0, the 0-simplices of X; the morphisms from x to y are given by all 1-simplices $f \in X_1$ such that $d_1(f) = x$

and $d_0(f) = y$; and identities are given by $s_0(x)$. Also, we need to explain how to compose morphisms: Two composable morphisms determine a map $I^2 = \Lambda_1^2 \to X$ which we can extend over Δ^2 and restrict to the new edge. We claim that this is indeed a category, for which we have to check that identities really are identities and that composition is associative (both follow from uniqueness):

(a) $s_0(f) \colon \Delta^2 \to X$ has $d_0 s_0(f) = f = d_1 s_0(f)$ and $d_2 s_0(f) = s_0 d_1(f) = s_0(x)$. Similarly, $s_1(f) \colon \Delta^2 \to X$ is a 2-simplex, witnessing that $\mathrm{id}_y \circ f = f$.

(b) Let f, g, h be composable 1-simplices. Consider the associated map $I^3 \to X$. It can be uniquely extended to a map $\Phi \colon \Delta^3 \to X$. The restriction of this map to $\Delta^{\{0,2\}}$ is gf. Therefore, the 2-simplex $d_2(\Phi)$ is a composition of gf and h, i.e., $\Phi_{|\Delta^{\{0,3\}}} = h \circ gf$. On the other hand, the 2-simplex $d_0(\Phi)$ yields hg, and thus the 2-simplex $d_1(\Phi)$ yields $hg \circ f$. Hence

$$h \circ gf = \Phi_{\Delta^{\{0,3\}}} = hg \circ f,$$

which shows the associativity of composition.

We claim that there is a preferred map $X \to N(\mathcal{C})$ given by the following construction: A map $\Delta^n \to X$ can be restricted along the spine and thus determines a sequence of composable morphisms of $\mathcal{C}$, so that we obtain a map $I^n \to N(\mathcal{C})$. This map can be (uniquely) extended over Δ^n and thus provides an association which maps n-simplices of X to n-simplices of $N(\mathcal{C})$. Using the fact that an n-simplex of X is determined by its restriction to the spine, it is now easily seen that the map $X \to N(\mathcal{C})$ satisfies the following two criteria:

(a) It is an isomorphism on 0- and 1-simplices.
(b) There is a commutative diagram

$$
\begin{array}{ccc}
X_n & \longrightarrow & N(\mathcal{C})_n \\
\downarrow & & \downarrow \\
X_1 \times_{X_0} \cdots \times_{X_0} X_1 & \longrightarrow & N(\mathcal{C})_1 \times_{N(\mathcal{C})_0} \cdots \times_{N(\mathcal{C})_0} N(\mathcal{C})_1
\end{array}
$$

where the vertical maps are bijections and where the lower fibre product is over the source and target maps and has n-many factors on both sides.

By (a) the lower map is a bijection, so the upper horizontal map is a bijection as well. This shows that the map $X \to N(\mathcal{C})$ is an isomorphism of simplicial sets, which shows (2) $\Rightarrow$ (3).

Next, we prove (1) $\Rightarrow$ (2) via induction over n. For $n = 2$, this is clear, since the 2-spine is the inner 2-horn. Therefore, we may assume that one can uniquely lift maps $I^k \to X$ to Δ^k for all k strictly smaller than n and consider a map $I^n \to X$ for which we wish to show that it extends uniquely to Δ^n. We will show that it extends uniquely to Λ_j^n for some $0 \leq j \leq n$, and then use (1) to deduce the claim. We first observe that $I^n \cap \Delta^{n\backslash\{n\}}$ is the spine I^{n-1} of this simplex, and likewise that $I^n \cap \Delta^{n\backslash\{0\}}$ is the spine as well. Thus, by the inductive hypothesis, there are unique maps $\Delta^{n\backslash\{\varepsilon\}} \to X$ extending the map from the spine to X for $\varepsilon = 0, n$. Since the intersection of these two faces is given by $\Delta^{n\backslash\{0,n\}}$, which intersects the spine again in a smaller spine, these two extensions agree on this intersection, by the inductive hypothesis.

Hence, we obtain a map

$$I^n \cup \Delta^{n \setminus \{0\}} \cup \Delta^{n \setminus \{n\}} = \Delta^{n \setminus \{0\}} \cup \Delta^{n \setminus \{n\}} \to X$$

where the union is in Δ^n. We claim that there exists a unique extension to the union

$$\Delta^{n \setminus \{0\}} \cup \Delta^{n \setminus \{n\}} \cup \Delta^{n \setminus \{1\}}.$$

For this, we claim that $\Delta^{n \setminus \{0\}} \cup \Delta^{n \setminus \{n\}}$ contains the spine of $\Delta^{n \setminus \{1\}}$: The edges from $i \to i+1$ for $2 \leq i \leq n - 1$ all lie in $\Delta^{n \setminus \{0\}}$, and the edge from $0 \to 2$ lies in $\Delta^{n \setminus \{n\}}$ because of $n \geq 3$. Hence, there is a unique map from $\Delta^{n \setminus \{1\}} \to X$ extending this map on the spine. It remains to be shown that this map agrees with the given one on

$$\left(\Delta^{n \setminus \{0\}} \cup \Delta^{n \setminus \{n\}} \right) \cap \Delta^{n \setminus \{1\}} = \Delta^{n \setminus \{0,1\}} \cup \Delta^{n \setminus \{1,n\}}.$$

On both of these simplices, the map is determined by its restriction to the spine, which shows the claim. Inductively, we find that there exists a unique extension of the map in question to a map $\Lambda^n_{n-1} \to X$. This map can now be uniquely extended to Δ^n by assumption (1).

In order to see that (2) $\Rightarrow$ (1), we consider an extension problem $\beta \colon \Lambda^n_i \to X$ for which we want to show that it extends uniquely to Δ^n. Clearly we may assume that $n \geq 3$, because $I^2 = \Lambda^2_1$. By assumption (2) and Exercise 13, there is an inclusion $I^n \to \Lambda^n_i$ and we can consider the restricted extension problem. This problem can be solved uniquely by assumption (2), so that we obtain a map $\alpha \colon \Delta^n \to X$. This map can in turn be restricted to Λ^n_i, and we want to show that this map is given by β. In order to do so, we may restrict the map to the faces of Λ^n_i, i.e., to the union of $\Delta^{n \setminus \{j\}}$ for $j \neq i$. It is easy to see that

$$\alpha_{| \Delta^{n \setminus \{0\}}} = \beta_{| \Delta^{n \setminus \{0\}}},$$

because the spine of that simplex is given by a subset of the big spine, and by its very definition $\alpha_{|I^n} = \beta_{|I^n}$. The same holds for

$$\alpha_{| \Delta^{n \setminus \{n\}}} = \beta_{| \Delta^{n \setminus \{n\}}},$$

provided that β is defined there. We need to show that

$$\alpha_{| \Delta^{n \setminus \{j\}}} = \beta_{| \Delta^{n \setminus \{j\}}}$$

under the assumption that $j \neq 0, n$. For this, we show that $\alpha_{| \Delta^{\{j-1,j+1\}}} = \beta_{| \Delta^{\{j-1,j+1\}}}$ (again, all the other edges of the spine are already contained in the big spine). Since $n \geq 3$, this edge is contained in $\Delta^{n \setminus \{\varepsilon\}}$ for ε either 0 or 1. Then we can induct that this determines the map from Λ^n_i, which completes the proof. $\square$

In fact, one can say slightly more:

Lemma 1.1.53

If $n \geq 4$ and $0 \leq j \leq n$ and $\mathcal{C}$ is a category, then every lifting problem

can be uniquely solved.

Proof It is the content of Exercise 19 to show that nerves of categories are 2-coskeletal, and we give another argument in Corollary 1.2.21. Hence, it suffices to recall that $\mathrm{sk}_2(\Lambda_j^n) \to \mathrm{sk}_2(\Delta^2)$ is an isomorphism for $n \geq 4$ and all $0 \leq j \leq n$. $\qquad\square$

Lemma 1.1.54

The nerve of a category $\mathcal{C}$ is a Kan complex if and only if $\mathcal{C}$ is a groupoid.

Proof In Lemma 1.1.49, we have seen already the direction that $N(\mathcal{C})$ being a Kan complex implies that $\mathcal{C}$ is a groupoid. Now, we need to show the other direction. Let us thus assume that $\mathcal{C}$ is a groupoid, and let us show that $N(\mathcal{C})$ is a Kan complex. By Theorem 1.1.52 we already know that we can (uniquely) lift all inner horns and all horns of dimension greater or equal to 4. Therefore, we only need to prove that we can lift outer 2-horns and outer 3-horns. By passing to opposite categories, it suffices to show that every extension problem

has a (unique) solution for $n = 2, 3$. For $n = 2$, such a map is given by two maps $f \colon x \to y$ and $g \colon x \to z$. We can then choose $f \circ g^{-1}$ for the other edge. In order to show the claim for the left outer 3-horn, we consider the restriction along the spine and obtain three composable maps f, g, and h. We find an extension to Δ^3 precisely if the edge $\Delta^{\{1,3\}} \to \Lambda_0^3 \to \mathcal{C}$ is given by the composite hg. Considering the 2-simplex $\Delta^{\{0,1,3\}} \to \Lambda_0^3 \to \mathcal{C}$, we find that this edge satisfies the condition that precomposition with f is given by hgf. Since f is an isomorphism, the claim follows. $\qquad\square$

1.2   ∞-Categories

In this section, we will work towards the definition of ∞-categories, taking Theorem 1.1.52 as a motivation. This naturally leads to two (a priori and a posteriori) different definitions that one could come up with, the more naive one being called a *composer* in this book. We will work out why the notion of an ∞-category as defined by Joyal behaves better than the notion of a composer and how this makes it the "correct" generalization of ordinary categories. Also, we will discuss examples of ∞-categories and some basic operations which may be performed on them. One "import tool" to construct ∞-categories are simplicially enriched categories, so we will take a little detour and discuss enriched categories to the extent necessary for our purposes.

Definition 1.2.1

A *composer* is a simplicial set which has the extension property for spine inclusions $I^n \to \Delta^n$.

Definition 1.2.2

In a composer (in fact, in a general simplicial set) we call 0-simplices *objects*, and a 1-simplex f is called a *morphism* from $d_1(f)$ (the source) to $d_0(f)$ (the target). We define the identity morphism of an object x to be $s_0(x)$. For composers, we define a composition of n-composable morphisms to be a choice of an extension to Δ^n, sometimes also just the restriction to the edge $\Delta^{\{0,n\}} \subseteq \Delta^n$.

From this definition, it becomes clear that the name "composer" originates from the fact that one *can* compose morphisms. In order to avoid associativity questions, a composer is equipped with an n-ary composition law.

Example The singular set of a topological space is a composer: Objects are the points, morphisms from x to y are paths. A composition of morphisms is any path which is homotopic relative endpoints to the concatenation of the paths.

Definition 1.2.3

Let X be a simplicial set. We call two 1-simplices f and g from x to y *equivalent* if there exists a 2-simplex $\sigma : \Delta^2 \to X$ which satisfies the following conditions:

(1) $\sigma_{|\Delta^{\{0,1\}}} = f$
(2) $\sigma_{|\Delta^{\{0,2\}}} = g$
(3) $\sigma_{|\Delta^{\{1,2\}}} = \mathrm{id}_y$

Observation 1.2.4

This relation is obviously reflexive, but again a priori neither transitive nor symmetric. Can you find a further lifting criterion for a composer so that the relation becomes in fact an equivalence relation?

Definition 1.2.5

Let X be a simplicial set. We define a category hX by means of generators and relations: The objects are given by X_0; morphisms are generated by X_1, i.e., for every 1-simplex $f: \Delta^1 \to X$ there is a morphism from $d_1(f)$ to $d_0(f)$; the free composites will be denoted by $f \star g$. Now we start to impose relations:

(1) The 1-simplex $s_0(x)$ is the identity of x.
(2) For every 2-simplex $\sigma: \Delta^2 \to X$ with boundary given by a triple (f, g, h), we impose the relation that $h = g \star f$.
(3) If $f \sim f'$, then $f \star g \sim f' \star g$ and $g' \star f \sim g' \star f'$.

The category hX is called the *homotopy category* of X.

Remark 1.2.6

This construction is obviously functorial, i.e., we have a functor $h: \mathrm{sSet} \to \mathrm{Cat}$.

Lemma 1.2.7

Let $f: X \to Y$ be a map of simplicial sets which induces an isomorphism $\mathrm{sk}_2(X) \to \mathrm{sk}_2(Y)$. Then the induced map $hX \to hY$ is an isomorphism of categories.

Proof The whole construction only refers to the 2-skeleton of X. In other words, the evident map $h(\mathrm{sk}_2(X)) \to hX$ induced by functoriality of h is an isomorphism. Apart from that, we need to use the fact that the following diagram commutes:

$$
\begin{array}{ccc}
h(\mathrm{sk}_2 X) & \xrightarrow{\;\cong\;} & h(\mathrm{sk}_2 Y) \\
{\scriptstyle\cong}\big\downarrow & & \big\downarrow{\scriptstyle\cong} \\
hX & \xrightarrow{\hspace{2cm}} & hY
\end{array}
$$

$\square$

Observation 1.2.8

In general, the morphisms in hX are formal composites of 1-simplices (with correct source and target). If X is a composer, then we see that the set of morphisms of hX is a quotient of the set of 1-simplices, and that any two composites of the same two morphisms will be identified. In particular, equivalent morphisms are identified: If we have a 2-simplex

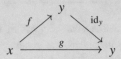

then we find that $g \sim \mathrm{id}_y \star f \sim f$.

Lemma 1.2.9

Suppose that a composer X has in addition the lifting property with respect to inner 3-horn inclusions. Let f and g be composable 1-simplices in X. Then

(1) there exists a composite of f and g;
(2) the relation "equivalence" of morphisms in the sense of Definition 1.2.3 is an equivalence relation;
(3) any two composites of f and g are equivalent in the sense of Definition 1.2.3; and
(4) given a 2-simplex σ with $\sigma_{\Delta^{\{0,1\}}} = \mathrm{id}_x$, $\sigma_{\Delta^{\{1,2\}}} = h$ and $\sigma_{\Delta^{\{0,2\}}} = h'$, then $h' \sim h$.

Proof (1) follows from the definition of a composer. For (2), we need to prove symmetry and transitivity. Let us first prove symmetry. For this purpose, let $f, g \colon x \to y$ be morphisms with $f \sim g$. Pick a 2-simplex σ with $\sigma_{\Delta^{\{0,1\}}} = f$, $\sigma_{\Delta^{\{0,2\}}} = g$, and $\sigma_{\Delta^{\{1,2\}}} = \mathrm{id}_y$. Together with $s_1(f)$ and $s_0(\mathrm{id}_y)$, this determines a map $\Lambda_1^3 \to X$:

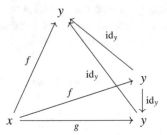

Since X has the extension property for inner 3-horns, there exists an extension to Δ^3, which can be restricted to the face $\Delta^{\{0,2,3\}}$. This 2-simplex witnesses that $g \sim f$.

In order to show transitivity, suppose that $f \sim g \sim h$. Pick 2-simplices σ and σ' witnessing these relations. Together with $s_0(\mathrm{id}_y)$, these define a map $\Lambda_2^3 \to X$:

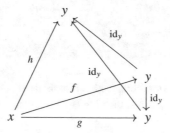

Extending to Δ^3 and then restricting to $\Delta^{\{0,1,3\}}$ shows that $f \sim h$. In order to prove (3), let $f \colon x \to y$ and $g \colon y \to z$ be composable morphisms. Choosing compositions h and h', together with $s_1(g)$, determines a map $\Lambda_1^3 \to X$:

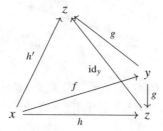

Extending to Δ^3 and then restricting to $\Delta^{\{0,1,3\}}$ shows that $h \sim h'$. In order to prove (4), we consider the map $\Lambda_2^3 \to X$ given by the diagram

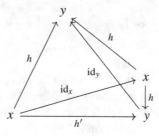

and extend to Δ^3. Restricting the result to $\Delta^{\{0,2,3\}}$ shows that $h' \sim h$. □

Lemma 1.2.10

Suppose that X is a composer with the inner 3-horn extension property. Then there is a category $\pi(X)$ with objects given by 0-simplices of X and morphisms given by

(continued)

Lemma 1.2.10 (continued)
equivalence classes (in the sense of Definition 1.2.3) of 1-simplices in X. Composition is defined via lifting along $I^2 \to \Delta^2$. The uniqueness of composition (up to equivalence) shows that composition in $\pi(X)$ is associative.

Proof It only remains to prove that composition is associative. For this, it suffices to show that if f, g, h are composable morphisms, then $h \circ (gf)$ is a composition of hg and f. Consider the map $\Lambda_1^3 \to X$ given by

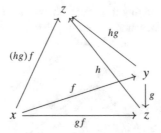

Extending to Δ^3 and then restricting to $\Delta^{\{0,2,3\}}$ shows that $(hg)f$ is a composition of gf and h. Since composition in X is unique up to equivalence, the associativity of composition in $\pi(X)$ follows. $\square$

Corollary 1.2.11
Let X be a composer which has the inner 3-horn lifting property. Then hX is isomorphic to $\pi(X)$. In particular, for composers with the additional inner 3-horn lifting condition, there is a very explicit description of the homotopy category of X.

Proof There is a canonical functor $hX \to \pi(X)$ which is constructed as follows: It is the identity on objects and induced by the identity on 1-simplices. Since all relations imposed in hX are fulfilled in $\pi(X)$, this in fact leads to a functor as needed. It suffices to prove that for any two objects $x, y \in X$, the canonical map

$$\mathrm{Hom}_{hX}(x, y) \to \mathrm{Hom}_{\pi(X)}(x, y)$$

is a bijection. In order to show this, we observe that there is a commutative diagram

$$X_1(x, y)$$

$$\mathrm{Hom}_{hX}(x, y) \longrightarrow \mathrm{Hom}_{\pi(X)}(x, y)$$

where $X_1(x, y)$ denotes the set of 1-simplices f with $d_1(f) = x$ and $d_0(f) = y$. Thanks to the definition of π and Observation 1.2.8, both maps from $X_1(x, y)$ are surjective. Therefore, the horizontal map is surjective. It remains to show that this map is injective as well. For this purpose, assume that you are given two morphisms $f, g \in X_1(x, y)$ with the same image in $\pi(X)$. This means that they are equivalent in the sense of Definition 1.2.3. But again by Observation 1.2.8, this means that $f \sim g$.   □

Remark 1.2.12
In the above corollary, X needed not really be a composer, but only a composition of 2 composable morphisms. In the sequel, we will call a simplicial set which has the extension property for the 2-spine (which is the inner 2-horn) a *weak composer*. In order to compose many morphisms at the same time, we compose inductively and obtain a well-defined "n-fold composition" up to equivalence (provided that the weak composer satisfies the extension property for inner 3-horns as well).

Now notice that $I^2 = \Lambda_1^2$. Therefore, we can reformulate the above statement as follows:

Corollary 1.2.13
Let X be a simplicial set which admits liftings for inner 2- and 3-horns. Then hX is isomorphic to $\pi(X)$.

Corollary 1.2.14
Let $\mathcal{C}$ be a category. Then $h(N(\mathcal{C}))$ is canonically isomorphic to $\mathcal{C}$.

Proof We have seen that $N(\mathcal{C})$ admits (unique) lifts for many horns, including the ones described in the previous corollary. Hence, it suffices to prove that $\pi(N(\mathcal{C})) \cong \mathcal{C}$. But we recall that the relation of "equivalence" for morphisms in $N(\mathcal{C})$ is the relation of "being equal".   □

Definition 1.2.15
A simplicial set is called an ∞-*category* if it has the extension property for *all* inner horn inclusions $\Lambda_j^n \to \Delta^n, n \geq 2, 0 < j < n$.

Definition 1.2.16
A functor between two ∞-categories is just a map of simplicial sets. In other words, the category of ∞-categories is the full subcategory of sSet on objects which are ∞-categories.

At a later stage of this book, we will see that, informally, an ∞-category is a composer in which the choice of a composition is unique up to a contractible space of choices: For each pair of composable morphisms in a composer X, there is a simplicial set of compositions of f and g, $\mathrm{Comp}_X(f, g)$, given by the pullback

$$\begin{array}{ccc} \mathrm{Comp}_X(f, g) & \longrightarrow & \underline{\mathrm{Hom}}(\Delta^2, X) \\ \downarrow & & \downarrow \\ \Delta^0 & \xrightarrow{\ (f,g)\ } & \underline{\mathrm{Hom}}(\Lambda_1^2, X) \end{array}$$

According to Exercise 32, the inner 3-horn lifting condition tells us that $\mathrm{Comp}_X(f, g)$ is connected, i.e., its $\pi_0^\Delta(-)$ vanishes. If we now demand that X has indeed the extension property for inner horn inclusions for all $\Lambda_j^n \to \Delta^n$, then we obtain the following two facts (which we will prove later):

(1) The simplicial set $\mathrm{Comp}_X(f, g)$ is a Kan complex.
(2) *All* simplicial homotopy groups $\pi_n^\Delta(\mathrm{Comp}_X(f, g))$ vanish.

Therefore, an ∞-category is a composer in which composition is well-defined up to a contractible space of choices.

Example A Kan complex, and thus the singular set $\mathcal{S}(X)$ of a space X, is an ∞-category. Also, the nerve of any category is an ∞-category.

In order to compare ∞-categories with composers, we state the following proposition. The proof will be given later, see Propositions 1.3.22 and 1.3.12.

Proposition 1.2.17
Every ∞-category is a composer. However, there are composers which are not ∞-categories.

Proposition 1.2.18
The pair of functors

$$h \colon \mathrm{sSet} \rightleftarrows \mathrm{Cat} \colon \mathrm{N}$$

are adjoint, N being the right adjoint to h.

Proof We need to specify unit and counit transformations. The counit is the isomorphism

$$h(\mathrm{N}(\mathcal{C})) \cong \mathcal{C}$$

of Corollary 1.2.14. In order to construct a natural map $X \to \mathrm{N}(hX)$, i.e., the unit of the adjunction, we find that there are canonical maps on 0- and 1-simplices: Recall that the objects of hX are the 0-simplices of X and that there is a canonical map from X_1 to the morphisms of hX (this map is a surjection if X is a composer). For the construction of the map on general n-simplices, we observe that an n-simplex $\Delta^n \to \mathrm{N}(hX)$ is uniquely determined by its restriction to the spine, i.e., consider the diagram

$$
\begin{array}{ccc}
\mathrm{Hom}_{\mathrm{sSet}}(\Delta^n, X) & \dashrightarrow & \mathrm{Hom}_{\mathrm{sSet}}(\Delta^n, \mathrm{N}(hX)) \\
\downarrow & & \downarrow \cong \\
\mathrm{Hom}_{\mathrm{sSet}}(I^n, X) & \longrightarrow & \mathrm{Hom}_{\mathrm{sSet}}(I^n, \mathrm{N}(hX))
\end{array}
$$

But then, simply use the bijection

$$
\mathrm{Hom}_{\mathrm{sSet}}(I^n, X) \simeq X_1 \times_{X_0} \cdots \times_{X_0} X_1
$$

and the corresponding bijection for $\mathrm{N}(hX)$ and the previous observation to obtain a map for general n-simplices. It is easily checked that these are in fact unit and counit of an adjunction.

Alternatively, you can also show that the map induced by h and the above counit,

$$
\mathrm{Hom}_{\mathrm{sSet}}(X, \mathrm{N}(\mathcal{C})) \to \mathrm{Fun}(hX, \mathcal{C}),
$$

is a bijection. This follows immediately from the definitions. $\qquad\square$

Lemma 1.2.19

Let $(F, G, \varepsilon, \eta)$ be an adjunction with $F \colon \mathcal{C} \to \mathcal{D}$ the left adjoint, G the right adjoint, $\varepsilon \colon FG \to \mathrm{id}$ the counit and $\eta \colon \mathrm{id} \to GF$ the unit. Then

(1) G is fully faithful if and only if ε is an isomorphism, and
(2) F is fully faithful if and only if η is an isomorphism.

Proof Consider the diagram

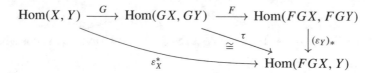

The fact that the right triangle commutes is part of the solution of Exercise 9. (In fact, this is how adjunctions with a binatural isomorphism τ are translated into adjunctions using unit and counit.) We now claim that the big diagram also commutes. Spelling this out, we need to check that for every morphism $f \colon X \to Y$, the diagram

$$
\begin{array}{ccc}
FGX & \xrightarrow{FG(f)} & FGY \\
\downarrow{\varepsilon_X} & & \downarrow{\varepsilon_Y} \\
X & \xrightarrow{\ \ f\ \ } & Y
\end{array}
$$

commutes. This is true, since $\varepsilon \colon FG \to \mathrm{id}$ is a natural transformation of functors. Hence, we see that G is fully faithful if and only if for all $X, Y \in \mathcal{D}$, the map ε_X^* is an isomorphism. By the Yoneda lemma, this is the case if and only if ε_X itself is an isomorphism for all X, which shows claim (1). The argument for (2) is similar.   □

Corollary 1.2.20
The nerve functor is fully faithful. Its essential image is described by Proposition 1.1.52.

Proof We proved in Proposition 1.2.18 that N is a right adjoint to h by constructing explicit unit and counit maps. Thus, by Lemma 1.2.19 it suffices to check that the counit of this adjunction is an isomorphism. But by construction, the counit is given by the canonical isomorphism $h(\mathrm{N}(\mathcal{C})) \to \mathcal{C}$.   □

Let us also record the following consequence, which Exercise 19 asks to prove from the definitions.

Corollary 1.2.21
The nerve of a category is 2-coskeletal.

Proof There is a commutative diagram

$$
\begin{array}{ccc}
\mathrm{Fun}(hX, \mathcal{C}) & \longrightarrow & \mathrm{Hom}_{\mathrm{sSet}}(X, \mathrm{N}(\mathcal{C})) \\
\downarrow & & \downarrow \qquad\qquad\searrow \\
\mathrm{Fun}(h(\mathrm{sk}_2 X), \mathcal{C}) & \longrightarrow & \mathrm{Hom}_{\mathrm{sSet}}(\mathrm{sk}_2 X, \mathrm{N}(\mathcal{C})) \longrightarrow \mathrm{Hom}_{\mathrm{sSet}}(X, \mathrm{cosk}_2 \mathrm{N}(\mathcal{C}))
\end{array}
$$

where the diagonal arrow is induced by the canonical map $\mathrm{N}(\mathcal{C}) \to \mathrm{cosk}_2 \mathrm{N}(\mathcal{C})$. By Proposition 1.2.18, the two left horizontal maps are bijections, and by Lemma 1.2.7 the remaining horizontal arrow is a bijection as well. Thus, the vertical map is a bijection if and only if the diagonal map is a bijection. Since X is arbitrary, the claim follows from Yoneda's lemma.   □

Definition 1.2.22

A morphism in an ∞-category is called an *equivalence* if its image in the homotopy category is an isomorphism.

Lemma 1.2.23

A morphism $f: x \to y$ in an ∞-category $\mathcal{C}$ is an equivalence if and only if there exist 2-simplices $\sigma^l: \Delta^2 \to \mathcal{C}$ and $\sigma^r: \Delta^2 \to \mathcal{C}$ such that

$$\sigma^l_{|\Lambda^2_0} = (f, \mathrm{id}) \text{ and } \sigma^r_{|\Lambda^2_2} = (f, \mathrm{id}).$$

Here, the notation (g, h) means that the second morphism h is the one from 0 to 2 in Δ^2, and the first morphism is the one from 0 to 1 in the first case, and the one from 1 to 2 in the second case.

Proof If a 2-simplex σ^l exists, then $\sigma^l_{|\Delta^{\{1,2\}}}$ is a left inverse of the image of f in $h\mathcal{C}$. Similarly, $\sigma^r_{|\Delta^{\{0,1\}}}$ is a right inverse of the image of f in $h\mathcal{C}$. For the converse, suppose that the image of f is an equivalence in $h\mathcal{C}$. This means that there exists a 1-simplex $g: y \to x$ such that $[fg]$ and $[gf]$ are the identity in $h\mathcal{C}$, i.e., that there is a 2-simplex η which witnesses that h is a composite of f and g, and that there is a further 2-simplex η' which witnesses that h is equivalent to the identity. We can use these two 2-simplices (plus a degenerate 2-simplex on g) to obtain a map $\Lambda^3_2 \to \mathcal{C}$ which can be extended due to $\mathcal{C}$ being an ∞-category:

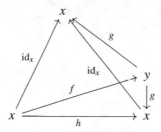

Restricting the resulting 3-simplex to the 2-simplex $\Delta^{\{0,1,3\}}$ yields a 2-simplex σ^l. The argument for σ^r is analogous. □

Definition 1.2.24

An ∞-category is called *∞-groupoid* if every morphism is an equivalence.

Definition 1.2.25

The maximal sub-groupoid of an ordinary category is the subcategory consisting of all isomorphisms and is denoted by $\mathcal{C}^\simeq \subseteq \mathcal{C}$. For an ∞-category $\mathcal{C}$, we define the maximal sub-∞-groupoid to be the pullback of simplicial sets

$$
\begin{array}{ccc}
\mathcal{C}^{\simeq} & \longrightarrow & \mathcal{C} \\
\downarrow & & \downarrow \\
N(h\mathcal{C}^{\simeq}) & \longrightarrow & N(h\mathcal{C})
\end{array}
$$

Remark 1.2.26

For an ordinary category $\mathcal{C}$, we have $N(\mathcal{C}^{\simeq}) = N(\mathcal{C})^{\simeq}$, since by definition there is a pullback

$$
\begin{array}{ccc}
N(\mathcal{C})^{\simeq} & \longrightarrow & N(\mathcal{C}) \\
\downarrow & & \downarrow \\
N(h(N(\mathcal{C}))^{\simeq}) & \longrightarrow & N(h(N(\mathcal{C}))) \\
\downarrow \cong & & \downarrow \cong \\
N(\mathcal{C}^{\simeq}) & \longrightarrow & N(\mathcal{C})
\end{array}
$$

where the right vertical map composite is the identity, and the lower vertical maps are induced by the canonical isomorphism $h(N(\mathcal{C})) \cong \mathcal{C}$.

Lemma 1.2.27

An n-simplex x of an ∞-category $\mathcal{C}$ belongs to the maximal sub-∞-groupoid if and only if all edges are equivalences.

Proof It suffices to observe that an n-simplex in $N(h\mathcal{C})$ is determined by its restriction to all edges, and that this n-simplex lies in $N(h\mathcal{C}^{\simeq})$ if and only if all edges are isomorphisms in $h\mathcal{C}$.   □

Corollary 1.2.28

The maximal sub-∞-groupoid of an ∞-category is in fact an ∞-groupoid, and it is the largest such ∞-groupoid which sits inside the given ∞-category.

Proof Let us first prove that $\mathcal{C}^{\simeq}$ is an ∞-category. For this, consider a lifting problem

$$
\begin{array}{ccc}
\Lambda^n_j & \longrightarrow & \mathcal{C}^{\simeq} \subseteq \mathcal{C} \\
\downarrow & \nearrow & \\
\Delta^n & &
\end{array}
$$

for $0 < j < n$. Since $\mathcal{C}$ is an ∞-category, this problem can be solved in $\mathcal{C}$. The claim follows if we can prove that if we are given a map $\Delta^n \to \mathcal{C}$ which induces a map

$$\Lambda^n_j \to \Delta^n \to \mathcal{C} \to N(h\mathcal{C})$$

whose image is contained in $N(h\mathcal{C}^\simeq)$, then already the map $\Delta^n \to N(h\mathcal{C})$ factors through $N(h\mathcal{C}^\simeq)$. For this, we again recall that an n-simplex of the nerve of a category is determined by its restriction to the spine. Since the spine is included in the inner horns Λ^n_j for all n, the claim follows. This formalizes that $\mathcal{C}^\simeq$ is the sub-∞-category consisting of the equivalences of $\mathcal{C}$. In particular, it follows that $h(\mathcal{C}^\simeq) = (h\mathcal{C})^\simeq$, so that $\mathcal{C}^\simeq$ is an ∞-groupoid. It is then clear that $\mathcal{C}^\simeq$ is the largest such ∞-groupoid sitting inside $\mathcal{C}$. □

Lemma 1.2.29

A Kan complex X is an ∞-groupoid.

Proof It suffices to show that for every morphism in X, there is a left and a right inverse σ^l and σ^r, respectively, as in Lemma 1.2.23. But since X is a Kan complex, the maps $(f, \mathrm{id}) \colon \Lambda^2_0 \to X$ and $(f, \mathrm{id}) \colon \Lambda^2_2 \to X$ can be extended to 2-simplices σ^l and σ^r. □

The converse is also true – this is a non-trivial and very important theorem in higher categories, whose proof will soon be our next aim. But before that, we will spend some time on more examples of ∞-categories, i.e., on how can we produce ∞-categories.

Definition 1.2.30

Let V be a category. A monoidal structure on V consists of the following data:

(1) A functor $- \otimes - \colon V \times V \to V$, called the *monoidal product*;
(2) a unit object $\mathbb{1} \in V$, together with natural isomorphisms $\eta_l \colon X \to \mathbb{1} \otimes X$, and $\eta_r \colon X \to X \otimes \mathbb{1}$, called *left unit* and *right unit*; and
(3) natural associativity isomorphisms $\alpha_{X,Y,Z} \colon (X \otimes Y) \otimes Z \to X \otimes (Y \otimes Z)$.

These data are of course required to satisfy axioms, such as the pentagon axiom.

Example In a category $\mathcal{C}$ which admits finite products, there is a cartesian monoidal structure given by the product bifunctor $(X, Y) \mapsto X \times Y$. The unit is given by the terminal object (a product over the empty set). Dually, a category with finite coproducts admits a cocartesian monoidal structure with $(X, Y) \mapsto X \sqcup Y$. The unit is given by the initial object (the coproduct over the empty set).

Explicit examples which we will care about are:

(1) the category $(\text{Set}, \times, *)$
(2) the category $(\text{Cat}, \times, [0])$
(3) the category $(\text{sSet}, \times, \Delta^0)$

Definition 1.2.31

Let $(V, \otimes_V, \mathbb{1}_V)$ and $(W, \otimes_W, \mathbb{1}_W)$ be monoidal categories. A *lax monoidal functor* consists of the following data:

(1) a functor $F: V \to W$;
(2) a natural map $\mathbb{1}_W \to F(\mathbb{1}_V)$; and
(3) a natural map $FX \otimes_W FY \to F(X \otimes_V Y)$.

Dually, an *oplax monoidal functor* consists of the following data:

(1) a functor $F: V \to W$;
(2) a natural map $F(\mathbb{1}_V) \to \mathbb{1}_W$; and
(3) a natural map $F(X \otimes_V Y) \to FX \otimes_W FY$.

The structure morphisms have to satisfy compatibility with respect to the associativity isomorphisms and the left/right unit isomorphisms. Precisely, the following diagrams are required to commute:

$$
\begin{array}{ccc}
(FX \otimes FY) \otimes FZ & \longrightarrow & F(X \otimes Y) \otimes FZ \\
\alpha_W \downarrow & & \downarrow \\
FX \otimes (FY \otimes FZ) & & F((X \otimes Y) \otimes Z) \\
\downarrow & & \downarrow F(\alpha_V) \\
FX \otimes F(Y \otimes Z) & \longrightarrow & F(X \otimes (Y \otimes Z))
\end{array}
$$

where α_W and α_V are the associativity isomorphism in $(W, \otimes, \mathbb{1})$ and $(V, \otimes, \mathbb{1})$, respectively. Likewise for the units:

$$
\begin{array}{ccc}
FX & \xrightarrow{F(\eta_r^V)} & F(X \otimes \mathbb{1}) \\
\eta_r^W \downarrow & & \uparrow \\
FX \otimes \mathbb{1} & \longrightarrow & FX \otimes F(\mathbb{1})
\end{array}
$$

where η_r^V and η_r^W are the right unit isomorphisms of V and W, respectively. Similarly, the diagram involving left units is required to commute as well. A lax monoidal (or an oplax monoidal) functor is called monoidal if the natural maps of (2) and (3) are isomorphisms.

Definition 1.2.32

Let F and G be lax monoidal functors between monoidal categories V and W. A natural transformation $\tau \colon F \to G$ is called lax monoidal if for all $X, Y \in V$, the diagrams

$$
\begin{array}{ccc}
FX \otimes FY & \longrightarrow & F(X \otimes Y) \\
\downarrow & & \downarrow \\
GX \otimes GY & \longrightarrow & G(X \otimes Y)
\end{array}
$$

and

$$
\begin{array}{ccc}
& \mathbb{1} & \\
\downarrow & & \searrow \\
F(\mathbb{1}) & \longrightarrow & G(\mathbb{1})
\end{array}
$$

commute. We let $\mathrm{Fun}^{\mathrm{lax}}(V, W)$ be the category whose objects are the lax monoidal functors and whose morphisms are lax monoidal transformations. Furthermore, we let

$$
\mathrm{Fun}^{\otimes}(V, W) \subseteq \mathrm{Fun}^{\mathrm{lax}}(V, W)
$$

be the full subcategory on monoidal functors.

Remark 1.2.33

We see, therefore, that the category MonCat of monoidal categories with monoidal functors is canonically a 2-category: The hom-category between V and W is given by $\mathrm{Fun}^{\mathrm{lax}}(V, W)$. Of course, in order for this to make sense, we need to observe that the identity of a monoidal category is canonically lax monoidal (in fact monoidal) and that the composition of two lax monoidal functors is canonically lax monoidal.

Definition 1.2.34

Let $(V, \otimes, \mathbb{1})$ be a monoidal category. Then a V-enriched category $\mathcal{C}$ consists of a set of objects, and for any two objects $x, y \in \mathcal{C}$ an object $\mathrm{Hom}_{\mathcal{C}}(x, y) \in V$, together with composition functors

$$
\mathrm{Hom}_{\mathcal{C}}(x, y) \otimes \mathrm{Hom}_{\mathcal{C}}(y, z) \to \mathrm{Hom}_{\mathcal{C}}(x, z),
$$

and furthermore for every object an identity id_x

$$
\mathbb{1} \to \mathrm{Hom}_{\mathcal{C}}(x, x)
$$

satisfying the obvious associativity and unitality conditions, namely that the following diagrams are required to commute:

$$(\text{Hom}_{\mathcal{C}}(x, y) \otimes \text{Hom}_{\mathcal{C}}(y, z)) \otimes \text{Hom}_{\mathcal{C}}(z, w) \longrightarrow \text{Hom}_{\mathcal{C}}(x, z) \otimes \text{Hom}_{\mathcal{C}}(z, w)$$

$$\alpha \downarrow \qquad\qquad \text{Hom}_{\mathcal{C}}(x, w) \uparrow$$

$$\text{Hom}_{\mathcal{C}}(x, y) \otimes (\text{Hom}_{\mathcal{C}}(y, z) \otimes \text{Hom}_{\mathcal{C}}(z, w)) \longrightarrow \text{Hom}_{\mathcal{C}}(x, y) \otimes \text{Hom}_{\mathcal{C}}(y, w)$$

where α denotes the associativity isomorphism of $\mathcal{V}$, and

$$\text{Hom}_{\mathcal{C}}(x, y) \xrightarrow{\eta_r} \text{Hom}_{\mathcal{C}}(x, y) \otimes \mathbb{1}$$
$$\xrightarrow{\text{id}} \text{Hom}_{\mathcal{C}}(x, y)$$

$$\text{Hom}_{\mathcal{C}}(x, y) \xrightarrow{\eta_l} \mathbb{1} \otimes \text{Hom}_{\mathcal{C}}(x, y)$$
$$\searrow \qquad \downarrow$$
$$\text{Hom}_{\mathcal{C}}(x, y)$$

Definition 1.2.35

A V-enriched functor between V-enriched categories $f \colon \mathcal{C} \to \mathcal{D}$ consists of a map on objects $x \mapsto f(x)$, and for each two objects a morphism $f_{x,y} \colon \text{Hom}_{\mathcal{C}}(x, y) \to \text{Hom}_{\mathcal{D}}(fx, fy)$ in V such that the diagrams

$$\text{Hom}_{\mathcal{C}}(x, y) \otimes \text{Hom}_{\mathcal{C}}(y, z) \longrightarrow \text{Hom}_{\mathcal{C}}(x, z)$$
$$f_{x,y} \otimes f_{y,z} \downarrow \qquad\qquad\qquad \downarrow f_{x,z}$$
$$\text{Hom}_{\mathcal{D}}(fx, fy) \otimes \text{Hom}_{\mathcal{D}}(fy, fz) \longrightarrow \text{Hom}_{\mathcal{D}}(fx, fz)$$

and

$$\mathbb{1} \xrightarrow{\text{id}_x} \text{Hom}_{\mathcal{C}}(x, x)$$
$$\text{id}_{fx} \searrow \qquad \downarrow f_{x,x}$$
$$\text{Hom}_{\mathcal{D}}(fx, fx)$$

commute.

Definition 1.2.36

Let V be a monoidal category. Then we let Cat_V be the category of V-enriched categories, i.e., its objects are V-enriched categories and its morphisms are V-enriched functors.

Example A Set-enriched category is just an ordinary category. A Cat-enriched category is a strict 2-category.

Example A Cat-enriched category with one object is the same datum as a monoidal category which is strictly associative (i.e., where the associativity isomorphisms are the identity). An enriched functor between two such categories is the same datum as a monoidal functor.

Lemma 1.2.37

If $\Phi\colon V \to V'$ is a lax monoidal functor between monoidal categories, then applying Φ to each hom-object yields a functor $\Phi_*\colon \mathrm{Cat}_V \to \mathrm{Cat}_{V'}$. In fact, this construction determines a 2-functor

$$\mathrm{MonCat} \to \mathrm{Cat}$$

from the 2-category of monoidal categories (with lax monoidal functors and monoidal transformations as morphisms and 2-morphisms) to the 2-category of categories (with functors and natural transformations as morphisms and 2-morphisms). In particular, a monoidal adjunction between V and V' determines an adjunction on the level of enriched categories.

Proof In order to construct this 2-functor, we first consider the map on the level of objects: It takes a monoidal category V to the category Cat_V of V-enriched categories. In order for this map to become a 2-functor, we need to construct for every pair V, W of monoidal categories a functor

$$\mathrm{Fun}^{\mathrm{lax}}(V, W) \to \mathrm{Fun}(\mathrm{Cat}_V, \mathrm{Cat}_W)$$

and then show that this construction is compatible with composition. For the construction of this functor, again we first consider its effect on objects: Given a lax monoidal functor $\Phi\colon V \to W$ and a V-enriched category $\mathcal{C}$, we consider the W-enriched category $\Phi_*(\mathcal{C})$, whose objects are the same as the objects of $\mathcal{C}$, and for $X, Y \in \mathrm{ob}(\mathcal{C})$ we define

$$\mathrm{Hom}_{\Phi_*(\mathcal{C})}(X, Y) = \Phi(\mathrm{Hom}_{\mathcal{C}}(X, Y)).$$

It is straightforward to show that $\Phi_*(\mathcal{C})$ is a W-enriched category: For instance, in order to show that composition satisfies the associativity constraint, one uses the compatibility of Φ with the associativity isomorphisms of V and W. Next, we need to explain the effect on morphisms. For this purpose, let $\tau\colon \Phi \to \Psi$ be a monoidal transformation. We wish to construct a natural transformation between Φ_* and Ψ_*. Concretely, we need to construct natural maps $\Phi_*(\mathcal{C}) \to \Psi_*(\mathcal{C})$ in Cat_W, i.e., natural W-enriched functors $\tau_*\colon \Phi_*(\mathcal{C}) \to \Psi_*(\mathcal{C})$. On objects, this functor is defined to be the identity, and on morphisms between X and Y we have to construct a map

$$\tau_*\colon \Phi(\mathrm{Hom}_{\mathcal{C}}(X, Y)) \to \Psi(\mathrm{Hom}_{\mathcal{C}}(X, Y)).$$

For this, we simply use the natural map $\tau_{\mathrm{Hom}_{\mathcal{C}}(X,Y)}$ given by the natural transformation $\tau\colon \Phi \to \Psi$. In order to see that this is compatible with composition, we use the fact that τ is a monoidal transformation, so that the diagram

$$\begin{array}{ccc}
\Phi(\mathrm{Hom}_{\mathcal{C}}(X, Y)) \otimes \Phi(\mathrm{Hom}_{\mathcal{C}}(Y, Z)) & \longrightarrow & \Phi(\mathrm{Hom}_{\mathcal{C}}(X, Y) \otimes \mathrm{Hom}_{\mathcal{C}}(Y, Z)) \\
\downarrow & & \downarrow \\
\Psi(\mathrm{Hom}_{\mathcal{C}}(X, Y)) \otimes \Psi(\mathrm{Hom}_{\mathcal{C}}(Y, Z)) & \longrightarrow & \Psi(\mathrm{Hom}_{\mathcal{C}}(X, Y) \otimes \mathrm{Hom}_{\mathcal{C}}(Y, Z))
\end{array}$$

commutes. Then we use naturality of τ to see that the diagram

$$\begin{array}{ccc}
\Phi(\mathrm{Hom}_{\mathcal{C}}(X, Y) \otimes \mathrm{Hom}_{\mathcal{C}}(Y, Z)) & \longrightarrow & \Phi(\mathrm{Hom}_{\mathcal{C}}(X, Z)) \\
\downarrow & & \downarrow \\
\Psi(\mathrm{Hom}_{\mathcal{C}}(X, Y) \otimes \mathrm{Hom}_{\mathcal{C}}(Y, Z)) & \longrightarrow & \Psi(\mathrm{Hom}_{\mathcal{C}}(X, Z))
\end{array}$$

also commutes. Glueing these two diagrams together, we find that the map τ_* is compatible with composition, and thus is in fact a W-enriched functor $\Phi_*(\mathcal{C}) \to \Psi_*(\mathcal{C})$ as needed.

Now it is also clear by definition that for two composable lax monoidal functors Ψ and Φ, we have $\Psi_*(\Phi_*(\mathcal{C})) = (\Psi \circ \Phi)_*(\mathcal{C})$, so that compatibility with composition is an immediate consequence.

To see the "in particular" part of the claim is now easy: A monoidal adjunction consists of lax monoidal functors Φ and Ψ and unit and counit transformations which are themselves monoidal transformations. By the previously established parts, these are sent to functors Φ_* and Ψ_* equipped with candidates for the unit and counit. The only thing to check is the triangle identities, but they follow from the fact that they hold for Φ and Ψ, and that the constructed functor preserves identities.   □

Definition 1.2.38

Let V be a monoidal category and $\mathcal{C}$ a V-enriched category. Then its *underlying category* $u\mathcal{C}$ is obtained via the lax monoidal functor

$$\mathrm{Hom}(\mathbb{1}, -)\colon V \to \mathrm{Set}.$$

In formulas, we have

$$u\mathcal{C} = \mathrm{Hom}_V(\mathbb{1}, -)_*(\mathcal{C}) \in \mathrm{Cat}_{\mathrm{Set}} = \mathrm{Cat}.$$

Definition 1.2.39

For the sake of simplicity, we will call a category which is enriched in simplicial sets a *simplicial category*. Here, sSet is viewed as a monoidal category via the cartesian product, and we will write Cat_Δ for $\mathrm{Cat}_{\mathrm{sSet}}$.

Remark 1.2.40

Note that there is some ambiguity in the above definition: Usually a simplicial category would rather refer to a simplicial object in categories, i.e., a functor $\Delta^{op} \to \mathrm{Cat}$. Luckily we have the following lemma about this:

Lemma 1.2.41

There is a canonical fully faithful embedding $\mathrm{Cat}_\Delta \to \mathrm{Fun}(\Delta^{op}, \mathrm{Cat})$ determined by the family of functors $(\mathrm{ev}_n)_* \colon \mathrm{Cat}_\Delta \to \mathrm{Cat}$, and the essential image can be characterized as those simplicial objects in categories whose underlying simplicial set of objects is constant.

Lemma 1.2.42

Suppose that $F \colon \mathcal{C} \to \mathcal{D}$ is left-adjoint to $G \colon \mathcal{D} \to \mathcal{C}$. Furthermore, suppose that

(1) $\mathcal{C}$ is cocomplete, and
(2) G is fully faithful.

Then $\mathcal{D}$ is cocomplete as well.
Dually, if $\mathcal{D}$ is complete and F is fully faithful, then $\mathcal{C}$ is complete as well.

Proof Assume that (1) and (2) hold. Let $X \colon I \to \mathcal{D}$ be a diagram and consider the object $F(\mathrm{colim}_I G(X_i))$, which exists since $\mathcal{C}$ is cocomplete. We wish to prove that it satisfies the universal property of a colimit:

$$\mathrm{Hom}_\mathcal{D}(F(\mathrm{colim}_I G(X_i)), Y) \cong \mathrm{Hom}_\mathcal{C}(\mathrm{colim}_I G(X_i), G(Y))$$

$$\cong \lim_I \mathrm{Hom}_\mathcal{C}(G(X_i), G(Y))$$

$$\cong \lim_I \mathrm{Hom}_\mathcal{C}(X_i, Y)$$

where the last bijection holds by full faithfulness of G.

Dually, let $X \colon I \to \mathcal{C}$ be a diagram and consider the object $G(\lim_I F(X_i))$, which exists since $\mathcal{D}$ is complete. Then we calculate that

$$\mathrm{Hom}_\mathcal{C}(Y, G(\lim_I F(X_i))) \cong \mathrm{Hom}_\mathcal{D}(F(Y), \lim_I F(X_i))$$

$$\cong \lim_I \mathrm{Hom}_\mathcal{D}(F(Y), F(X_i))$$

$$\cong \lim_I \mathrm{Hom}_\mathcal{C}(Y, X_i),$$

where the last bijection uses the fact that F is fully faithful. □

Proposition 1.2.43
The category Cat *is bicomplete, and the functor* $\mathrm{ob}(-)\colon \mathrm{Cat} \to \mathrm{Set}$ *preserves limits and colimits.*

Proof The case of limits can be done by hand: Given a diagram $\mathcal{C}\colon I \to \mathrm{Cat}$, sending $i \in I$ to $\mathcal{C}_i$, we define its limit $\lim_I \mathcal{C}_i$ as follows:

(1) $\mathrm{ob}(\lim_I \mathcal{C}_i) = \lim_I \mathrm{ob}(\mathcal{C}_i)$, and
(2) for any two objects $\{X_i\}_{i \in I}$ and $\{Y_i\}_{i \in I}$ we have

$$\mathrm{Hom}_{\lim_I \mathcal{C}_i}(\{X_i\}, \{Y_i\}) = \lim_I \mathrm{Hom}_{\mathcal{C}_i}(X_i, Y_i).$$

In order to see that this is in fact a category, we use the fact that limits commute with products (when defining composition) and that a canonical map exists from a limit indexed over $I \times I$ to the limit indexed over the restriction to I along the diagonal. (This is one reason why the situation is more complicated with colimits.) By definition then, the functor $\mathrm{ob}(-)\colon \mathrm{Cat} \to \mathrm{Set}$ commutes with limits (as it must do, since it is right-adjoint to the discrete category functor $d\colon \mathrm{Set} \to \mathrm{Cat}$).

For the proof of the existence of colimits, we make use of Lemma 1.2.42: We consider the adjunction (h, N) of functors between sSet and Cat of Proposition 1.2.18. We have seen that

(1) sSet is cocomplete (in fact, bicomplete by Corollary 1.1.24), and
(2) $\mathrm{N}\colon \mathrm{Cat} \to \mathrm{sSet}$ is fully faithful, see Corollary 1.2.20.

It follows that Cat is cocomplete and that a colimit of a diagram $\mathcal{C}\colon I \to \mathrm{Cat}$ is given by $h(\mathrm{colim}_I \mathrm{N}(\mathcal{C}_i))$. With this, we calculate that the object functor commutes with colimits as follows:

$$\mathrm{ob}(\mathop{\mathrm{colim}}_I \mathcal{C}_i) \cong \mathrm{ob}(h(\mathop{\mathrm{colim}}_I \mathrm{N}(\mathcal{C}_i)))$$

$$\cong (\mathop{\mathrm{colim}}_I \mathrm{N}(\mathcal{C}_i))_0 \cong \mathop{\mathrm{colim}}_I (\mathrm{N}(\mathcal{C}_i)_0)$$

$$\cong \mathop{\mathrm{colim}}_I \mathrm{ob}(\mathcal{C}_i)$$

□

Lemma 1.2.44
Let $\mathcal{C}' \subseteq \mathcal{C}$ *be a full subcategory of a category* $\mathcal{C}$ *and let* $X\colon I \to \mathcal{C}'$ *be a diagram. If* X *has a (co)limit in* $\mathcal{C}$ *which happens to lie in* $\mathcal{C}'$, *then this is also a (co)limit in* $\mathcal{C}'$.

Proof Obvious from the universal property and the fact that the inclusion is full. □

Corollary 1.2.45

The category Cat_Δ *is bicomplete.*

Proof Consider an I-shaped diagram of simplicial categories $\mathcal{C}_i$. By Lemma 1.2.41, this gives rise to an I-shaped diagram in $\mathrm{Fun}(\Delta^{\mathrm{op}}, \mathrm{Cat})$, whose associated I-shaped diagram of simplicial sets of objects in constant, i.e., for all $i \in I$, we find that $\mathrm{ob}(\mathcal{C}_i)$ is a constant simplicial set. We wish to show that in this case $\mathrm{ob}(\mathrm{colim}_I \, \mathcal{C}_i)$ is also a constant simplicial set. By Proposition 1.2.43, we have

$$\mathrm{ob}(\underset{I}{\mathrm{colim}} \, \mathcal{C}_i) \cong \underset{I}{\mathrm{colim}} \, \mathrm{ob}(\mathcal{C}_i),$$

and the latter is a colimit in simplicial sets, over constant simplicial sets. Since the constant functor $c \colon \mathrm{Set} \to \mathrm{sSet}$ admits a right adjoint (ev_0), it preserves colimits. Therefore, the colimit over constant simplicial sets is itself constant (on the colimit of the sets involved). The argument for limits works the same. □

Remark 1.2.46

Nothing is special about Δ here. In fact, the same argument holds to prove that for any small category $\mathcal{C}$, there is a fully faithful inclusion $\mathrm{Cat}_{\mathcal{P}(\mathcal{C})} \subseteq \mathrm{Fun}(\mathcal{C}^{\mathrm{op}}, \mathrm{Cat})$ with essential image given by those functors whose presheaf of objects is constant. It follows completely analogously that $\mathrm{Cat}_{\mathcal{P}(\mathcal{C})}$ is bicomplete. Here, we always use the pointwise (cartesian) monoidal structure on $\mathcal{P}(\mathcal{C})$.

Example The nerve functor $\mathrm{N} \colon \mathrm{Cat} \to \mathrm{sSet}$ is monoidal: It preserves products since it is a right adjoint. Hence, given a 2-category, we obtain a simplicial category by applying the nerve functor to all hom-categories. Furthermore, for a 2-category $\mathcal{C}$ we have $u(\mathrm{N}_*(\mathcal{C})) = u\mathcal{C}$, simply because

$$\mathrm{Hom}_{\mathrm{sSet}}(\Delta^0, -)_* \circ \mathrm{N}_* = (\mathrm{Hom}_{\mathrm{sSet}}(\Delta^0, -) \circ \mathrm{N})$$

$$= \mathrm{Hom}_{\mathrm{sSet}}(\Delta^0, \mathrm{N}(-))$$

$$= \mathrm{Hom}_{\mathrm{Cat}}(h\Delta^0, -),$$

and $h\Delta^0$ is the unit of the monoidal structure on Cat given by cartesian product.

Lemma 1.2.47

The functors c: Set $\to$ sSet and π_0, ev_0: sSet $\to$ Set are canonically monoidal.

Proof We claim that every functor F: $(V, \times, *) \to (W, \times, *)$ is canonically oplax monoidal. The oplax monoidal structure maps are given by

1. $F(*) \to *$, i.e., the unique map to the terminal object of W, and
2. $F(X \times Y) \to F(X) \times F(Y)$ given by the effect of F on the two projections

$$X \leftarrow X \times Y \to Y.$$

It hence suffices to check that the canonical oplax structure maps in our examples are isomorphisms. For c and ev_0, this follows directly from the definitions. Only the functor π_0 requires an actual argument. We want to check that the map

$$\pi_0(X \times Y) \to \pi_0(X) \times \pi_0(Y)$$

is a bijection. By definition of π_0 of a simplicial set, Definition 1.1.9, we have a commutative square

$$
\begin{array}{ccc}
(X \times Y)_0 & \xrightarrow{\ \cong\ } & X_0 \times Y_0 \\
\downarrow & & \downarrow \\
\pi_0(X \times Y) & \longrightarrow & \pi_0(X) \times \pi_0(Y)
\end{array}
$$

in which the vertical maps are bijections. Since the top horizontal map is a bijection (this is the monoidality of ev_0), it follows that the oplax monoidal structure map is surjective. In order to see that it is injective, it suffices to check that generators of the relations can be lifted. Since these are given by 1-simplices, and since ev_1 is also monoidal, the claim follows. Also, it follows directly that the map $\pi_0(*) \to *$ is an isomorphism. $\square$

Definition 1.2.48

We obtain the following functors:

(1) $c = c_*$: Cat $\to$ Cat_Δ, which sends a category to the simplicially enriched category with constant simplicial enrichment;
(2) $\pi = (\pi_0)_*$: $\mathrm{Cat}_\Delta \to$ Cat, called the *homotopy category* of a simplicial category; and
(3) $u = (\mathrm{ev}_0)_*$: $\mathrm{Cat}_\Delta \to$ Cat, called the *underlying category*.

Notice that $\mathrm{ev}_0 = \mathrm{Hom}_{\mathrm{sSet}}(\Delta^0, -)$, therefore it is the same underlying category as in Definition 1.2.38.

Definition 1.2.49

A simplicial functor $f\colon \mathcal{C} \to \mathcal{D}$ between simplicial categories is called a *weak equivalence* if it induces

(1) a weak equivalence on all hom-simplicial sets (weakly fully faithful), and
(2) an essentially surjective functor $\pi(\mathcal{C}) \to \pi(\mathcal{D})$ (weakly essentially surjective).

Lemma 1.2.50

Every functor between cartesian monoidal categories is canonically oplax monoidal. Every natural transformation between two such functors is also canonically oplax monoidal. In particular, the adjunctions (π_0, c) and (c, ev_0) are monoidal adjunctions.

Proof In the proof of Lemma 1.2.47, we have seen that every functor $F\colon (V, \times, *) \to (W, \times, *)$ is canonically oplax monoidal. So let $\tau\colon F \to G$ be a natural transformation. We wish to show that the diagram

$$
\begin{array}{ccc}
F(X \times Y) & \longrightarrow & F(X) \times F(Y) \\
\downarrow{\scriptstyle \tau_{X \times Y}} & & \downarrow{\scriptstyle \tau_X \times \tau_Y} \\
G(X \times Y) & \longrightarrow & G(X) \times G(Y)
\end{array}
$$

commutes. For this, it suffices to check that each of the two following squares commutes:

$$
\begin{array}{ccccc}
F(X) & \longleftarrow & F(X \times Y) & \longrightarrow & F(Y) \\
\downarrow{\scriptstyle \tau_X} & & \downarrow{\scriptstyle \tau_{X \times Y}} & & \downarrow{\scriptstyle \tau_Y} \\
G(X) & \longleftarrow & G(X \times Y) & \longrightarrow & G(Y)
\end{array}
$$

But this follows from naturality of τ.

It now suffices to show that an oplax monoidal transformation between monoidal functors is also a monoidal transformation. But this follows from the general fact that if a square which both vertical maps being isomorphisms commutes, then the square with the inverse vertical maps also commutes. $\qquad\square$

Corollary 1.2.51

The two functors $c\colon \mathrm{Cat} \to \mathrm{Cat}_\Delta$ and $\pi\colon \mathrm{Cat}_\Delta \to \mathrm{Cat}$ form an adjoint pair, with π being left-adjoint to the constant functor c. Similarly, the two functors $u\colon \mathrm{Cat}_\Delta \to \mathrm{Cat}$ and $c\colon \mathrm{Cat} \to \mathrm{Cat}_\Delta$ form an adjoint pair, with c being left-adjoint to the underlying functor.

Proof This follows from Lemmas 1.2.37 and 1.2.50.   □

Observation 1.2.52

The adjunction gives rise to a canonical functor $\mathcal{C} \to c\pi(\mathcal{C})$ of simplicial categories.

Definition 1.2.53

Given a simplicial category $\mathcal{C}$ and two objects $x, y \in \mathcal{C}$, we say that a morphism from x to y is a morphism in the underlying category $u\mathcal{C}$. In other words, it is a 0-simplex of $\text{Hom}_{\mathcal{C}}(x, y)$. Such a morphism is called an *equivalence* if its image in $\pi(\mathcal{C})$ is an isomorphism.

Next, we wish to extend the notion of the nerve of a category to simplicially enriched categories. For this purpose, we need a version of the category $[n]$ which is well suited for simplicially enriched categories.

Definition 1.2.54

Let J be a finite non-empty linearly ordered set, and let i, j be elements of this set. We let $P_{i,j}$ be the following set of subsets of J:

$$P_{i,j} = \{I \subseteq J : i, j \in I \text{ and } k \in I \Rightarrow i \leq k \leq j\}$$

In words, $P_{i,j}$ consists of all subsets of $[i, j] \subseteq J$ which contain i and j.

$P_{i,j}$ is partially ordered by inclusion: $I \leq I' \Leftrightarrow I \subseteq I'$. Notice that $P_{i,j}$ is only non-empty if $i \leq j$.

Observation 1.2.55

Given a triple $i \leq j \leq k$ in J, there is a canonical map of partially ordered sets

$$P_{i,j} \times P_{j,k} \to P_{i,k},$$

given by sending (I, I') to $I \cup I'$. This clearly defines an associative binary operation.

As a consequence, we obtain the following definition.

Definition 1.2.56

Let J be a non-empty linearly ordered set. Then the following defines a simplicially enriched category $\mathfrak{C}[\Delta^J] \in \text{Cat}_\Delta$. Objects of $\mathfrak{C}[\Delta^J]$ are given by the elements of J. Furthermore,

$$\mathrm{Hom}_{\mathcal{C}[\Delta^J]}(i, j) = \begin{cases} \emptyset & \text{if } j < i \\ \mathrm{N}(P_{i,j}) & \text{if } i \le j. \end{cases}$$

Composition is defined via the previous observation.

Lemma 1.2.57

For every $n \ge 1$, we have that $\mathrm{N}(P_{0,n})$ is isomorphic to $(\Delta^1)^{n-1}$. Furthermore, $P_{i,j} \cong P_{0,j-i}$.

Proof The latter statement simply follows by choosing an isomorphism $J \cong [n]$ for some n, and then considering the unique order-preserving isomorphism of $[i, j]$ with $[0, j - i]$. In order to show that $\mathrm{N}(P_{0,n}) \cong (\Delta^1)^{n-1}$, it suffices to find an isomorphism of posets

$$P_{0,n} \cong [1] \times \cdots \times [1],$$

where the latter product has $n-1$ factors. This simply comes from the following construction: A subset $I \subseteq [n]$ containing $\{0, n\}$ is determined by checking which of the elements $1, \ldots, n - 1$ is contained in I. We label an element contained in I with a 1 and elements not contained in I with a 0. This constructs a map of posets $P_{0,n} \to [1] \times \cdots \times [1]$, which is clearly an isomorphism. □

Lemma 1.2.58

Let $\mathcal{C}$ be a category with initial or terminal object. Then $\mathrm{N}(\mathcal{C})$ is contractible, i.e., its identity map is homotopic to the constant map at the initial or terminal object.

Proof Let us consider the case where $\mathcal{C}$ has an initial object $\emptyset$. (The other case follows from the fact that a simplicial set X is contractible if and only if X^{op} is contractible.) We claim that the identity functor of $\mathcal{C}$ admits a natural transformation from the constant functor with value $\emptyset$, which is simply given on an object $X \in \mathcal{C}$ by the unique map $\emptyset \to X$. The relevant diagrams commute by the uniqueness of the maps. We obtain a functor $\mathcal{C} \times [1] \to \mathcal{C}$ whose restriction to 0 is the constant map at $\emptyset$ and whose restriction to 1 is the identity of $\mathcal{C}$. Applying the nerve functor N, we obtain a simplicial homotopy

$$\mathrm{N}(\mathcal{C}) \times \Delta^1 \to \mathrm{N}(\mathcal{C})$$

from the constant map at the vertex given by $\emptyset$ to the identity of $\mathrm{N}(\mathcal{C})$. □

Corollary 1.2.59
For $i \leq j$, the simplicial set $\mathrm{Hom}_{\mathfrak{C}[\Delta^J]}(i, j)$ is contractible.

Proof We have just seen that $\mathrm{Hom}_{\mathfrak{C}[\Delta^J]}(i, j) \cong \Delta^{j-i-1} = \mathrm{N}([1] \times \cdots \times [1])$. The category $[1] \times \cdots \times [1]$ has both an initial and a terminal object, so we can apply Lemma 1.2.58. □

Remark 1.2.60
The functor $|-|: \mathrm{sSet} \to \mathrm{Top}$ preserves products. In order to see this, it suffices to check that $|\Delta^n \times \Delta^m| \cong \Delta^n_{\mathrm{top}} \times \Delta^m_{\mathrm{top}}$, which is a concrete calculation. Hence, a simplicial homotopy induces a homotopy of geometric realizations. It follows that every contractible simplicial set is weakly contractible (i.e., its geometric realization is contractible).

Lemma 1.2.61
There is a unique isomorphism $\pi(\mathfrak{C}[\Delta^n]) \cong [n]$ which is the identity on objects. By adjunction we obtain a canonical functor $\mathfrak{C}[\Delta^n] \to c[n]$, and this functor is a weak equivalence of simplicial categories.

Proof Everything follows from Corollary 1.2.59, which calculates $\pi(\mathfrak{C}[\Delta^n])$ to be $[n]$. It follows that the induced functor $\mathfrak{C}[\Delta^n] \to c[n]$ is bijective on objects (and thus weakly essentially surjective) and a weak equivalence on hom-simplicial sets. □

Lemma 1.2.62
The association $J \mapsto \mathfrak{C}[\Delta^J]$ extends to a functor

$$\mathrm{Lin.or.Set} \to \mathrm{Cat}_\Delta.$$

In particular, we obtain a functor

$$\Delta \to \mathrm{Cat}_\Delta \qquad [n] \mapsto \mathfrak{C}[\Delta^n]$$

which is a cosimplicial object in simplicially enriched categories.

Proof We need to show that every map $J \to J'$ of linearly ordered sets induces a simplicially enriched functor $\mathfrak{C}[\Delta^J] \to \mathfrak{C}[\Delta^{J'}]$. For this, it suffices to explain how such a map $J \to J'$

induces, for every $i \leq j$ in J, a map of posets $P_{i,j} \to P_{f(i),f(j)}$. This map is simply given by sending a subset I to its image $f(I)$. It is easy to see that this produces in fact a functor.   □

Definition 1.2.63

Let $\mathcal{C}$ be a simplicially enriched category. Then we define its *simplicial nerve* (or the homotopy-coherent nerve) as the following simplicial set:

$$N(\mathcal{C})_n = \mathrm{Hom}_{\mathrm{Cat}_\Delta}(\mathfrak{C}[\Delta^n], \mathcal{C})$$

Lemma 1.2.64

If $\mathcal{C}$ is an ordinary category, viewed as a simplicially enriched category $c\mathcal{C}$ via the constant functor, then there is an isomorphism

$$N(\mathcal{C}) \cong N(c\mathcal{C}).$$

Proof This follows from:

$$N(c\mathcal{C})_n = \mathrm{Hom}_{\mathrm{Cat}_\Delta}(\mathfrak{C}[\Delta^n], c\mathcal{C})$$

$$\cong \mathrm{Hom}_{\mathrm{Cat}}(\pi(\mathfrak{C}[\Delta^n]), \mathcal{C})$$

$$\cong \mathrm{Hom}_{\mathrm{Cat}}([n], \mathcal{C}) = N(\mathcal{C})_n$$

The only thing that needs further justification is the isomorphism $\pi(\mathfrak{C}[\Delta^n]) \cong [n]$, which we proved in Lemma 1.2.61.   □

Remark 1.2.65

The analog of such a statement with $\pi(\mathcal{C})$ and $u\mathcal{C}$ is wrong!

Observation 1.2.66

Let $\mathcal{C}$ be a simplicially enriched category. Let us now investigate the simplicial set $N(\mathcal{C})$ in more detail. For this purpose, we recall that

$$N(\mathcal{C})_n = \mathrm{Hom}_{\mathrm{Cat}_\Delta}(\mathfrak{C}[\Delta^n], \mathcal{C}).$$

(continued)

1.2.66 (continued)

Unravelling the simplicial categories $\mathfrak{C}[\Delta^0]$ and $\mathfrak{C}[\Delta^1]$, we find:

(1) $\mathfrak{C}[\Delta^0]$ is a simplicial category with a single object, whose morphism-simplicial set is given by $N(P_{0,0})$ which is also given by Δ^0.
(2) $\mathfrak{C}[\Delta^1]$ is a simplicial category with two objects, 0 and 1, and all morphism-simplicial sets are given by Δ^0.

In other words, both $\mathfrak{C}[\Delta^0]$ and $\mathfrak{C}[\Delta^1]$ are given by $c[0]$ and $c[1]$, where we view $[0]$ and $[1]$ as categories and then apply the constant functor $c \colon \mathrm{Cat} \to \mathrm{Cat}_\Delta$. As a consequence, we obtain

(1) $N(\mathcal{C})_0 = \mathrm{Hom}_{\mathrm{Cat}_\Delta}(\mathfrak{C}[\Delta^0], \mathcal{C}) \cong \mathrm{Hom}_{\mathrm{Cat}}([0], u\mathcal{C})$, and
(2) $N(\mathcal{C})_1 = \mathrm{Hom}_{\mathrm{Cat}_\Delta}(\mathfrak{C}[\Delta^1], \mathcal{C}) \cong \mathrm{Hom}_{\mathrm{Cat}}([1], u\mathcal{C})$,

by Corollary 1.2.51. In words, objects of $N(\mathcal{C})$ are given by the objects of $\mathcal{C}$, and morphisms of $N(\mathcal{C})$ are given by the morphisms (i.e., the 0-simplices of the hom-simplicial sets) of $\mathcal{C}$.

Let us go one step further and analyze the simplicial category $\mathfrak{C}[\Delta^2]$, whose objects are given by $\{0, 1, 2\}$. All endomorphism-simplicial sets are given by Δ^0. Furthermore,

$$\mathrm{Hom}_{\mathfrak{C}[\Delta^2]}(0, 1) = \Delta^0 = \mathrm{Hom}_{\mathfrak{C}[\Delta^2]}(1, 2).$$

However, in order to analyze the hom-simplicial set from 0 to 2, we have to investigate $N(P_{0,2})$. By definition, $P_{0,2}$ is the partially ordered set of subsets of $\{0 < 1 < 2\}$ which contain 0 and 2. There are precisely two such subsets, so that we obtain $P_{0,2} = [1]$. In particular,

$$\mathrm{Hom}_{\mathfrak{C}[\Delta^2]}(0, 2) = N(P_{0,2}) = \Delta^1.$$

Subsequently, we find that a 2-simplex in $N(\mathcal{C})$ consists of the following data: objects X, Y, and Z (associated with the three objects 0, 1, and 2); a morphism $f \colon X \to Y$ (associated with the unique morphism from 0 to 1 in $\mathfrak{C}[\Delta^2]$) and a morphism $g \colon Y \to Z$ (associated with the unique morphism from 1 to 2 in $\mathfrak{C}[\Delta^2]$); a 1-simplex $\Delta^1 \to \mathrm{Hom}_\mathcal{C}(X, Z)$ (associated with the hom-simplicial set between 0 and 2 in $\mathfrak{C}[\Delta^2]$) whose restriction to 0 is given by gf and whose restriction to 1 is given by some other morphism.

Informally, a 2-simplex thus consists of the data of two composable morphisms $X \to Y$ and $Y \to Z$, a further morphism $X \to Z$ and a homotopy between the composite of the first two morphisms to the last morphism.

Lemma 1.2.67

The category Cat_Δ *of simplicially enriched categories admits all small colimits. Hence, there exists a unique colimit-preserving functor*

$$\mathfrak{C}[-]\colon \mathrm{sSet} \to \mathrm{Cat}_\Delta$$

which sends Δ^n *to* $\mathfrak{C}[\Delta^n]$. *This functor is automatically left-adjoint to the simplicial nerve functor.*

Proof The existence of colimits was dealt with in Corollary 1.2.45. The rest of the proof is formal, and we have seen the argument several times. □

Fact 1.2.68 Given a simplicial set X, consider sub-simplicial sets $A_i \subseteq X$. Then the union $A = \bigcup A_i$ is also a sub-simplicial set of X. In this situation, we find that $\mathfrak{C}[A]$ is the sub-simplicial category of $\mathfrak{C}[X]$ generated by the $\mathfrak{C}[A_i]$. Recall that a sub-simplicial category of a simplicial category $\mathcal{C}$ is determined by a subset of the objects and, for any two such objects, a sub-simplicial set of the hom-simplicial set in $\mathcal{C}$.

This fact is, e.g., shown in [Joy07, Corollary 1.15]. It builds fundamentally on the fact that the functor $\mathfrak{C}[-]\colon \mathrm{sSet} \to \mathrm{Cat}_\Delta$ sends monomorphisms of simplicial sets to monomorphisms of simplicial categories.

Lemma 1.2.69

Let $0 < j < n$ *and consider the horn* Λ^n_j. *We have that* $\mathfrak{C}[\Lambda^n_j]$ *is the sub-simplicial category of* $\mathfrak{C}[\Delta^n]$ *with the following properties:*

(1) The objects of $\mathfrak{C}[\Lambda^n_j]$ *are given by the vertices of* Λ^n_j *and, therefore, by all objects of* $\mathfrak{C}[\Delta^n]$.

(2) The hom-simplicial sets are given as follows:

$$\mathrm{Hom}_{\mathfrak{C}[\Lambda^n_j]}(i, k) = \mathrm{Hom}_{\mathfrak{C}[\Delta^n]}(i, k)$$

unless $(i, k) = (0, n)$, *and*

$$\mathrm{Hom}_{\mathfrak{C}[\Lambda^n_j]}(0, n) \subseteq \mathrm{Hom}_{\mathfrak{C}[\Delta^n]}(0, n) = \mathrm{N}(P_{0,n})$$

is given by the sub-simplicial set of $(\Delta^1)^{n-1}$ *obtained by deleting the interior and the bottom j-face.*

Proof Let us first show that the candidate for $\mathfrak{C}[\Lambda_j^n]$ is in fact a sub-simplicial category of $\mathfrak{C}[\Delta^n]$. We know that $\Lambda_j^n = \bigcup_{i \neq j} \Delta^{[n] \setminus \{i\}}$. So let us first describe the sub-simplicial category

$$\mathfrak{C}[\Delta^{[n] \setminus \{i\}}] \subseteq \mathfrak{C}[\Delta^n].$$

The objects of this subcategory are given by all objects except for the object corresponding to $i \in [n]$. Now assume $k, l \in [n] \setminus \{i\}$ with $k \leq l$. If $l < i$, then there is an obvious isomorphism

$$P_{k,l}^{[n] \setminus \{i\}} \cong P_{k,l}^{[n]}$$

where the superscript indicates in which linearly ordered set to perform the construction $P_{k,l}$ of Definition 1.2.54. Thus we obtain an equality of hom-simplicial sets

$$\mathrm{Hom}_{\mathfrak{C}[\Delta^{[n] \setminus \{i\}}]}(k, l) = \mathrm{Hom}_{\mathfrak{C}[\Delta^n]}(k, l).$$

Likewise, there is such an isomorphism for $k > i$. Now let $(k, l) \neq (0, n)$, i.e., either $k \neq 0$ or $l \neq n$. Assume first that $k \neq 0$. Then we have the following chain on inclusions of simplicial sets:

$$\mathrm{Hom}_{\mathfrak{C}[\Delta^{[n] \setminus \{0\}}]}(k, l) \subseteq \mathrm{Hom}_{\mathfrak{C}[\Lambda_j^n]}(k, l) \subseteq \mathrm{Hom}_{\mathfrak{C}[\Delta^n]}(k, l)$$

where the composition is an equality by the previous arguments. Therefore, we find that both inclusions must in fact be equalities. If $l \neq n$, the same argument works. Summing up, we find that

$$\mathrm{Hom}_{\mathfrak{C}[\Lambda_j^n]}(k, l) = \mathrm{Hom}_{\mathfrak{C}[\Delta^n]}(k, l)$$

unless $(k, l) = (0, n)$.

Next, let us determine $\mathrm{Hom}_{\mathfrak{C}[\Lambda_j^n]}(0, n)$. By Fact 1.2.68, we have for every $i \neq j$ and $i \neq 0, n$ inclusions as follows:

$$\mathrm{Hom}_{\mathfrak{C}[\Delta^{[n] \setminus \{i\}}]}(0, n) \subseteq \mathrm{Hom}_{\mathfrak{C}[\Lambda_j^n]}(0, n) \subseteq \mathrm{Hom}_{\mathfrak{C}[\Delta^n]}(0, n).$$

For $0 < i < n$, we find that

$$P_{0,n}^{[n] \setminus \{i\}} \subseteq P_{0,n}^{[n]}$$

is a sub-poset, consisting precisely of those $I \in P_{k,l}^{[n]}$ for which $i \notin I$. In other words, it is given by the map of posets

$$[1]^{\times (n-2)} \to [1]^{\times (n-1)}$$

which includes a 0 at the ith spot of $[1]^{\times(n-1)}$. Upon taking the nerve, this yields the map $(\Delta^1)^{n-2} \subseteq (\Delta^1)^{n-1}$, which is the zero-vertex of Δ^1 in the ith coordinate. It follows that for every $i \neq j$, the face $(\Delta^1)^{n-2} \subseteq (\Delta^1)^{n-1}$ with the ith entry being 0 is contained in $\mathrm{Hom}_{\mathfrak{C}[\Lambda^n_j]}(0, n)$. Let us call the face with the ith entry being 0 the *bottom i-face*. Then we find that, a priori, the bottom j-face is not contained in $\mathrm{Hom}_{\mathfrak{C}[\Lambda^n_j]}(0, n)$, as promised in the statement of the lemma. It remains to show that the top k-face is contained in $\mathrm{Hom}_{\mathfrak{C}[\Lambda^n_j]}(0, n)$ for all $0 < k < n$. For this, we consider the diagram

$$
\begin{array}{ccc}
\mathrm{Hom}_{\mathfrak{C}[\Lambda^n_j]}(0, k) \times \mathrm{Hom}_{\mathfrak{C}[\Lambda^n_j]}(k, n) & \longrightarrow & \mathrm{Hom}_{\mathfrak{C}[\Lambda^n_j]}(0, n) \\
\Big\downarrow{\scriptstyle\cong} & & \Big\downarrow \\
\mathrm{Hom}_{\mathfrak{C}[\Delta^n]}(0, k) \times \mathrm{Hom}_{\mathfrak{C}[\Delta^n]}(k, n) & \longrightarrow & \mathrm{Hom}_{\mathfrak{C}[\Delta^n]}(0, n)
\end{array}
$$

which encodes the composition in the respective categories. This diagram must commute, as $\mathfrak{C}[\Lambda^n_j]$ is a sub-simplicial category of $\mathfrak{C}[\Delta^n]$. Since the left vertical map is an isomorphism, it hence suffices to show that the top i-face is contained in the image of the lower horizontal map. This map is induced by the map of posets

$$
P_{0,k} \times P_{k,n} \to P_{0,n}
$$

which is induced by sending (I, I') to $I \cup I'$. Since both I and I' contain the element k, after identifying these posets with cubes as in Lemma 1.2.57, we obtain the map

$$
(\Delta^1)^{n-2} \cong (\Delta^1)^{k-1} \times (\Delta^1)^{n-k-1} \to (\Delta^1)^{n-1}
$$

which is induced by inserting a 1 in the kth slot. This shows that the top k-face is contained in $\mathrm{Hom}_{\mathfrak{C}[\Lambda^n_j]}(0, n)$. Since we already know that this punctured cube gives rise to a sub-simplicial category, and that $\mathfrak{C}[\Lambda^n_j]$ is contained in it and itself contains the sub-simplicial categories determined by the i-faces of Δ^n for $i \neq j$, the lemma is proved. $\square$

Lemma 1.2.70

The coherent nerve of a simplicial category $\mathfrak{C}$ is a composer. Furthermore, it is an ∞-category if all hom-simplicial sets are Kan complexes, i.e., if it is in fact enriched in the symmetric monoidal category of Kan complexes.

Proof We consider the extension problem

which, by adjunction, is equivalent to the extension problem

In order to show that the latter can be solved, it suffices to show that the map $\mathfrak{C}[I^n] \to \mathfrak{C}[\Delta^n]$ has a retraction. To see this, we first claim that $\mathfrak{C}[I^n] \cong c[n]$. This follows simply from the fact that it is true for $n = 1$ and subsequent induction (using the fact that $I^n \cong I^{n-1} \amalg_{\Delta^0} I^1$ and that the functor $c \colon \mathrm{Cat} \to \mathrm{Cat}_\Delta$ preserves colimits). In Lemma 1.2.61, we have seen that there is a unique simplicial functor $\mathfrak{C}[\Delta^n] \to c[n]$ which is the identity on objects. It follows that the composition

$$c[n] \cong \mathfrak{C}[I^n] \to \mathfrak{C}[\Delta^n] \to c[n]$$

is a functor which is the identity on objects, and therefore an isomorphism.

In order to show that, in the case of a Kan enrichment, $N(\mathcal{C})$ is an ∞-category, we need to solve the extension problem

$$
\begin{array}{ccc}
\Lambda^n_j & \xrightarrow{\hspace{1cm}} & N(\mathcal{C}) \\
\downarrow & \nearrow & \\
\Delta^n & &
\end{array}
$$

for $0 < j < n$.

By adjunction, we need to argue why every simplicially enriched functor $\mathfrak{C}[\Lambda^n_j] \to \mathcal{C}$ extends to a simplicially enriched functor $\mathfrak{C}[\Delta^n] \to \mathcal{C}$, provided that $0 < j < n$. For this purpose, we use our analysis of $\mathfrak{C}[\Lambda^n_j]$ in Lemma 1.2.69.

In order to solve the extension problem which we are interested in, it thus suffices to prove that there exists an extension in the diagram

$$
\begin{array}{ccc}
\mathrm{Hom}_{\mathfrak{C}[\Lambda^n_j]}(0, n) & \xrightarrow{\quad f \quad} & \mathrm{Hom}_{\mathcal{C}}(f(0), f(n)) \\
\downarrow & \nearrow & \\
\mathrm{Hom}_{\mathfrak{C}[\Delta^n]}(0, n) & &
\end{array}
$$

From the description of Lemma 1.2.69 and Fact 1.1.48, it follows that the vertical map is an anodyne map of simplicial sets. (We will also give an independent argument in Lemma 1.3.31.) Hence, the dotted arrow exists. It now suffices to show that this construction in fact produces a simplicial functor $\mathfrak{C}[\Delta^n] \to \mathcal{C}$. To see this, one uses that $\mathrm{Hom}_{\mathfrak{C}[\Delta^n]}(0, 0) = \Delta^0 = \mathrm{Hom}_{\mathfrak{C}[\Delta^n]}(n, n)$, so that no new relations for functoriality are to be checked: Since the diagram

$$\operatorname{Hom}_{\mathfrak{C}[\Lambda_j^n]}(0, k) \times \operatorname{Hom}_{\mathfrak{C}[\Lambda_j^n]}(k, n) \longrightarrow \operatorname{Hom}_{\mathfrak{C}[\Lambda_j^n]}(0, n)$$

$$\Big\downarrow \cong \qquad\qquad\qquad\qquad \Big\downarrow$$

$$\operatorname{Hom}_{\mathfrak{C}[\Delta^n]}(0, k) \times \operatorname{Hom}_{\mathfrak{C}[\Delta^n]}(k, n) \longrightarrow \operatorname{Hom}_{\mathfrak{C}[\Delta^n]}(0, n)$$

commutes, any composition which factors through the object k, with $0 < k < n$, is already contained in the sub-simplicial category $\mathfrak{C}[\Lambda_j^n]$. □

Observation 1.2.71

By applying the nerve, we find that a strict 2-category gives rise to a simplicial category, so that its coherent nerve is a composer. Moreover, a strict $(2, 1)$-category gives rise to a simplicial category where all hom-simplicial sets are nerves of groupoids and thus Kan complexes by Lemma 1.1.54. In particular, a strict $(2, 1)$-category gives rise to an ∞-category in our sense.

Definition 1.2.72

Consider the simplicial category with CW-complexes as objects and hom-simplicial sets given by the singular set of the mapping space. (Notice that the singular set commutes with products.) The simplicial nerve of this category yields an ∞-category which we call the ∞-category of spaces and denote by Spc.

Observation 1.2.73

Objects of Spc are CW-complexes, and morphisms are given by points in the space $\mathrm{map}(X, Y)$, i.e., by a continuous map from X to Y. The homotopy category is what one would expect: Morphisms are homotopy classes of maps.

Remark 1.2.74

We would like to have a "purely simplicial" model of the ∞-category of spaces (what a perverse thing to say—but it comes from the fact that we wish to think of Kan complexes/spaces as ∞-groupoids and later want to have spaces and ∞-categories on equal footing), i.e., a model where we directly construct a simplicial category whose objects are Kan complexes. For this purpose, we will need to show that for a simplicial set K and a Kan complex X, the simplicial set of maps $\underline{\operatorname{Hom}}(K, X)$ is again a Kan complex. The requirements for showing this are also needed on the way of showing that ∞-groupoids are Kan complexes, and we will start to develop these tools in the next section.

Lemma 1.2.75

The product of ∞-categories is an ∞-category, and the coproduct of ∞-categories is also an ∞-category.

Proof For products, one can solve the extension problem in every ∞-category individually, which provides an extension for the product. For coproducts, we observe that both Λ_j^n and Δ^n are connected. Therefore, an extension problem for a coproduct of ∞-categories is in fact an extension problem for a single one.   □

Definition 1.2.76

A sub-∞-category $\mathcal{C}'$ of an ∞-category $\mathcal{C}$ is a sub-simplicial set determined by a subset $X \subseteq \mathcal{C}_0$ of 0-simplices and a subset $S \subseteq \mathcal{C}_1$ of 1-simplices between objects lying in X, such that S contains identities and is closed under compositions and equivalences. Then an n-simplex of $\mathcal{C}$ belongs to $\mathcal{C}'$ if and only if the edges of its restriction to the spine I^n are contained in S. A subcategory is called *full* (on objects $X \subseteq \mathcal{C}_0$) if S equals the set of all 1-simplices whose boundary lies in X.

Lemma 1.2.77

A sub-∞-category of an ∞-category is itself an ∞-category. Its homotopy category is the subcategory of $h\mathcal{C}$ on the image of the morphisms lying in S. The diagram

$$
\begin{array}{ccc}
\mathcal{C}' & \longrightarrow & \mathcal{C} \\
\downarrow & & \downarrow \\
\mathrm{N}(h\mathcal{C}') & \longrightarrow & \mathrm{N}(h\mathcal{C})
\end{array}
$$

is a pullback. Furthermore, for any subcategory $(h\mathcal{C})' \subseteq h\mathcal{C}$ of the homotopy category, this pullback defines a sub-∞-category of $\mathcal{C}$ with X and S given by the preimage of objects and morphisms along the canonical map $\mathcal{C} \to \mathrm{N}(h\mathcal{C})$.

Proof Let $\mathcal{C}$ be an ∞-category and let $\mathcal{D} \subseteq h\mathcal{C}$ be a subcategory of its homotopy category. Let $\mathcal{C}'$ be the pullback $\mathrm{N}(\mathcal{D}) \times_{\mathrm{N}(h\mathcal{C})} \mathcal{C}$. Since pullbacks preserve monomorphisms, $\mathcal{C}'$ is a sub-simplicial set of $\mathcal{C}$. Let us spell out what it means for an n-simplex of $\mathcal{C}$ to lie in $\mathcal{C}'$: It means precisely that the induced n-simplex of $\mathrm{N}(h\mathcal{C})$ lies in $\mathrm{N}(\mathcal{D})$. This is the case if and only if the spine, and thus every edge of the simplex, lies in $\mathcal{D}$. Now observe that a subset $S \subseteq \mathcal{C}_1$ which contains all identities of objects in X and is closed under compositions is in fact the preimage of the 1-morphisms of the nerve of a subcategory of $h\mathcal{C}$. Notice that in the above definition, we have that $(\mathcal{C}')_1 = S$.   □

Corollary 1.2.78

There is a bijective correspondence between sub-∞-categories of $\mathcal{C}$ and subcategories of $h\mathcal{C}$ induced by taking the homotopy category. Full sub-∞-categories of $\mathcal{C}$ correspond precisely to full subcategories of $h\mathcal{C}$.

Definition 1.2.79

A *natural transformation* between two functors $f, g \colon \mathcal{C} \to \mathcal{D}$ is a simplicial map $\mathcal{C} \times \Delta^1 \to \mathcal{D}$ which restricts appropriately to the given functors.

Observation 1.2.80

Functors and natural transformations of functors between X and Y are precisely the 0- and 1-simplices of the hom-simplicial set $\underline{\mathrm{Hom}}(X, Y)$. As in ordinary category theory, we would like to have an ∞-category of functors between two ∞-categories. Also, one would expect $\mathrm{N}(\mathrm{Fun}(\mathcal{C}, \mathcal{D})) = \mathrm{Fun}(\mathrm{N}\mathcal{C}, \mathrm{N}\mathcal{D})$. Since one also has $\mathrm{N}(\mathrm{Fun}(\mathcal{C}, \mathcal{D})) = \underline{\mathrm{Hom}}(\mathrm{N}\mathcal{C}, \mathrm{N}\mathcal{D})$ (due to the nerve functor being fully faithful), one might hope that the hom-simplicial set is again an ∞-category. This will turn out to be true, and like in the case of Kan complexes, it is not a triviality. One of the objectives of the next sections is to prove this fact.

1.3 Anodyne Maps and Fibrations

The goal of this section is to set up some combinatorial notions which will allow us to prove that ∞-groupoids, i.e., ∞-categories in which every morphism is invertible, are in fact Kan complexes. These methods are central to our definition of the ∞-category of spaces and of the ∞-category of ∞-categories. This section can be seen as one of the technical hearts of the theory, and we try to give a concise treatment. We start with the notion of saturated sets and discuss a special case of what is called the small object argument of Quillen. Then we will use these findings to prove that certain natural operations which one can perform in simplicial sets preserve ∞-categories and ∞-groupoids, most notably Theorem 1.3.37 and its consequences. From this, we shall obtain, for any two objects of an ∞-category, an ∞-category of morphisms between them in a canonical way. It will be the content of the following sections to show that these ∞-categories of morphisms are in fact ∞-groupoids.

Definition 1.3.1

A map of simplicial sets $X \to Y$ is an (inner, left, right) *fibration* if it satisfies the right lifting property (RLP) with respect to (inner, left, right) horn inclusions as depicted

by the diagram

Definition 1.3.2

A map of simplicial sets $A \to B$ is an (inner, left, right) *anodyne map*, if it satisfies the left lifting property (LLP) with respect to (inner, left, right) fibrations as depicted by the diagram

---

Definition 1.3.3

Let S be a set of morphisms in a category $\mathcal{C}$. We let $\chi_R(S)$ be the set of morphisms having the RLP with respect to S, and we let $\chi_L(S)$ be the set of morphisms having the LLP with respect to S. Furthemore, we let $\chi(S) = \chi_L(\chi_R(S))$, i.e., $\chi(S)$ is the set of morphisms which have the LLP with respect to morphisms having the RLP with respect to S.

Example Let S be the set of (inner, left, right) horn inclusions. Then the (inner, left, right) fibrations are given by $\chi_R(S)$, and the (inner-, left-, right-) anodyne maps are given by $\chi(S)$.

Next, we want to introduce the notion of saturated sets of maps of simplicial sets. For this purpose, we need the following definition.

Definition 1.3.4

Let $\mathcal{C}$ be a category. A morphism $f : A \to B$ is a *retract* of a morphism $f' : A' \to B'$ if there is a commutative diagram

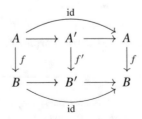

Definition 1.3.5

A *saturated set* of morphisms of simplicial sets is a set of morphisms T which is closed under taking pushouts (along arbitrary maps), arbitrary coproducts, countable compositions (i.e., colimits along $\mathbb{N}$), and retracts. Given an arbitrary set of morphisms S, we call the smallest saturated set containing S the *saturated closure* of S, and we denote it by $\overline{S}$.

Remark 1.3.6

Let us state more precisely what the conditions of Definition 1.3.5 mean: Being closed under arbitrary coproducts means that given a family $\{f_i\}_{i \in I}$ such that each f_i is an element of T, then also $\coprod_{i \in I} f_i$ is an element of T; and being closed under pushouts means that given a map $f \colon A \to B$ belonging to T and any other map $\varphi \colon A \to A'$, then the map f' in the following pushout diagram also belongs to T:

$$
\begin{array}{ccc}
A & \xrightarrow{\ \varphi\ } & A' \\
\downarrow{\scriptstyle f} & & \downarrow{\scriptstyle f} \\
B & \longrightarrow & B'
\end{array}
$$

The set T is called closed under countable compositions if for all diagrams $X \colon \mathbb{N} \to$ sSet the following condition holds: If, for every $i \in \mathbb{N}$, the map

$$X(i) \to X(i+1)$$

is contained is S, then the map $X(0) \to \operatorname{colim}_{\mathbb{N}} X$ is also contained in S. Finally, T is called closed under retracts if for f and f' as in Definition 1.3.4, if f' is an element of T, then so is f.

Remark 1.3.7

Notice that the intersection of saturated sets is again saturated. Therefore, in order to see that the saturated closure exists, it suffices to show that there is a saturated set containing S. One can, e.g., simply take the set of all morphisms, which is obviously saturated and contains S.

Example The set of monomorphisms in sSet is a saturated set.

Lemma 1.3.8

Given a set of morphisms S, we have that $\chi_L(S)$ is a saturated set. In particular, $\chi(S)$ is a saturated set.

Proof Let $\alpha\colon A \to B$ be a morphism in $\chi_L(S)$ and let $\varphi\colon A \to A'$ be an arbitrary morphism. Consider the pushout $B' = A' \amalg_A B$. We wish to show that the canonical map $A' \to B'$ is contained in $\chi_L(S)$. So let $f\colon X \to Y$ be a map in S and consider the diagram

$$
\begin{array}{ccccc}
A & \longrightarrow & A' & \dashrightarrow & X \\
\downarrow & & \downarrow & \nearrow & \downarrow \\
B & \longrightarrow & B' & \longrightarrow & Y
\end{array}
$$

where we wish to construct the dashed map. By assumption, we have the dotted map, and hence obtain the dashed map by the universal property of a pushout. Likewise, suppose that $A \to B$ is a retract of $A' \to B'$ and that $A' \to B'$ is contained in $\chi_L(S)$. In order to show that then $A \to B$ is also contained in $\chi_L(S)$, we consider a map $f\colon X \to Y$ in S and the diagram

$$
\begin{array}{ccccccc}
A & \longrightarrow & A' & \longrightarrow & A & \longrightarrow & X \\
\downarrow & & \downarrow & & \downarrow & \nearrow & \downarrow \\
B & \longrightarrow & B' & \longrightarrow & B & \longrightarrow & Y
\end{array}
$$

where we wish to show that the dashed arrow exists. Again, the dotted arrow exists, which we may restrict to B along the map $B \to B'$.

Suppose now that a family $\{f_i\colon A_i \to B_i\}_{i \in I}$ of elements of $\chi_L(S)$ is given. We then want to show that $\coprod_{i \in I} f_i \in \chi_L(S)$ as well. For this purpose, consider a lifting problem

$$
\begin{array}{ccc}
\coprod_{i \in I} A_i & \longrightarrow & X \\
{\scriptstyle \coprod_{i \in I} f_i} \downarrow & \nearrow & \downarrow {\scriptstyle \varphi} \\
\coprod_{i \in I} B_i & \longrightarrow & Y
\end{array}
$$

with $\varphi\colon X \to Y$ being an element of S. Then the dashed arrow exists simply by the universal property of coproducts and the assumption that each f_i is an element of $\chi_L(S)$.

It remains to show that for any diagram $A\colon \mathbb{N} \to \mathcal{C}$, where each map $A_i \to A_{i+1}$ is contained in $\chi_L(S)$, the map $A_0 \to A = \operatorname{colim}_i X_i$ is also contained in $\chi_L(S)$. This follows simply from the universal property of colimits: We consider again $f\colon X \to Y$ in S and a diagram

$$
\begin{array}{ccc}
A_0 & \longrightarrow & X \\
\downarrow & \nearrow & \downarrow \\
A & \longrightarrow & Y
\end{array}
$$

where we need to show existence of the dashed arrow. Since this is a map out of a colimit, it suffices to show that there exist compatible maps $A_i \to X$ which make everything commute. But this can be done inductively, using the fact that each map $A_i \to A_{i+1}$ is contained in $\chi_L(S)$. $\hfill\square$

The following result is known as the *small object argument*. It is a very useful tool for constructing non-trivial factorizations of maps.

Proposition 1.3.9

Let S be a set of morphisms $\{A_i \to B_i\}_{i \in I}$ of simplicial sets such that, for every $i \in I$, the simplicial set A_i has only finitely many non-degenerate simplices. If f is an arbitrary morphism, then f can be factored into a map contained in $\overline{S}$ followed by a map which has the RLP with respect to S.

Proof First, consider the set Θ_S which consists of triples (α_i, u_i, v_i), where $\alpha_i : A_i \to B_i$ is an element of S, and where $u_i : A_i \to X$ and $v_i : B_i \to Y$ are maps such that the diagram

$$
\begin{array}{ccc}
A_i & \xrightarrow{\ u_i\ } & X \\
\downarrow{\scriptstyle \alpha_i} & & \downarrow{\scriptstyle f} \\
B_i & \xrightarrow{\ v_i\ } & Y
\end{array}
$$

commutes. We obtain a commutative diagram

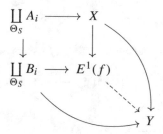

where $E^1(f)$ is defined to be the pushout. Since $\overline{S}$ is closed under small coproducts and pushouts, we can deduce that the map $X \to E^1(f)$ is contained in $\overline{S}$.

Doing the same with the map $E^1(f) \to Y$ instead of the map $f : X \to Y$, we obtain a sequence

$$
X \to E^1(f) \to E^2(f) \to \cdots \to Y.
$$

Let us define $E^\omega(f) = \mathrm{colim}_k E^k(f)$, where ω is used as the notation for the first limit ordinal, so that we obtain a factorization

$$
X \to E^\omega(f) \to Y.
$$

By construction, every map $E^k(f) \to E^{k+1}(f)$ is contained in $\overline{S}$ because $\overline{S}$ is closed under pushouts and coproducts. Hence, since $\overline{S}$ is saturated, the map $X \to E^\omega(f)$ is also contained in $\overline{S}$. It remains to show that the map $E^\omega(f) \to Y$ is contained in $\chi_R(S)$. For this purpose, let us consider a map $A_i \to B_i$ in S and a diagram

$$
\begin{array}{ccc}
A_i & \longrightarrow & E^\omega(f) \\
\downarrow & \nearrow & \downarrow \\
B_i & \longrightarrow & Y
\end{array}
$$

where we wish to show the existence of the dashed arrow, which would make both triangles commute. We now claim that the canonical map

$$
\operatorname*{colim}_{k \in \mathbb{N}} \operatorname{Hom}_{\mathrm{sSet}}(A_i, E^k(f)) \to \operatorname{Hom}_{\mathrm{sSet}}(A_i, E^\omega(f))
$$

is a bijection. (In fact, such A_i are *compact*, i.e., $\operatorname{Hom}_{\mathrm{sSet}}(A_i, -)$ commutes with general filtered colimits.) This follows from the fact that there are only finitely many non-degenerate simplices in A_i, and that any simplicial map is determined on the non-degenerate simplices. Hence, we find a $k \in \mathbb{N}$ such that the given map $A_i \to E^\omega(f)$ factors into a composition

$$
A_i \to E^k(f) \to E^\omega(f).
$$

Now the diagram

$$
\begin{array}{ccc}
A_i & \longrightarrow E^k(f) \longrightarrow & E^\omega(f) \\
\downarrow & & \downarrow \\
B_i & \longrightarrow & Y
\end{array}
$$

commutes, and thus by the very definition of $E^{k+1}(f)$, there exists a commutative diagram

$$
\begin{array}{cccc}
A_i & \longrightarrow E^k(f) & \longrightarrow & E^\omega(f) \\
\downarrow & \downarrow & \nearrow & \downarrow \\
B_i & \longrightarrow E^{k+1}(f) & \longrightarrow & Y
\end{array}
$$

which solves our lifting problem. $\square$

Remark 1.3.10

The simplicial sets A_i for a small object argument in this book will have only finitely many non-degenerate simplices. In general, we would have to find a regular cardinal κ which is larger that the cardinality of any of the A_i's appearing in S (which ensures

(continued)

1.3.10 (continued)

that all A_i are indeed κ-compact). Moreover, we would have to define a saturated set to be closed under colimits indexed over arbitrary ordinals, not just over the ordinal ω. Then we could continue the above inductive process, namely considering pushouts for successor ordinals as before, and taking a colimit for limit ordinals as before. At some point, one would have defined $E^\kappa(f)$ which sits in a factorization $X \to E^\kappa(f) \to Y$, and a similar argument as before would show that the first map is contained in $\overline{S}$ (with the new definition of saturated sets) and that the latter map is contained in $\chi_R(S)$. We chose to not deal with colimits over ordinals in this text, since we will not need it. But it is of course useful to know that the small object argument does not depend on such size issues.

Remark 1.3.11

The factorization as described above is functorial: Whenever given a commutative diagram

$$
\begin{array}{ccc}
X & \xrightarrow{\ f\ } & Y \\
\downarrow & & \downarrow \\
X' & \xrightarrow{\ f'\ } & Y'
\end{array}
$$

then the small object argument in fact provides a commutative diagram

$$
\begin{array}{ccccc}
X & \longrightarrow & E^\omega(f) & \longrightarrow & Y \\
\downarrow & & \downarrow & & \downarrow \\
X' & \longrightarrow & E^\omega(f') & \longrightarrow & Y'
\end{array}
$$

With the help of the small object argument, we can now give a proof of Proposition 1.2.17. (Note that it is also possible to give a direct proof.)

Proposition 1.3.12

The saturated set generated by spine inclusions is not equal to the inner-anodyne maps. Moreover, there exists a composer which is not an ∞-category.

Proof Consider the inclusion $\Lambda_1^3 \to \Delta^3$. By the small object argument, we can factor this map into a composition

$$\Lambda_1^3 \to X \to \Delta^3$$

such that the first map is in the saturated class generated by the spine inclusions, and the last map satisfies the RLP with respect to spine inclusions, and hence with respect to the saturated set generated by the spine inclusions. It follows that X is a composer, because the map $\Delta^3 \to \Delta^0$ also has the RLP with respect to spine inclusions. If the saturated set generated by the spine inclusions equaled the inner-anodyne maps, it would follow that the map $X \to \Delta^3$ is an inner fibration (since it has the RLP with respect to inner-anodyne maps). We will show that this is not the case. For this purpose,

we claim that the lifting problem

$$
\begin{array}{ccc}
\Lambda_1^3 & \longrightarrow & X \\
\downarrow & \nearrow & \downarrow \\
\Delta^3 & = & \Delta^3
\end{array}
$$

does not have a solution. This is readily proved via induction over the filtration on X obtained from the small object argument, using the following observation: Given a pushout

$$
\begin{array}{ccc}
I^n & \longrightarrow & A \\
\downarrow & & \downarrow \\
\Delta^n & \longrightarrow & B
\end{array}
$$

of simplicial sets with $n \geq 3$, it follows that the image of $\Lambda_j^n \subseteq \Delta^n$ in B is not contained in A.

If X is an ∞-category, then the map $X \to \Delta^3$ would indeed be an inner fibration, see Exercise 51. But we have shown that it is not.   □

Remark 1.3.13

The proposition also follows from a different argument: Combining Exercise 43 and Lemma 1.2.70, we see that there exists a composer X which is not an ∞-category. Now the map $X \to \Delta^0$ is contained in $\chi_R(\{I^n \to \Delta^n\}_{n \in \mathbb{N}})$, but it is not an inner fibration. This shows that the saturated closure of the spine inclusions cannot be given by the inner-anodyne maps, since otherwise any map that satisfies the RLP with respect to the spine inclusions would also be an inner fibration.

Notice that the sheer fact that the saturated closure of the spine inclusions is not equal to the inner-anodyne maps does not formally imply the existence of a composer which is not an ∞-category. However, the converse holds: The existence of a composer which is not an ∞-category shows that the saturated set generated by spine inclusions cannot contain all inner-anodyne maps.

Remark 1.3.14

We will show in Proposition 1.3.22 that spine inclusions are inner-anodyne, so that Proposition 1.3.12 can be restated as saying that the saturated set generated by spine inclusions is strictly contained in the inner-anodyne maps.

Lemma 1.3.15

Consider a set of morphisms $S = \{A_i \to B_i\}_{i \in I}$ *such that all* A_i *have only finitely many non-degenerate simplices. Then the saturated closure* $\overline{S}$ *of* S *is given by* $\chi(S)$.

Proof Obviously, we have $S \subseteq \chi(S)$. In Lemma 1.3.8, we have shown that $\chi(S)$ is itself saturated, so we have $\overline{S} \subseteq \chi(S)$. In order to prove the converse, consider a map $f : x \to y$ with $f \in \chi(S)$. By the small object argument from Proposition 1.3.9, we find a factorization of this map,

$$
\begin{array}{ccc}
x & \xrightarrow{\ j\ } & z \\
{\scriptstyle f}\downarrow & \nearrow{\scriptstyle \alpha} & \downarrow{\scriptstyle p} \\
y & =\!=\!= & y
\end{array}
$$

where $j \in \overline{S}$ and p satisfies the RLP with respect to S. Since $f \in \chi(S)$, it follows that there exists a dashed arrow α, which makes the diagram commute. Hence, we have a commutative diagram

$$
\begin{array}{ccccc}
x & =\!=\!= & x & =\!=\!= & x \\
{\scriptstyle f}\downarrow & & {\scriptstyle j}\downarrow & & \downarrow{\scriptstyle f} \\
y & \xrightarrow{\ \alpha\ } & z & \xrightarrow{\ p\ } & y
\end{array}
$$

which shows that f is a retract of j. Since $j \in \overline{S}$, the same applies for f. $\qquad\square$

Corollary 1.3.16

(Inner-, left-, right-) anodyne maps are precisely the saturated closure of the (inner, left, right) horn inclusions. In particular, all of these are monomorphisms.

As an extension of Fact 1.1.48, we also note the following fact.

Fact 1.3.17 A monomorphism is anodyne if and only if its geometric realization is a weak equivalence. This is part of the existence of the Kan–Quillen model structure on simplicial sets [GJ09, Theorem 11.3], which is equivalent to the Quillen model structure on topological spaces.

The following corollary is the relative version of Lemma 1.2.70.

Corollary 1.3.18
Let $F: \mathcal{C} \to \mathcal{D}$ be a morphism of simplicial categories. Assume that for all objects $X, Y \in \mathcal{C}$, the induced map $\mathrm{Hom}_{\mathcal{C}}(X, Y) \to \mathrm{Hom}_{\mathcal{D}}(FX, FY)$ is a Kan fibration. Then $NF: N\mathcal{C} \to N\mathcal{D}$ is an inner fibration.

Corollary 1.3.19
Let $F: \mathcal{C} \to \mathcal{D}$ be a functor between ordinary categories. Then the induced functor $NF: N\mathcal{C} \to N\mathcal{D}$ is an inner fibration.

Proof Given any map between sets $A \to B$, then the induced map of constant simplicial sets $cA \to cB$ is a Kan fibration. This follows from the fact that, for all $n \geq 1$ and all $0 \leq j \leq n$, the map $\pi_0(\Lambda_j^n) \to \pi_0(\Delta^n)$ is an isomorphism. Then use the fact that $N(\mathcal{C}) = N(c\mathcal{C})$ and the previous corollary to complete the proof. $\square$

Remark 1.3.20
In fact, more holds true: In Exercise 51, it is shown that a map of simplicial sets $X \to N(\mathcal{D})$, where $\mathcal{D}$ is an ordinary category, is an inner fibration if and only if X is an ∞-category.

For what comes next, we have to recall the definition of the S-horn Λ_S^n of Definition 1.1.37, so that we have

$$\Lambda_S^n = \bigcup_{s \notin S} \Delta^{[n] \setminus \{s\}}.$$

Lemma 1.3.21
Let $S \subseteq [n]$ be a non-empty subset. Then the map $\Lambda_S^n \to \Delta^n$ is

(1) anodyne, provided that $S \neq [n]$,
(2) left-anodyne, provided that $\{n\} \notin S$,
(3) right-anodyne, provided that $\{0\} \notin S$,
(4) inner-anodyne, provided that S is not the complement of an interval, i.e., there are $a < b < c \in [n]$ with $a, c \notin S$ but $b \in S$.

Proof Let $S \subseteq [n]$ be a non-empty subset, let $s \in S$ and let $S' = S \setminus \{s\}$. Consider the pushout diagram

$$
\begin{array}{ccc}
\Delta^{[n]\setminus\{s\}} \cap \Lambda^n_S & \longrightarrow & \Delta^{[n]\setminus\{s\}} \\
\downarrow & & \downarrow \\
\Lambda^n_S & \longrightarrow \Lambda^n_{S'} \longrightarrow & \Delta^n
\end{array}
$$

Notice that the upper horizontal arrow is a generalized horn inclusion $\Lambda^{[n]\setminus\{s\}}_{S'} \subseteq \Delta^{[n]\setminus\{s\}}$. It follows that the inclusion $\Lambda^n_S \to \Delta^n$ is contained in the smallest saturated set containing the inclusions $\Lambda^{[n]\setminus\{s\}}_{S'} \subseteq \Delta^{[n]\setminus\{s\}}$ and $\Lambda^n_{S'} \subseteq \Delta^n$.

We will prove the lemma by induction over the size of S (for arbitrary $[n]$). Let us first consider (1). If S contains only one element, say $S = \{i\}$, then Λ^n_S is a horn inclusion, and hence anodyne. If S contains more than one element, then S' still contains at least one element and is smaller than S, so that $\Lambda^n_{S'} \subseteq \Delta^n$ and $\Lambda^{[n]\setminus\{s\}}_{S'} \subseteq \Delta^n$ is anodyne by induction.

For (2), suppose that $S = \{i\}$. Then the horn inclusion is left-anodyne, because $i \neq n$. If S contains more than one element, then S' is smaller and still does not contain n. The statement for (3) is similar.

In order to show (4), we first note that if $S = \{i\}$ then $0 < i < n$, else its complement is an interval, so that the inclusion $\Lambda^n_S \to \Delta^n$ is inner-anodyne. If S contains more than one element, then we claim that there exists an element s in S such that $S \setminus \{s\}$ is again not the complement of an interval: By assumption, there are $a < b < c$ such that $b \in S$ and $a, c \notin S$; and also by assumption, we have $S \neq \{b\}$, so we can choose some other element $s \in S \setminus \{b\}$. Then $S' = S \setminus \{s\}$ is again not the complement of an interval (because $b \in S'$). □

Proposition 1.3.22
The spine inclusions $I^n \to \Delta^n$ are inner-anodyne.

Proof The spine inclusion $i_n : I^n \to \Delta^n$ can be factored as follows:

$$
I^n \xrightarrow{f_n} \Delta^{[n]\setminus\{0\}} \cup I^n \xrightarrow{g_n} \Delta^n
$$

We will show by induction on n that both f_n and g_n are inner-anodyne. The induction start $n = 1$ and $n = 2$ is obvious. Now let $n \geq 3$ and consider the pushout diagram

$$
\begin{array}{ccc}
I^{[n]\setminus\{0\}} & \longrightarrow & \Delta^{[n]\setminus\{0\}} \\
\downarrow & & \downarrow \\
I^n & \longrightarrow & \Delta^{[n]\setminus\{0\}} \cup I^n
\end{array}
$$

where the upper composite is inner-anodyne by induction. Hence f_n is inner-anodyne as a pushout of an inner-anodyne map. It remains to show that g_n is also inner-anodyne. In this case, we consider the pushout diagram

$$
\begin{array}{ccc}
\Delta^{[n]\setminus\{0,n\}} \cup I^{[n]\setminus\{n\}} & \xrightarrow{\; g_{n-1}\;} & \Delta^{[n]\setminus\{n\}} \\
\downarrow & & \downarrow \\
\Delta^{[n]\setminus\{0\}} \cup I^n & \longrightarrow & \Delta^{[n]\setminus\{0\}} \cup \Delta^{[n]\setminus\{n\}} & \longrightarrow & \Delta^n
\end{array}
$$

where the upper horizontal map is inner-anodyne by induction. Therefore, the left lower horizontal map is also inner-anodyne. The right lower horizontal map is given by $\Lambda^n_{[n]\setminus\{0,n\}} \subseteq \Delta^n$, which is inner-anodyne by Lemma 1.3.21, part (4), because $\{0, n\}$ is not an interval. □

Corollary 1.3.23
Every ∞-category is a composer.

Proof Obvious from Proposition 1.3.22.   □

Definition 1.3.24
A *trivial fibration* is a map which has the RLP with respect to the boundary inclusions $\partial\Delta^n \to \Delta^n$ for $n \geq 0$.

Definition 1.3.25
Let J be the nerve of the category consisting of two objects with a unique isomorphism between them.

Observation 1.3.26
The category with two objects and a unique isomorphism between them is a classifier for isomorphisms in a category. In other words, the functor corepresented by this category is the functor which assigns to a category its set of isomorphisms.

Given a morphism f in an ∞-category X, one can thus wonder under what circumstances its classifying map $\Delta^1 \to X$ extends over J. It is easy to see that if this is the case, then f is an equivalence (see Exercise 61). The converse turns out to be true as well: It is yet another application of the fact that ∞-groupoids are Kan complexes. (Note that the map $\Delta^1 \to J$ is anodyne.)

Definition 1.3.27

A *Joyal fibration* between ∞-categories is an inner fibration which in addition has the RLP with respect to the map $\Delta^0 \to J$.

Construction 1.3.28   Let $f \colon X \to Y$ and $i \colon A \to B$ be maps of simplicial sets. Then there is a commutative diagram

$$
\begin{array}{ccc}
X^B & \longrightarrow & X^A \\
\downarrow & & \downarrow \\
Y^B & \longrightarrow & Y^A
\end{array}
$$

and therefore, there is an induced map

$$
\langle f, i \rangle \colon X^B \to X^A \times_{Y^A} Y^B.
$$

Construction 1.3.29   Dually, for morphisms $i \colon A \to B$ and $g \colon S \to T$, we obtain a commutative diagram

$$
\begin{array}{ccc}
A \times S & \longrightarrow & A \times T \\
\downarrow & & \downarrow \\
B \times S & \longrightarrow & B \times T
\end{array}
$$

and therefore, there is an induced map

$$
i \boxtimes g \colon A \times T \amalg_{A \times S} B \times S \to B \times T.
$$

Lemma 1.3.30

The following two lifting problems are equivalent:

$$
\begin{array}{ccc}
S & \longrightarrow & X^B \\
\downarrow & \nearrow & \downarrow \\
T & \longrightarrow & X^A \times_{Y^A} Y^B
\end{array}
\qquad
\begin{array}{ccc}
A \times T \amalg_{A \times S} B \times S & \longrightarrow & X \\
\downarrow & \nearrow & \downarrow \\
B \times T & \longrightarrow & Y
\end{array}
$$

Proof This follows immediately from spelling out all definitions and the universal properties of pullbacks, pushouts, and the fact that the functor $(-)^A$ is a right adjoint of the functor $A \times -$. $\square$

The crucial technical lemma about the maps $i \boxtimes g$ is the following.

Lemma 1.3.31
In the above notation, with i and g monomorphisms, we have that

(1) $i \boxtimes g$ is inner-anodyne if i or g is inner-anodyne;
(2) $i \boxtimes g$ is left-anodyne if i or g is left-anodyne,
(3) $i \boxtimes g$ is right-anodyne if i or g is right-anodyne,
(4) $i \boxtimes g$ is anodyne if i or g is anodyne.

In order to prove this lemma, we will need four preliminary steps in the form of Lemma 1.3.32 to Lemma 1.3.35.

Lemma 1.3.32
Let S and T be two sets of morphisms whose domains are all compact. Then $\overline{S} \boxtimes T \subseteq \overline{S \boxtimes T} \subseteq \overline{S} \boxtimes \overline{T}$. In particular, $\overline{\overline{S} \boxtimes T} = \overline{\overline{S} \boxtimes \overline{T}} = \overline{S \boxtimes T}$.

Proof The very first inclusion is obvious. In order to see the second inclusion, we let $\mathcal{F} = \chi_R(S \boxtimes T)$. Then $\overline{S \boxtimes T} = \chi_L(\mathcal{F})$ by Lemma 1.3.15. Now consider the set $\mathcal{A} = \{f : f \boxtimes T \in \chi_L(\mathcal{F})\}$. By Lemma 1.3.30 we have that

$$\mathcal{A} = \chi_L(\langle \mathcal{F}, T \rangle),$$

thus $\mathcal{A}$ is a saturated set by Lemma 1.3.8. Since $S \subseteq \mathcal{A}$ by definition of $\mathcal{A}$, it follows that $\overline{S} \subseteq \mathcal{A}$. Thus $\overline{S} \boxtimes T \subseteq \overline{S \boxtimes T}$. Next, let us consider the set $\mathcal{B} = \{f : \overline{S} \boxtimes f \in \chi_L(\mathcal{F})\}$. As before, we get that

$$\mathcal{B} = \chi_L(\langle \mathcal{F}, \overline{S} \rangle)$$

so that $\mathcal{B}$ is also a saturated set. We see that $T \subseteq \mathcal{B}$ by the previous step, so that $\overline{T} \subseteq \mathcal{B}$ as well. This proves the first part. For the second part, we argue as follows: Since $\overline{S} \boxtimes \overline{T}$ is saturated and contains $\overline{S} \boxtimes \overline{T}$ and $\overline{S} \boxtimes T$, we find that

$$\overline{\overline{S} \boxtimes T} \subseteq \overline{\overline{S} \boxtimes \overline{T}} \subseteq \overline{S \boxtimes T}.$$

On the other hand, $S \boxtimes T \subseteq \overline{S} \boxtimes T \subseteq \overline{S} \boxtimes \overline{T}$, so the other inclusion holds as well.   □

Lemma 1.3.33
For $0 < j < n$, the inclusion $\Lambda_j^n \to \Delta^n$ is a retract of the map

$$\Lambda_j^n \times \Delta^2 \amalg_{\Lambda_j^n \times \Lambda_1^2} \Delta^n \times \Lambda_1^2 \to \Delta^n \times \Delta^2.$$

Proof For the first part, consider the maps

$$[n] \xrightarrow{s} [n] \times [2] \xrightarrow{r} [n]$$

where

$$s(i) = \begin{cases} (i, 0) & \text{if } i < j, \\ (i, 1) & \text{if } i = j, \\ (i, 2) & \text{if } i > j, \end{cases}$$

and where

$$r(i, k) = \begin{cases} i & \text{if } i < j \text{ and } k = 0, \\ i & \text{if } i > j \text{ and } k = 2, \\ j & \text{else.} \end{cases}$$

We now have to show that

(1) $rs = \text{id}$,
(2) $s(\Lambda^n_j) \subseteq \Lambda^n_j \times \Delta^2 \cup \Delta^n \times \Lambda^2_1$, and
(3) $r(\Lambda^n_j \times \Delta^2 \cup \Delta^n \times \Lambda^2_1) \subseteq \Lambda^n_j$.

(1) is an immediate check. In order to see (2), we observe that in fact $s(\Lambda^n_j) \subseteq \Lambda^n_j \times \Delta^2$: For this, it suffices to see that composing s with the projection $[n] \times [2] \to [n]$ is the identity. For (3), we need to show two things:

(a) $r(\Lambda^n_j \times \Delta^2) \subseteq \Lambda^n_j$, and
(b) $r(\Delta^n \times \Lambda^2_1) \subseteq \Lambda^n_j$.

In order to prove (a), consider a k-simplex of Λ^n_j, i.e., $f : [k] \to [n]$ such that there exists an $m \in [n] \setminus j$ which is not in the image of f, and an arbitrary k-simplex $\alpha : [k] \to [2]$ of Δ^2. Then the composite

$$[k] \to [n] \times [2] \xrightarrow{r} [n]$$

is easily seen to send $i \in [k]$ to either $f(i)$ or j. Hence its image is contained in the image of $f \cup \{j\}$. In particular, m is not in the image of this composite, and thus represents a k-simplex of Λ^n_j. In order to see (b), consider again a general k-simplex $\beta : [k] \to [n]$ of Δ^n and a k-simplex $f : [k] \to [2]$ of Λ^2_1, i.e., where either 0 or 2 is not in the image of f. For the sake of definiteness, let us say that 2 is not in the image. (The other case works similarly.) We find that the composite

$$[k] \xrightarrow{(\beta, f)} [n] \times [2] \xrightarrow{r} [n]$$

sends $i \in [k]$ to $\beta(i)$ if $\beta(i) < j$ and to j else. Thus the image is contained in $\{0, \ldots, j\}$. Now since $0 < j < n$, we see that n is not in the image, so that the above composite represents a k-simplex of Λ_j^n. In the case that 0 is not in the image of f, we find that 0 is not in the image by a similar argument. This proves the lemma. $\square$

The following lemma corresponds to [Lur09, Proposition 2.3.2.1].

Lemma 1.3.34

The following sets of morphisms all generate the set of inner-anodyne maps:

(1) the inner horn inclusions $S_1 = \{\Lambda_j^n \to \Delta^n\}$ for all $n \geq 2$;

(2) the maps $S_2 = \{(K \to L) \boxtimes (i \colon \Lambda_1^2 \to \Delta^2)\}$ for all monomorphisms $K \to L$;

(3) the maps $S_3 = \{(\partial \Delta^n \to \Delta^n) \boxtimes (i \colon \Lambda_1^2 \to \Delta^2)\}$ for all $n \geq 0$;

(4) the maps $S_4 = \{(K \to L) \boxtimes (\Lambda_j^n \to \Delta^n)\}$ for all monomorphisms $K \to L$ and all inner horns.

Proof Let us first introduce some more notation by denoting the set of monomorphisms by T_2 and the set of boundary inclusions by T_3. Hence, we have $S_2 = T_2 \boxtimes i$, $S_3 = T_3 \boxtimes i$, and $S_4 = T_2 \boxtimes S_1$. In Exercise 59, it is shown that $\overline{T_2} = \overline{T_3}$,. Therefore, it follows from Lemma 1.3.32 that

$$\overline{S_3} = \overline{T_3 \boxtimes i} = \overline{\overline{T_3} \boxtimes i} = \overline{\overline{T_2} \boxtimes i} = \overline{T_2 \boxtimes i} = \overline{S_2}.$$

Tautologically, we have that $\overline{S_2} \subseteq \overline{S_4}$, since $S_2 \subseteq S_4$. We also find that $\overline{S_1} \subseteq \overline{S_2} = \overline{T_2 \boxtimes i}$, since any inner horn inclusion is a retract of a map in S_2 by Lemma 1.3.33. Now notice that $T_2 \boxtimes T_2 = T_2$, so that we additionally obtain

$$\overline{S_4} = \overline{T_2 \boxtimes S_1} \subseteq \overline{T_2 \boxtimes \overline{T_2 \boxtimes i}} = \overline{T_2 \boxtimes T_2 \boxtimes i} = \overline{T_2 \boxtimes i} = \overline{S_2},$$

so that $\overline{S_2} = \overline{S_4}$. The lemma is proved once we can show that $\overline{S_3} \subseteq \overline{S_1}$, for which it suffices to show that $S_3 \subseteq \overline{S_1}$. For this purpose,

let $m \geq 0$ and consider the inclusion

$$\Delta^m \times \Lambda_1^2 \cup \partial \Delta^m \times \Delta^2 \subseteq \Delta^m \times \Delta^2.$$

If $m = 0$, then this map is given by $\Lambda_1^2 \to \Delta^2$ and thus is inner-anodyne. So let us assume $m \geq 1$. We will construct a filtration of $\Delta^m \times \Delta^2$ as follows: For $0 \leq i \leq j < m$, consider the $(m+1)$-simplices of $\Delta^m \times \Delta^2$ given by

$$\sigma_{i,j}(k) = \begin{cases} (k, 0) & \text{if } 0 \leq k \leq i, \\ (k-1, 1) & \text{if } i+1 \leq k \leq j+1, \\ (k-1, 2) & \text{if } j+2 \leq k \leq m+1. \end{cases}$$

For $0 \leq i \leq j \leq m$, consider the $(m + 2)$-simplices of $\Delta^m \times \Delta^2$ given by

$$\tau_{i,j}(k) = \begin{cases} (k, 0) & \text{if } 0 \leq k \leq i, \\ (k - 1, 1) & \text{if } i + 1 \leq k \leq j + 1, \\ (k - 2, 2) & \text{if } j + 2 \leq k \leq m + 2. \end{cases}$$

We observe that the non-degenerate k-simplices of $\Delta^m \times \Delta^2$ correspond to paths of the length k in the grid $[m] \times [2]$ which do not "take a break at any point", i.e., they are precisely the injective maps $[k] \to [m] \times [2]$. We claim that

(1) the simplices $\tau_{i,j}$ are the non-degenerate $(m + 2)$-simplices of $\Delta^m \times \Delta^2$: Necessarily, the paths corresponding to non-degenerate $(m + 2)$-simplices have to start at $(0, 0)$ and end at $(m, 2)$ in order for an injective map $[m + 2] \to [m] \times [2]$ to exist;
(2) the simplices $\sigma_{i,j}$ are non-degenerate;
(3) the simplex $\sigma_{i,j}$ is a face of $\tau_{i,j}$ and of $\tau_{i,j+1}$ but of no other of the τ's which we just constructed;
(4) the simplices $\sigma_{i,j}$ and $\tau_{i,j}$ are not contained in $\Delta^m \times \Lambda_1^2 \cup \partial \Delta^m \times \Delta^2$.

(1) and (2) are obvious from the previous observation. (3) follows immediately from the definitions, and (4) is also straightforward: The projection $[m] \times [2] \to [m]$ sends the simplices in question to surjections, so that they are not contained in $\partial \Delta^m \times \Delta^2$; likewise, the projection $[m] \times [2] \to [2]$ sends the simplices in question to surjections as well, so that they are not contained in $\Delta^m \times \Lambda_1^2$ either.

Let us now inductively define simplicial sets $X(i, j)$ for $0 \leq i \leq j < m$ as follows:

$$X(0, 0) = \Delta^m \times \Lambda_1^2 \cup \partial \Delta^m \times \Delta^2$$

For fixed j, we inductively define, for $i \leq j < m$,

$$X(i + 1, j) = X(i, j) \cup \sigma_{i,j},$$

and we set

$$X(0, j + 1) = X(j + 1, j).$$

Then we define $Y(0, 0) = X(0, m)$, again inductively define, for $i \leq j \leq m$,

$$Y(i + 1, j) = Y(i, j) \cup \tau_{i,j},$$

and set

$$Y(0, j + 1) = Y(j + 1, j).$$

Since the $\tau_{i,j}$ are the non-degenerate $(m+2)$-simplices of $\Delta^m \times \Delta^2$, we find that $Y(0, m+1) = \Delta^m \times \Delta^2$. In order to complete the proof of the lemma, we have to show the following statements:

(A) The simplicial set $X(i, j) \cap \sigma_{i,j}$ is an inner horn.
(B) The simplicial set $Y(i, j) \cap \tau_{i,j}$ is an inner horn.

From this, it follows that all maps $X(i, j) \to X(i+1, j)$ and all maps $Y(i, j) \to Y(i+1, j)$ are inner-anodyne: In either case, they are pushouts of inner horn inclusions, which is why their composite $X(0, 0) \to Y(0, m+1)$ is also inner-anodyne, which concludes the lemma.

• *Proof of (A):* W e have to prove that for all $0 \le i \le j < m$, the map

$$[m + 1] \xrightarrow{\sigma_{i,j}} [m] \times [2]$$

sends all m-dimensional faces to $X(i, j)$, except for one inner m-dimensional face. We claim that the $(i + 1)$-face of $\sigma_{i,j}$ is the one which is not contained in $X(i, j)$. Notice that $0 < i+1 < m+1$, so that $\sigma_{i,j} \cap X(i, j)$ is an inner horn. There are only two faces of $\sigma_{i,j}$ which do not lie in $\partial\Delta^m \times \Delta^2 \subseteq X(0, 0)$, namely the ones which compose the horizontal edges adjacent to the unique vertical edge. These are $d_i\sigma_{i,j}$ and $d_{i+1}\sigma_{i,j}$, therefore it suffices to check that $d_i\sigma_{i,j}$ is contained in $X(i, j)$ and that $d_{i+1}\sigma_{i,j}$ is not contained in $X(i, j)$. The former statement is readily seen by observing that $d_i\sigma_{i-1,j} = d_i\sigma_{i,j}$, provided that $i > 0$ and that $d_0\sigma_{0,j}$ is contained in $\Delta^m \times \Lambda_1^2$, since 0 is not in the image after projecting to [2]. It remains to prove the latter statement, namely that $d_{i+1}\sigma_{i,j}$ is not contained in $X(i, j)$. We have already seen that $d_{i+1}\sigma_{i,j}$ is not contained in $\partial\Delta^m \times \Delta^2$. Hence, we need to show that

(1) $d_{i+1}\sigma_{i,j}$ is not contained in $\Delta^m \times \Lambda_1^2$,
(2) $d_{i+1}\sigma_{i,j}$ is not contained in $\sigma_{k,j}$ for $k < i$, and
(3) $d_{i+1}\sigma_{i,j}$ is not contained in $\sigma_{k,l}$ for $k \le l < j$.

(1) follows from the fact that $0 < i + 1 < m + 1$, i.e., after projecting $[m] \times [2] \to [2]$, 0 and 2 are in the image of $d_{i+1}\sigma_{i,j}$. For (2), we observe that the path corresponding to $d_{i+1}\sigma_{i,j}$ runs through the spot $(i, 0)$, which is not the case for $\sigma_{k,j}$ with $k < i$, and hence not for any face of it either. Likewise, if $i < j$ then $d_{i+1}\sigma_{i,j}$ runs through the spot $(j, 1)$, which is not the case for any $\sigma_{k,l}$ with $l < j$. If $i = j$, then $d_{i+1}\sigma_{i,j}$ runs again through $(i, 0)$, which is not the case for $\sigma_{k,l}$ with $k \le l < j = i$, so that $k < i$ as in the first case.

• *Proof of (B):* Again, we claim that $d_{i+1}\tau_{i,j}$ is the only face which is not contained in $Y(i, j)$.

(a) We first consider the case $i < j$.
Observe again that $d_\ell\tau_{i,j}$ is contained in $\partial\Delta^m \times \Delta^2$ unless $\ell \in \{i, i+1, j+1, j+2\}$, and that $d_{j+1}\tau_{i,j} = \sigma_{i,j-1}$ and $d_{j+2}\tau_{i,j} = \sigma_{i,j}$, so that both are contained in $Y(i, j)$. Likewise, if $i > 0$, then $d_i\tau_{i,j} = d_i\tau_{i-1,j}$, so that this face is also contained in $Y(i, j)$. If $i = 0$, then $d_i\tau_{i,j}$ is contained in $\Delta^m \cup \Lambda_1^2$, since its projection to [2] does not have 0 in its image. We now show that $d_{i+1}\tau_{i,j}$ is not contained in $Y(i, j)$, which implies

that $Y(i, j) \cap \tau_{i,j}$ is again an inner horn (since $0 < i + 1 < m + 2$). As in case (A), we need to see that

(1) $d_{i+1}\tau_{i,j}$ is not contained in $\tau_{k,j}$ for $k < i$, and
(2) $d_{i+1}\tau_{i,j}$ is not contained in $\tau_{k,l}$ for $k \leq l < j$.

As before, $d_{i+1}\tau_{i,j}$ runs through the spot $(i, 0)$, which is not the case for $\tau_{k,j}$ if $k < i$. Likewise, $d_{i+1}\tau_{i,j}$ runs through the spot $(j, 1)$, which is not the case for $\tau_{k,l}$ if $l < j$.

(b) Finally, we consider the case $i = j$. We claim that $d_{i+1}\tau_{i,i}$ is again the only face which is not contained in $Y(i, i)$. If $i = j = 0$, then $d_0\tau_{0,0}$ is contained in $\Delta^m \times \Lambda_1^2$, because its projection to $[2]$ misses 0. If $i > 0$, then we have that $d_i\tau_{i,i} = d_i\tau_{i-1,i}$ so $d_i\tau_{i,i}$ is contained in $Y(i, i)$ in all cases. Likewise, we have that $d_{i+2}\tau_{i,i} = \sigma_{i,i}$. If $l < i$ or $l > i + 2$, then the projection of $d_l\tau_{i,i}$ to $[m]$ is not surjective and thus $d_l\tau_{i,i}$ is contained in $\partial\Delta^m \times \Delta^2$. It remains to show that

(1) $d_{i+1}\tau_{i,i}$ is not contained in $\tau_{k,j}$ for $k < i$, and
(2) $d_{i+1}\tau_{i,i}$ is not contained in $\tau_{k,l}$ for $k \leq l < j$.

Note that $k < i$ in both cases and that $d_{i+1}\tau_{i,i}$ runs through the spot $(i, 0)$. But $\tau_{k,l}$ does not run through $(i, 0)$, no matter what l is.

$\square$

Lemma 1.3.35

The following sets of morphisms all generate the set of left-anodyne maps:

(1) the left horn inclusions $S_1 = \{\Lambda_j^n \to \Delta^n\}$ for all $n \geq 1$ and $0 \leq j < n$;
(2) the maps $S_2 = \{(K \to L) \boxtimes (i: \{0\} \to \Delta^1)\}$ for all monomorphisms $K \to L$;
(3) the maps $S_3 = \{(\partial\Delta^n \to \Delta^n) \boxtimes (i: \{0\} \to \Delta^1)\}$ for all $n \geq 0$;
(4) the maps $S_4 = \{(K \to L) \boxtimes (\Lambda_j^n \to \Delta^n)\}$ for all monomorphisms $K \to L$, all $n \geq 1$ and $0 \leq j < n$.

Proof As in the proof of Lemma 1.3.34, only the following two statements do not follow from previous considerations:

(1) The maps in S_3 are left-anodyne.
(2) The map $\Lambda_j^n \to \Delta^n$ is a retract of the pushout product map

$$\Delta^n \times \{0\} \amalg_{\Lambda_j^n \times \{0\}} \Lambda_j^n \times \Delta^1 \to \Delta^n \times \Delta^1.$$

Both statements are included in Exercise 62. Here, we give the following hints: For (1), consider a similar filtration of $\Delta^n \times \Delta^1$ starting with the domain of the pushout product by adding the missing non-degenerate $(n + 1)$-simplices in $\Delta^n \times \Delta^1$, and then run the same argument as in Lemma 1.3.34. For (2), consider the maps

$$[n] \xrightarrow{s} [n] \times [1] \xrightarrow{r} [n],$$

where s is the inclusion $k \mapsto (k, 1)$ and

$$r(k, i) = \begin{cases} k & \text{if } k \neq j+1 \text{ and } i = 0, \\ j & \text{if } (k, i) = (j+1, 0), \\ k & \text{if } i = 1. \end{cases}$$

□

Corollary 1.3.36

The following sets of morphisms all generate the set of right-anodyne maps:

(1) the right horn inclusions $S_1 = \{\Lambda_j^n \to \Delta^n\}$ for all $n \geq 1$ and $0 < j \leq n$;

(2) the maps $S_2 = \{(K \to L) \boxtimes (i \colon \{1\} \to \Delta^1)\}$ for all monomorphisms $K \to L$;

(3) the maps $S_3 = \{(\partial \Delta^n \to \Delta^n) \boxtimes (i \colon \{1\} \to \Delta^1)\}$ for all $n \geq 0$;

(4) the maps $S_4 = \{(K \to L) \boxtimes (\Lambda_j^n \to \Delta^n)\}$ for all monomorphisms $K \to L$, all $n \geq 1$ and $0 < j \leq n$.

Proof We observe that a map is right-anodyne if and only if its opposite if left-anodyne, because a map is a left fibration if and only if its opposite is a right fibration according to Exercise 51. Then the first part follows from the fact that $(\Lambda_j^n)^{\mathrm{op}} \cong \Lambda_{n-j}^n$ and that all other morphisms of the involved simplicial sets are "self-opposite". □

Proof of Lemma 1.3.31 Part (1) follows from the equality $\overline{S_4} = \overline{S_1}$ of Lemma 1.3.34, Part (2) follows from the equality $\overline{S_4} = \overline{S_1}$ of Lemma 1.3.35, and Part (3) follows from the equality $\overline{S_4} = \overline{S_1}$ of Corollary 1.3.36. In order to see Part (4), we simply observe that anodyne maps are generated (as a saturated set) by left- and right-anodyne maps. Hence, it suffices to show that for i being a left-anodyne (respectively right-anodyne) map and g being a monomorphism, $i \boxtimes g$ is anodyne. Since left-anodyne (respectively right-anodyne) maps are anodyne, this follows from statement (2) and (3), respectively. □

Theorem 1.3.37

Let $f \colon X \to Y$ be a (inner, left, right) fibration and let $i \colon A \to B$ be a monomorphism. Then

(1) the map $\langle f, i \rangle$ is a (inner, left, right) fibration.

(2) If additionally i is (inner-, left-, right-) anodyne, then the map $\langle f, i \rangle$ is a trivial fibration.

Proof We prove (1) first. For this purpose, consider a lifting problem

$$
\begin{array}{ccc}
S & \longrightarrow & X^B \\
{\scriptstyle g}\downarrow & \nearrow & \downarrow{\scriptstyle \langle f,i\rangle} \\
T & \longrightarrow & X^A \times_{Y^A} Y^B
\end{array}
$$

where the map $g\colon S \to T$ is (inner-, left-, right-) anodyne. By Lemma 1.3.30, solving this problem is equivalent to solving the lifting problem

$$
\begin{array}{ccc}
A \times T \amalg_{A\times S} B \times S & \longrightarrow & X \\
{\scriptstyle i\boxtimes g}\downarrow & \nearrow & \downarrow{\scriptstyle f} \\
B \times T & \longrightarrow & Y
\end{array}
$$

By Lemma 1.3.31, we know that $i \boxtimes g$ is (inner-, left-, right-) anodyne, and by assumption f is an (inner, right, left) fibration.

In order to prove (2), we need to consider again a lifting problem as above, where now g is a monomorphism. Since i is (inner-, left-, right-) anodyne, we see again that we can solve this lifting problem by Lemma 1.3.31. $\qquad\square$

Corollary 1.3.38

Let K and X be simplicial sets. If X is an ∞-category, then so is X^K. If X is a Kan complex, then so is X^K.

Proof This is a special case of Theorem 1.3.37, part (1), where A is empty, $B = K$ and $Y = \Delta^0$. $\qquad\square$

Definition 1.3.39

Let $\mathcal{C}$ and $\mathcal{D}$ be ∞-categories. We define the ∞-category of functors from $\mathcal{C}$ to $\mathcal{D}$ by $\mathrm{Fun}(\mathcal{C}, \mathcal{D}) = \underline{\mathrm{Hom}}(\mathcal{C}, \mathcal{D})$. It is an ∞-category due to Corollary 1.3.38.

Observation 1.3.40

Notice again that the 0- and 1-simplices of $\mathrm{Fun}(\mathcal{C}, \mathcal{D})$ are functors and natural transformations as defined in Definition 1.2.16 and Definition 1.2.79.

If additionally $\mathcal{D}$ is a Kan complex, then $\mathrm{Fun}(\mathcal{C}, \mathcal{D})$ is itself a Kan complex and hence an ∞-groupoid in particular. This, together with the (to be proven) fact that ∞-groupoids are Kan complexes, suggests that in the functor category, a morphism which is pointwise an equivalence is in fact an equivalence. This will turn out to be true, but

(continued)

1.3.40 (continued)

even more complicated to prove then showing that Kan complexes are precisely the ∞-groupoids, see Theorem 2.2.1.

We are now in the position to give a new definition of the ∞-category of spaces, which is purely simplicial.

Definition 1.3.41

We consider the simplicial category Kan with objects given by Kan complexes X and with hom-simplicial sets from X to Y given by the internal hom-simplicial set $\underline{\mathrm{Hom}}(X, Y)$. We let $\widehat{\mathrm{Spc}} = \mathrm{N}(\mathrm{Kan})$ be its coherent nerve.

Observation 1.3.42

There is a canonical functor $\mathrm{Spc} \to \widehat{\mathrm{Spc}}$. For the construction of this simplicial functor, recall that Spc is the coherent nerve of the simplicial category of CW-complexes as in Definition 1.2.72. We claim that sending a CW-complex X to its singular complex $S(X)$ induces the required simplicial functor. For this, we need to show that there is a canonical map of simplicial sets

$$S(\mathrm{map}(X, Y)) \to \underline{\mathrm{Hom}}(S(X), S(Y))$$

which is compatible with composition. This map is constructed as follows: By adjunction it suffices to construct a map

$$S(\mathrm{map}(X, Y)) \times S(X) \to S(Y).$$

Using that S (as a right adjoint) commutes with products, this is equivalently provided by a canonical map

$$S(\mathrm{map}(X, Y) \times X) \to S(Y).$$

Now we use the continuous evaluation map $\mathrm{map}(X, Y) \times X \to Y$ and apply the functor S to it. It is not hard to see that the map obtained in this way is compatible with composition, and hence determines a functor $\mathrm{Spc} \to \widehat{\mathrm{Spc}}$ as claimed.

Observation 1.3.43

Furthermore, we claim that the map

$$S(\mathrm{map}(X, Y)) \to \underline{\mathrm{Hom}}(S(X), S(Y))$$

(continued)

1.3.43 (continued)

is a weak equivalence, so that the simplicial functor from CW-complexes to Kan complexes is a weak equivalence in the sense of Definition 1.2.49. At a later stage, we will prove that weak equivalences of Kan enriched simplicial categories induce equivalences of ∞-categories upon applying the coherent nerve, so that both ∞-categories of spaces which we have defined are in fact equivalent.

Corollary 1.3.44

A simplicial set $\mathcal{C}$ is an ∞-category if and only if the canonical map $\mathcal{C}^{\Delta^2} \to \mathcal{C}^{\Lambda^2_1}$ is a trivial fibration. In particular, for an ∞-category $\mathcal{C}$ the fibre over a fixed diagram $\Lambda^2_1 \to \mathcal{C}$ is a contractible Kan complex.

Proof The "only if" part follows from the fact that $\Lambda^2_1 \to \Delta^2$ is inner-anodyne, $\mathcal{C} \to *$ is an inner fibration, and Theorem 1.3.37. In order to see the converse, we wish to show that $\mathcal{C} \to *$ is an inner fibration if $\mathcal{C}^{\Delta^2} \to \mathcal{C}^{\Lambda^2_1}$ is a trivial fibration.

For showing that $\mathcal{C} \to *$ is an inner fibration, by Lemma 1.3.34 it suffices to show that it admits the extension property for maps in S_2. By adjunction, this holds if and only if $\mathcal{C}^{\Delta^2} \to \mathcal{C}^{\Lambda^2_1}$ satisfies the extension property for all monomorphisms, which is the case if and only if it is a trivial fibration.

For the "in particular" part, consider two composable morphisms f and g in $\mathcal{C}$, and view them as a map $\Lambda^2_1 \to \mathcal{C}$. Then in the pullback diagram

$$
\begin{array}{ccc}
\mathrm{Comp}_{\mathcal{C}}(f, g) & \longrightarrow & \mathrm{Fun}(\Delta^2, \mathcal{C}) \\
\downarrow & & \downarrow \\
\Delta^0 & \longrightarrow & \mathrm{Fun}(\Lambda^2_1, \mathcal{C})
\end{array}
$$

the right vertical map is a trivial fibration, therefore the left vertical map is a trivial fibration as well. In particular, for any two composable morphisms, the simplicial set $\mathrm{Comp}_{\mathcal{C}}(f, g)$ of compositions is a contractible Kan complex. $\square$

Remark 1.3.45

The same line of arguments shows that for an ∞-category $\mathcal{C}$, the map $\mathcal{C}^{\Delta^n} \to \mathcal{C}^{I^n}$ is a trivial fibration for all $n \geq 2$, because the maps $I^n \to \Delta^n$ are inner-anodyne.

Lemma 1.3.46
Any trivial fibration $X \to Y$ admits a section, and the collection of these sections assembles into a contractible Kan complex.

Proof For any simplicial set A, the map $\mathrm{Hom}(A, X) \to \mathrm{Hom}(A, Y)$ is again a trivial fibration. In particular, this is the case for $A = Y$. Then the fibre over id_Y is a contractible Kan complex whose zero-simplices are sections of p.   □

Due to the previous lemma, we can choose a section of the above trivial fibration and obtain the composite

$$\mathcal{C}^{\Lambda^2_1} \to \mathcal{C}^{\Delta^2} \to \mathcal{C}^{\Delta^{\{0,2\}}}$$

as functors of ∞-categories (i.e., maps of simplicial sets). Since the first category is equivalent to $\mathcal{C}^{\Delta^1} \times_{\mathcal{C}} \mathcal{C}^{\Delta^1}$, where the maps are target and source, we obtain a functor which encodes composition in the ∞-category $\mathcal{C}$:

$$\mathcal{C}^{\Delta^1} \times_{\mathcal{C}} \mathcal{C}^{\Delta^1} \to \mathcal{C}^{\Delta^1}$$

Definition 1.3.47
Let $\mathcal{C}$ be an ∞-category and let $x, y \in \mathcal{C}$ be objects. Then we define the mapping ∞-category between x and y to be the pullback

$$
\begin{array}{ccc}
\mathrm{map}_{\mathcal{C}}(x, y) & \longrightarrow & \mathrm{Fun}(\Delta^1, \mathcal{C}) \\
\downarrow & & \downarrow \\
* & \xrightarrow{\ (x,y)\ } & \mathcal{C} \times \mathcal{C}
\end{array}
$$

where the right vertical map is source and target (i.e., evaluation at 0 and 1).

Notice that it is easy to see that $\mathrm{map}_{\mathcal{C}}(x, y)$ is an ∞-category, since the map $\mathrm{Fun}(\Delta^1, \mathcal{C}) \to \mathcal{C} \times \mathcal{C}$ is an inner fibration by Theorem 1.3.37. In fact, more is true, but we have to defer the proof of the following proposition to a later stage of the book, see Corollary 2.2.4.

Proposition 1.3.48
If $\mathcal{C}$ is an ∞-category, then for all objects $x, y \in \mathcal{C}$, $\mathrm{map}_{\mathcal{C}}(x, y)$ is an ∞-groupoid.

Therefore, we obtain a functor

$$\mathrm{map}_{\mathcal{C}}(x, y) \times \mathrm{map}_{\mathcal{C}}(y, z) \to \mathrm{Fun}(\Delta^1, \mathcal{C}) \times_{\mathcal{C}} \mathrm{Fun}(\Delta^1, \mathcal{C}) \to \mathrm{Fun}(\Delta^1, \mathcal{C})$$

which makes the diagram

$$
\begin{array}{ccc}
\mathrm{map}_{\mathcal{C}}(x, y) \times \mathrm{map}_{\mathcal{C}}(y, z) & \longrightarrow & \mathrm{Fun}(\Delta^1, \mathcal{C}) \\
\downarrow & & \downarrow \\
* & \xrightarrow{\ (x,z)\ } & \mathcal{C} \times \mathcal{C}
\end{array}
$$

commute. We hence obtain a functor

$$\mathrm{map}_{\mathcal{C}}(x, y) \times \mathrm{map}_{\mathcal{C}}(y, z) \to \mathrm{map}_{\mathcal{C}}(x, z)$$

which we refer to as the composition in the ∞-category $\mathcal{C}$.

1.4 Joins and Slices

In this section, we discuss a construction in simplicial sets which mimics the notion of joins and slices in ordinary category theory. The section is of a similar technical flavour as the previous section, although it is of notably less technical effort. The tools developed here are a central building block for the proof of Joyal's lifting theorem which we will discuss in the next chapter.

Definition 1.4.1

Let $\mathcal{C}$ and $\mathcal{D}$ be categories. Then the *join* $\mathcal{C} \star \mathcal{D}$ is given by the following category:

$$\mathrm{Ob}(\mathcal{C} \star \mathcal{D}) = \mathrm{Ob}(\mathcal{C}) \amalg \mathrm{Ob}(\mathcal{D})$$

Hom-sets in this category are given by

$$
\mathrm{Hom}_{\mathcal{C}\star\mathcal{D}}(x, y) = \begin{cases}
\mathrm{Hom}_{\mathcal{C}}(x, y) & \text{if } x, y \in \mathcal{C}, \\
\mathrm{Hom}_{\mathcal{D}}(x, y) & \text{if } x, y \in \mathcal{D}, \\
* & \text{if } x \in \mathcal{C}, y \in \mathcal{D}, \\
\emptyset & \text{if } x \in \mathcal{D}, y \in \mathcal{C}.
\end{cases}
$$

Remark 1.4.2

Notice that the join is not symmetric.

Definition 1.4.3

Let $\mathcal{C}$ be a category, and let $x \in \mathcal{C}$ be an object. Then there are categories $\mathcal{C}_{/x}$ and $\mathcal{C}_{x/}$ of objects over and under x, called *slice categories*. Objects are given by morphisms $y \to x$ for $\mathcal{C}_{/x}$ and $x \to y$ for $\mathcal{C}_{x/}$. Morphisms between two such objects are given by commutative triangles.

Remark 1.4.4

If $\mathcal{C}$ is a cocomplete category and x an object of $\mathcal{C}$, then the slice $\mathcal{C}_{x/}$ of objects under x is again cocomplete. However, the forgetful functor $\mathcal{C}_{x/} \to \mathcal{C}$ does not preserve all colimits, see Lemma 1.4.14.

Definition 1.4.5

Given a linearly ordered set J, we define the set of *cuts* of J, denoted by $\mathrm{Cut}(J)$, as decompositions of $J = J_1 \sqcup J_2$ into two disjoint pieces J_1 and J_2 such that $x < y$ whenever $x \in J_1$ and $y \in J_2$. The half-empty cuts $(\emptyset, J)$ and $(J, \emptyset)$ are allowed.

Lemma 1.4.6

Given linearly ordered sets J and J' with a map $\alpha \colon J \to J'$, and given $(J_1', J_2') \in \mathrm{Cut}(J')$, then there exists a unique cut $(J_1, J_2) \in \mathrm{Cut}(J)$ such that α restricts to maps $\alpha_1 \colon J_1 \to J_1'$ and $\alpha_2 \colon J_2 \to J_2'$.

Proof We need to define $J_i = \alpha^{-1}(J_i')$. The only thing to check is that this is in fact a cut of J. But this follows from the fact that α preserves the order. $\qquad\square$

Observation 1.4.7

This result implies that $\mathrm{Cut}(-)$ is a contravariant functor from linearly ordered sets to sets. In fact, $\mathrm{Cut}(-)$ is representable by [1].

Definition 1.4.8

Let X and Y be simplicial sets. We define their join $X \star Y$ to be the simplicial set given as follows: For a finite linearly ordered set J, we set

$$(X \star Y)(J) = \coprod_{(J_1, J_2) \in \mathrm{Cut}(J)} X(J_1) \times Y(J_2)$$

where we declare that $X(\emptyset) = * = Y(\emptyset)$. Let $\alpha\colon J \to J'$ be a morphism of linearly ordered sets and consider a cut of J'. Then for the associated cut of J as in Lemma 1.4.6, there is a map

$$X(J'_1) \times Y(J'_2) \to X(J_1) \times Y(J_2).$$

This provides a unique map

$$(X \star Y)(J') \to (X \star Y)(J)$$

which restricts to the above map on each cut of J', making $X \star Y$ a simplicial set.

Example Given two ordinary categories $\mathcal{C}$ and $\mathcal{D}$, we have

$$N(\mathcal{C} \star \mathcal{D}) \cong N(\mathcal{C}) \star N(\mathcal{D}).$$

In particular, $\Delta^n \star \Delta^m = \Delta^{n+1+m}$ holds. In order to prove this claim, we construct a map of simplicial sets

$$N(\mathcal{C} \star \mathcal{D}) \to N(\mathcal{C}) \star N(\mathcal{D})$$

by observing that any n-simplex in $N(\mathcal{C} \star \mathcal{D})$ determines a cut of $[n]$: At some point, one jumps from morphisms in $\mathcal{C}$ to morphisms in $\mathcal{D}$. It is easy to see that this map is an isomorphism.

Lemma 1.4.9

Given a simplicial set X, the join construction determines a functor $X \star -\colon \mathrm{sSet} \to \mathrm{sSet}_{X/}$. Likewise, it yields a functor $- \star X\colon \mathrm{sSet} \to \mathrm{sSet}_{X/}$.

Proof We need to show that for every $Y \in \mathrm{sSet}$, the simplicial set $X \star Y$ is equipped with a map $X \to X \star Y$. This is obviously the case by the right half-empty cut inclusion

$$X(J) \times Y(\emptyset) \subseteq \coprod_{(J_1, J_2) \in \mathrm{Cut}(J)} X(J_1) \times Y(J_2) = (X \star Y)(J).$$

Furthermore, given a morphism $Y \to Y'$ of simplicial sets, we immediately see that the canonical diagram

$$
\begin{array}{ccc}
X & \longrightarrow & X \star Y \\
& \searrow & \downarrow \\
& & X \star Y'
\end{array}
$$

commutes, and functoriality is also obvious. $\qquad\square$

Lemma 1.4.10
Let K be a simplicial set equipped with a map $p\colon K \to \Delta^1$. Then there is a functorial factorization into a composite

$$K \to K_0 \star K_1 \overset{c}{\to} \Delta^1,$$

where $K_i = p^{-1}(i)$ and the map c is the map $K_0 \star K_1 \to \Delta^0 \star \Delta^0 \cong \Delta^1$.

Proof We need to construct the map $K \to K_0 \star K_1$. For any $n \geq 0$, we have that

$$\mathrm{Hom}_{\mathrm{sSet}}(\Delta^n, \Delta^1) = \mathrm{Hom}([n], [1]) = \mathrm{Cut}([n]).$$

Thus, for every n-simplex $x\colon \Delta^n \to K$, the composite $px\colon \Delta^n \to K \to \Delta^1$ determines a cut $([i], [j]) \in \mathrm{Cut}([n])$, so that the map $px\colon [n] \to [1]$ sends the first i points to 0 and the rest to 1. By definition, this determines a point of $K_0([i]) \times K_1([j])$ which in turn determines an n-simplex of $K_0 \star K_1$.

It is then easy to see that this in fact determines a map of simplicial sets $K \to K_0 \star K_1$ and that this construction is functorial in $\mathrm{sSet}_{/\Delta^1}$. Concretely, for a commutative triangle

the induced diagram

$$
\begin{array}{ccccc}
K & \longrightarrow & K_0 \star K_1 & \longrightarrow & \Delta^1 \\
\downarrow & & \downarrow & \nearrow & \\
K' & \longrightarrow & K_0' \star K_1' & &
\end{array}
$$

commutes as well. $\qquad\qquad\qquad\qquad\qquad\qquad\qquad\qquad\qquad\qquad\qquad\qquad\square$

Corollary 1.4.11
Given a map $\varphi\colon K \to X \star Y$ over Δ^1, i.e., a morphism in $\mathrm{sSet}_{/\Delta^1}$, there is a factorization into $K \to K_0 \star K_1 \overset{f \star g}{\longrightarrow} X \star Y$.

Proof We observe that the factorization provided by Lemma 1.4.10 for the map $X \star Y \to \Delta^1$ is given by $X \star Y \to X \star Y \to \Delta^1$. Since this factorization is functorial, we obtain a commutative diagram

$$
\begin{array}{ccc}
K & \xrightarrow{\varphi} & X \star Y \\
\downarrow & & \downarrow{\scriptstyle\cong} \\
K_0 \star K_1 & \xrightarrow{f \star g} & X \star Y
\end{array}
$$

from which the claim follows. $\qquad\square$

Proposition 1.4.12

X and Y are ∞-categories if and only if $X \star Y$ is an ∞-category.

Proof Consider a map

$$\Lambda_j^n \to X \star Y$$

for $n \geq 2$ and $0 < j < n$. We want to show that this map extends over Δ^n. We can post-compose the map with the canonical map $X \star Y \to \Delta^1$ and obtain a factorization

$$\Lambda_j^n \to (\Lambda_j^n)_0 \star (\Lambda_j^n)_1 \to X \star Y.$$

There are several possibilities for what the first map is. First recall that any map $\Lambda_j^n \to \Delta^1$ factors uniquely over Δ^n (since Δ^1 is the nerve of a category and Λ_j^n is an inner horn). There are three cases to be considered:

(1) The map $\Lambda_j^n \to \Delta^1$ is constant at 0.
(2) The map $\Lambda_j^n \to \Delta^1$ is constant at 1.
(3) The map $\Lambda_j^n \to \Delta^1$ is not constant.

In the first case, we find that the map $\Lambda_j^n \to X \star Y$ factors through $X \to X \star Y$ and can therefore be extended over Δ^n if and only if X is an ∞-category: If an extension of the composite

$$\Lambda_j^n \to X \to X \star Y$$

to Δ^n exists, then the composite $\Delta^n \to X \star Y \to \Delta^1$ is constant at 0, so that the map $\Delta^n \to X \star Y$ in fact factors through the inclusion $X \to X \star Y$. Similarly, in the second case we find that the map $\Lambda_j^n \to X \star Y$ factors through $Y \to X \star Y$, and thus can be extended over Δ^n if and only if Y is an ∞-category.

Lastly, let us consider the case where the map $\Lambda^n_j \to \Delta^1$ is not constant. Observe that this map factors uniquely through a non-constant map $\Delta^n \to \Delta^1$. The non-constant maps correspond precisely to the non-half-empty cuts of $[n]$, so there is a $0 \le k < n$ such that the map $\Delta^n \to \Delta^1$ is isomorphic to the canonical map $\Delta^k \star \Delta^\ell \to \Delta^1$. It follows that $(\Lambda^n_j)_0$ consists of all those m-simplices of Λ^n_j which are represented by maps $[m] \to [n]$ whose image is contained in $\{0, \dots, k\}$ (so that it lies in the fibre over 0) and such that there exists a number different from j which is not in the image of $[m] \to [n]$ (so that it lies in the horn). Since $k < n$, we can be sure that n does not lie in the image of the map $[m] \to [n]$ representing an m-simplex of $(\Lambda^n_j)_0$. In other words, we find that $(\Lambda^n_j)_0 \cong \Delta^k$. Likewise, we find that $(\Lambda^n_j)_1 \cong \Delta^\ell$. The factorization of Corollary 1.4.11 hence reads as

$$\Lambda^n_j \to \Delta^k \star \Delta^\ell \to X \star Y,$$

which is the desired extension. The proof of the proposition easily follows.   □

Now, we come to the construction of slice categories of ∞-categories. For this, we observe that if the functor $S \star -\colon \mathrm{sSet} \to \mathrm{sSet}_{S/}$ admits a right adjoint $\mathrm{sSet}_{S/} \to \mathrm{sSet}$,

$$(p\colon S \to X) \mapsto X_{p/},$$

then we obtain that a map from $Y \to X_{p/}$ is the same thing as a map $S \star Y \to X$ in $\mathrm{sSet}_{S/}$. Specializing this result to $Y = \Delta^n$, we obtain a simplicial set.

Definition 1.4.13

For $p\colon S \to X$, the association $n \mapsto \mathrm{Hom}_{\mathrm{sSet}_{S/}}(S \star \Delta^n, X)$ determines a simplicial set which we call $X_{p/}$.

Lemma 1.4.14

If an ordinary category $\mathcal{C}$ is (co)complete and $x \in \mathcal{C}$ is an object, then $\mathcal{C}_{x/}$ and $\mathcal{C}_{/x}$ are (co)complete as well. The forgetful map $\mathcal{C}_{x/} \to \mathcal{C}$ preserves limits and connected colimits (i.e., colimits indexed over connected categories), and the forgetful map $\mathcal{C}_{/x} \to \mathcal{C}$ preserves colimits and connected limits as well.

Proof We first observe that $\mathcal{C}_{/x} \cong (\mathcal{C}^{\mathrm{op}}_{x/})^{\mathrm{op}}$, therefore it suffices to treat the case of colimits. First, we show that the forgetful map $\mathcal{C}_{/x} \to \mathcal{C}$ preserves colimits. For this purpose, consider a diagram $F\colon I \to \mathcal{C}_{/x}$. The colimit of the underlying diagram $I \to \mathcal{C}_{/x} \to \mathcal{C}$ canonically comes with a map to x, and it is easy to see that this produces a colimit of F.

The case of colimits of a diagram $F\colon I \to \mathcal{C}_{x/}$ is slightly more complicated. Note that any such diagram is equivalently given by a diagram $G\colon I^{\triangleleft} \to \mathcal{C}$ whose restriction to the cone point is given by the object x. We observe that there are canonical functors $\Delta^0 \to I^{\triangleleft} \leftarrow I$ which are inclusions. In particular, we have a canonical map

$$x = \operatorname*{colim}_{\Delta^0} G_{|\Delta^0} \to \operatorname*{colim}_{I^{\triangleleft}} G.$$

We claim that this morphism is a colimit of F. For the proof of this claim, suppose that a functor $\bar{F}\colon I^{\triangleright} \to \mathcal{C}_{/x}$ is given, i.e., a compatible family of maps $F(i) \to (x \to y)$ in $\mathcal{C}_{x/}$. We wish to show that there exists a unique map

$$\operatorname*{colim}_{I^{\triangleleft}} G \to y$$

which is compatible with both maps from x. But this becomes clear from the observation that $G_{|I} = F$, so that for $i \in I$, there is a canonical map $G(i) = F(i) \to y$, and for the cone point, we have $G(*) = x$, so that there is a canonical map to y as well. These are compatible since F takes values in the slice $\mathcal{C}_{x/}$. This shows that F admits a colimit, namely the above-mentioned map

$$x \to \operatorname*{colim}_{I^{\triangleleft}} G.$$

To finish the proof of the lemma, we need to show that if I is connected, then the canonical map

$$\operatorname*{colim}_{I} F \to \operatorname*{colim}_{I^{\triangleleft}} G$$

is an isomorphism, so that the functor $\mathcal{C}_{x/} \to \mathcal{C}$ preserves connected colimits. We first construct a canonical map in the other direction, for which it is sufficient to construct a map $G(j) \to \operatorname*{colim}_{I} F$ for $j \in I^{\triangleleft}$, compatible in j. If $j \subset I$, then $G(j) = F(j)$, so that there is a canonical map to $\operatorname*{colim}_{I} F$. Therefore, we need to construct a map $x = G(*) \to \operatorname*{colim}_{I} F$. For this, we choose any object $i \in I$ and get a map

$$x = G(*) \to G(i) = F(i) \to \operatorname*{colim}_{I} F$$

as wanted. Next, we need to show that these maps assemble into a map

$$\operatorname*{colim}_{I^{\triangleleft}} G \to \operatorname*{colim}_{I} F.$$

In other words, we need to show that for any morphism in $I^{\triangleleft}$, the corresponding triangle commutes. For this, it suffices to treat morphisms of the form $* \to j$ for some $j \in I$. (For morphisms in I, it holds by construction.) Concretely, we need to show that for any two objects $i, j \in I$, the two maps

$$x = G(*) \to G(i) = F(i) \to \operatorname*{colim}_{I} F$$

and

$$x = G(*) \to G(j) = F(j) \to \operatorname*{colim}_I F$$

are the same maps. This is were the assumption that I is connected enters: Namely, we can find a sequence of morphisms connecting i to j in I. By induction on the length, we may assume that the length is one, so that there is in fact a map $i \to j$ in I. In this case, it follows from the property of colimits that the triangle

$$
\begin{array}{ccc}
F(i) & \longrightarrow & \operatorname*{colim}_I F \\
\downarrow & \nearrow & \\
F(j) & &
\end{array}
$$

commutes. On the other hand, by assumption on F, the triangle

$$
\begin{array}{ccc}
x & \longrightarrow & F(i) \\
& \searrow & \downarrow \\
& & F(j)
\end{array}
$$

also commutes, so that the claim is shown.

By construction, the composite

$$\operatorname*{colim}_I F \to \operatorname*{colim}_{I^\triangleleft} G \to \operatorname*{colim}_I F$$

is the identity. The other composite provides a map

$$\operatorname*{colim}_{I^\triangleleft} G \to \operatorname*{colim}_I F \to \operatorname*{colim}_{I^\triangleleft} G$$

whose restriction to $G(j) \to \operatorname*{colim}_{I^\triangleleft} G$ for $j \in I$ is the canonical map, and whose restriction to the cone point is given by $x \to G(i) \to \operatorname*{colim}_{I^\triangleleft} G$. This map is the canonical map $G(*) \to \operatorname*{colim}_{I^\triangleleft} G$, so that the above composite is also the identity. $\square$

> **Lemma 1.4.15**
> *The functors $S \star -$ and $- \star S$: sSet $\to$ sSet$_{S/}$ preserve colimits.*

Proof It suffices to check that the functors preserve coequalizers, which are calculated in sSet by Lemma 1.4.14, and that they preserve coproducts. For the latter, we recall that the coproduct $(S \to A) \amalg (S \to B)$ in sSet$_{S/}$ is given by the canonical map to the pushout

$S \to A \amalg_S B$ in sSet (see again Lemma 1.4.14 for the description of colimits in such slices). Then we observe that

$$(S * (\coprod_{i \in I} A_i))_n = S_n \amalg \coprod_{i \in I}(A_i)_n \amalg \coprod_{k+l=n-1} S_k \times \coprod_{i \in I}(A_i)_l,$$

whereas

$$\coprod_{i \in I}(S * A_i)_n = \coprod_{i \in I} \left(S_n \amalg (A_i)_n \amalg \coprod_{k+l=n-1} S_k \times (A_i)_l \right)/ \sim$$

where the relation identifies $\coprod_I S_n$ with S_n. For coequalizers, the statement follows similarly from the explicit description of the simplices of the join. □

Corollary 1.4.16
The functors $S \star -$ and $- \star S$: sSet $\to$ sSet preserve pushouts.

Proof The functors $S \star -$ and $- \star S$: sSet $\to$ sSet$_{S/}$ preserve all colimits by Lemma 1.4.15 and the forgetful functor sSet$_{S/} \to$ sSet preserves connected colimits by Lemma 1.4.14. Since pushouts are connected colimits, the claim follows. □

Corollary 1.4.17
The functor $(p: S \to X) \mapsto X_{p/}$ is right-adjoint to $S \star -$: sSet $\to$ sSet$_{S/}$. Likewise, the functor $(p: S \to X) \mapsto X_{/p}$ is right-adjoint to $- \star S$: sSet $\to$ sSet$_{S/}$.

Proof By definition, the adjunction property holds for representables. By Lemma 1.4.15, the functors $S \star -$ and $- \star S$ preserve colimits, so that the adjunction bijection extends from representables to all simplicial sets. □

Example Let $\mathcal{C}$ be an ∞-category and $x \in \mathcal{C}$ an object, which we view as a functor $x: \Delta^0 \to \mathcal{C}$. We obtain slices $\mathcal{C}_{x/}$ and $\mathcal{C}_{/x}$. For a general simplicial set K, we will write $K^{\lhd} = \Delta^0 \star K$ and $K^{\rhd} = K \star \Delta^0$ and call these constructions cone and cocone over K.

Observation 1.4.18
Let us explicitly spell out the unit and counit of the slice/join adjunction. For a fixed simplicial set S, the counit of the adjunction is given by a natural map as follows: Let

---

(continued)

1.4.18 (continued)

$p: S \to X$ be an object of $\mathrm{sSet}_{S/}$, so we obtain the slice $X_{p/}$. Then the counit is the map $S \star X_{p/} \to X$ in $\mathrm{sSet}_{S/}$ given by

$$S_n \sqcup \mathrm{Hom}_{\mathrm{sSet}_{S/}}(S \star \Delta^n, X) \sqcup \coprod_{k+l=n-1} S_k \times \mathrm{Hom}_{\mathrm{sSet}_{S/}}(S \star \Delta^l, X) \to X_n,$$

which is given by p in the first component, induced by precomposition with $\Delta^n \to S \star \Delta^n$ on the second component, and induced by precomposition with $\Delta^k \star \Delta^l \to S \star \Delta^l$ for each k-simplex of S on the last component.

Likewise, the unit is the map $X \to (S \star X)_{\mathrm{can}/}$, where $\mathrm{can}: S \to S \star X$ is the canonical map. It is given by joining an n-simplex of X with S, which yields a map

$$X_n \cong \mathrm{Hom}_{\mathrm{sSet}}(\Delta^n, X) \to \mathrm{Hom}_{\mathrm{sSet}_{S/}}(S \star \Delta^n, S \star X).$$

Definition 1.4.19

Let $\mathcal{C}$ be an ordinary category. We define a new category $\mathrm{Tw}(\mathcal{C})$, the *twisted arrow category* of $\mathcal{C}$, as follows: Objects are the morphisms of $\mathcal{C}$, and a morphism in $\mathrm{Tw}(\mathcal{C})$ from $f': x' \to y'$ to $f: x \to y$ is given by a commutative diagram

$$
\begin{array}{ccc}
x & \xrightarrow{f} & y \\
\downarrow{\alpha} & & \uparrow{\beta} \\
x' & \xrightarrow{f'} & y'
\end{array}
$$

We also write that the pair (α, β) is a morphism from f' to $f = \beta f' \alpha$. Composition is obtained by glueing such diagrams together.

Lemma 1.4.20

The slice construction induces a functor $\mathrm{Tw}(\mathrm{sSet}) \to \mathrm{sSet}$. *In particular, for*

$$A \xrightarrow{i} B \xrightarrow{\varphi} X \xrightarrow{f} Y$$

there is an induced map

$$X_{\varphi/} \to X_{\varphi i/} \times_{Y_{f\varphi i/}} Y_{f\varphi/}.$$

The same holds true for the other slice.

Proof The objects of Tw(sSet) are given by maps $p \colon S \to X$ of simplicial sets, and such an object is sent to $X_{/p}$. We need to show how this is functorial in morphisms of the twisted arrow category, i.e., we need to produce a canonical map

$$X_{\varphi/} \to Y_{f\varphi i/},$$

since the pair (i, f) is a morphism from φ to $f\varphi i$ in Tw(sSet).

First, we construct maps $X_{\varphi/} \to X_{\varphi i/}$ and $X_{\varphi/} \to Y_{f\varphi/}$ which correspond to the morphisms

$$
\begin{array}{ccc}
A & \xrightarrow{\varphi i} & X \\
\downarrow{\scriptstyle i} & & \| \\
B & \xrightarrow{\varphi} & X
\end{array}
\qquad\qquad
\begin{array}{ccc}
B & \xrightarrow{f\varphi} & Y \\
\| & & \uparrow{\scriptstyle f} \\
B & \xrightarrow{\varphi} & X
\end{array}
$$

in Tw(sSet). Using those constructions, we similarly obtain maps

$$X_{\varphi i/} \to Y_{f\varphi i/} \leftarrow Y_{f\varphi/},$$

and we will then show that the diagram

$$
\begin{array}{ccc}
X_{\varphi/} & \longrightarrow & Y_{f\varphi/} \\
\downarrow & & \downarrow \\
X_{\varphi i/} & \longrightarrow & Y_{f\varphi i/}
\end{array}
$$

commutes. This is already part of functoriality in the twisted arrow category, because the pair (i, f) satisfies

$$(i, \mathrm{id}) \circ (\mathrm{id}, f) = (i, f) = (\mathrm{id}, f) \circ (i, \mathrm{id}),$$

as the following diagrams show:

$$
\begin{array}{ccc}
A & \xrightarrow{f\varphi i} & Y \\
\| & & \uparrow{\scriptstyle f} \\
A & \xrightarrow{\varphi i} & X \\
\downarrow{\scriptstyle i} & & \| \\
B & \xrightarrow{\varphi} & X
\end{array}
\qquad\qquad
\begin{array}{ccc}
A & \xrightarrow{f\varphi i} & Y \\
\downarrow{\scriptstyle i} & & \| \\
B & \xrightarrow{f\varphi} & Y \\
\| & & \uparrow{\scriptstyle f} \\
B & \xrightarrow{\varphi} & X
\end{array}
$$

The map $X_{\varphi/} \to X_{\varphi i/}$ is adjoint to a map $A \star X_{\varphi/} \to X$ under A, which we define as the canonical composite

$$A \star X_{\varphi/} \to B \star X_{\varphi/} \to X$$

consisting of the map induced by i and the counit of the adjunction. Likewise, the map $X_{\varphi/} \to Y_{f\varphi/}$ is adjoint to a map $B \star X_{\varphi/} \to Y$ under B, which we define as the composite

$$B \star X_{\varphi/} \to X \to Y$$

consisting of the counit, followed by f. In order
to see that the diagram

$$
\begin{array}{ccc}
X_{\varphi/} & \longrightarrow & Y_{f\varphi/} \\
\downarrow & & \downarrow \\
X_{\varphi i/} & \longrightarrow & Y_{f\varphi i/}
\end{array}
$$

commutes, we observe that both composites are adjoint to the map

$$A \star X_{\varphi/} \to B \star X_{\varphi/} \to X \to Y.$$

It is then easy to see that this construction is functorial in $\mathrm{Tw}(\mathrm{sSet})$. For the other slice, the argument is similar. $\square$

Lemma 1.4.21
The slice/join adjunction induces a bijection of lifting problems between diagrams of the kind

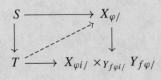

and diagrams of the kind

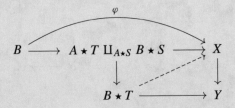

Proof Exercise 70. □

An analogue of Lemma 1.3.31 and Theorem 1.3.37 holds for joins and slices instead of product- and hom-simplicial sets.

Lemma 1.4.22

Let $i: A \to B$ and $g: S \to T$ be monomorphisms. Then the induced map

$$i \hat{\star} g: A \star T \amalg_{A \star S} B \star S \to B \star T$$

is a monomorphism. In addition,

(1) it is inner-anodyne if i is right-anodyne or g is left-anodyne;
(2) it is left-anodyne if i is left-anodyne;
(3) it is right-anodyne if g is right-anodyne.

Proof For (1), let us prove the case where i is right-anodyne. We claim that the set which contains all monomorphisms $i: A \to B$ such that the map $i \hat{\star} g$ is inner-anodyne (for any monomorphism $g: S \to T$) is a saturated class: This is because it is the set of all monomorphisms which has the LLP with respect to morphisms of the form

$$X_{\varphi/} \to X_{\varphi i/} \times_{Y_{f\varphi i/}} Y_{f\varphi/}$$

for an inner fibration $f: X \to Y$ and an arbitrary map $\varphi: T \to X$. It hence suffices to show that the horn inclusions $\Lambda^n_j \to \Delta^n$ for $0 < j \le n$ are in this set. We leave it as an exercise to see that the set of monomorphisms $g: S \to T$ such that the map $(\Lambda^n_j \to \Delta^n) \hat{\star} g$ is inner-anodyne is also a saturated set. It hence suffices to discuss the case of g being the boundary inclusions $\partial \Delta^m \to \Delta^m$. In this case, we have to check that

$$\Lambda^n_j \star \Delta^m \cup \Delta^n \star \partial \Delta^m \to \Delta^n \star \Delta^m$$

is inner-anodyne. But this follows from Exercise 71, which shows that the former is given by Λ^{n+1+m}_j, which is now an inner horn because of $j \le n < n + 1 + m$.

The case where g is left-anodyne follows from a similar calculation, based on the fact that

$$\partial \Delta^m \star \Delta^n \cup \Delta^m \star \Lambda^n_j \to \Delta^{m+1+n}$$

is isomorphic to the inclusion $\Lambda^{m+1+n}_{m+1+j} \to \Delta^{m+1+n}$ and $0 \le j < n$, so that this is again an inner horn.

Let us now prove (2). Using the same reduction arguments as before, it suffices to treat the case where $i: \Lambda^n_j \to \Delta^n$ with $0 \le j < n$ and $g: \partial \Delta^m \to \Delta^m$ is the boundary inclusion. Then we get, as before, that the map $i \hat{\star} g$ is given by $\Lambda^{n+1+m}_j \to \Delta^{n+1+m}$ with $0 \le j < n$, which is clearly a left-anodyne map. The case (3) is similar. □

Theorem 1.4.23

Let $A \xrightarrow{i} B \xrightarrow{\varphi} X \xrightarrow{f} Y$ be composable maps and assume that i is a monomorphism and that f is an inner fibration.

(1) The induced map

$$X_{\varphi/} \to X_{\varphi i/} \times_{Y_{f\varphi i/}} Y_{f\varphi/}$$

is a left fibration.

(2) If the map $f : X \to Y$ is a left fibration, then the induced map

$$X_{/\varphi} \to X_{/\varphi i} \times_{Y_{/f\varphi i}} Y_{/f\varphi}$$

is a left fibration.

(3) If the map $i : A \to B$ is right-anodyne, then the map

$$X_{\varphi/} \to X_{\varphi i/} \times_{Y_{f\varphi i/}} Y_{f\varphi/}$$

is a trivial fibration.

(4) If the map $f : X \to Y$ is a trivial fibration, then the map

$$X_{\varphi/} \to X_{\varphi i/} \times_{Y_{f\varphi i/}} Y_{f\varphi/}$$

is a trivial fibration.

Proof Again we consider a general lifting problem

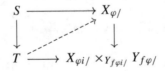

This lifting problem is equivalent (by Lemma 1.4.21, or rather by Exercise 70) to the following lifting problem:

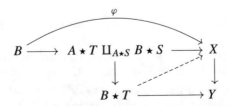

In order to prove (1), we thus need to check that the left vertical map is inner-anodyne provided that $S \to T$ is left-anodyne. But this follows from Lemma 1.4.22 part (1). For proving (2), we need to check that the map

$$S \star B \amalg_{S \star A} T \star A \to T \star B$$

is left-anodyne provided that the map $S \to T$ is left-anodyne. But this is the content of Lemma 1.4.22, part (2). In order to prove (3), we observe that if $S \to T$ is a monomorphism and $A \to B$ is right-anodyne, then the map $A \star T \amalg_{A \star S} B \star S \to B \star T$ is inner-anodyne, again by Lemma 1.4.22, part (1). Lastly, in order to prove (4), we only need to use the fact that the map $A \star T \amalg_{A \star S} B \star S \to B \star T$ is always a monomorphism and that trivial fibrations satisfy the RLP with respect to monomorphisms. □

Let us spell out some explicit special cases in the following corollary.

Corollary 1.4.24

Suppose that we are given maps $A \to B \to X \to Y$ as before, where $A \to B$ is a monomorphism and $X \to Y$ is an inner fibration.

(1) *If $Y = *$, so that X is an ∞-category, we get that $X_{\varphi/} \to X_{\varphi i/}$ is a left fibration and that $X_{/\varphi} \to X_{/\varphi i}$ is a right fibration. In particular, if $A = \emptyset$, then the map $X_{\varphi/} \to X$ is a left fibration and $X_{/\varphi} \to X$ is a right fibration. In particular, $X_{\varphi/}$ and $X_{/\varphi}$ are ∞-categories if X is.*

(2) *If $Y = *$, so that X is an ∞-category, we get that $X_{\varphi/} \to X_{\varphi i/}$ is a trivial fibration if $A \to B$ is right-anodyne, and that $X_{/\varphi} \to X_{/\varphi i}$ is a trivial fibration if $A \to B$ is left-anodyne.*

(3) *If $f: X \to Y$ is a trivial fibration, consider the case where $A = \emptyset$. Then the map $X_{\varphi/} \to X \times_Y Y_{f\varphi/}$ is a trivial fibration. Furthermore, the map $X \times_Y Y_{f\varphi/} \to Y_{f\varphi/}$ is a pullback of $X \to Y$ and therefore also a trivial fibration.*

Yet another special case of this result is the following: Suppose that $\mathcal{C}$ is an ∞-category and that $f: x \to y$ is a morphism in $\mathcal{C}$. We can consider the situation $\Delta^0 \to \Delta^1 \to \mathcal{C} \to *$ and obtain maps

$$\mathcal{C}_{/x} \leftarrow \mathcal{C}_{/f} \to \mathcal{C}_{/y}$$

which correspond to the two restrictions of f to Δ^0 (likewise for the other slice). Since the inclusion $\{0\} \to \Delta^1$ is left-anodyne, it follows from Corollary 1.4.24, part (2), that the map $\mathcal{C}_{f/} \to \mathcal{C}_{x/}$ is a trivial fibration. Therefore, we can choose a section in order to obtain a composite

$$\mathcal{C}_{/x} \to \mathcal{C}_{/f} \to \mathcal{C}_{/y},$$

which we can informally think of as the functor of post-composition with f. The same works for the other slice in order to obtain a functor $\mathcal{C}_{y/} \to \mathcal{C}_{x/}$ which we can think of as precomposition with f.

Joyal's Theorem, Applications, and Dwyer–Kan Localizations

2

2.1 Joyal's Special Horn Lifting Theorem

In this section, we start out with the notion of a conservative functor, i.e., a functor which detects whether a given morphism is an equivalence. Afterwards, we will show Joyal's lifting theorem, which states that conservative inner fibrations are characterized by a lifting property with respect to the so-called special horns, i.e., those horns where a particular edge is sent to an equivalence. From this, we will deduce that ∞-groupoids are Kan complexes, which is one of the central results in the theory.

Definition 2.1.1

A functor $F\colon \mathcal{C} \to \mathcal{D}$ between ∞-categories is called *conservative* if it detects equivalences, i.e., if whenever $f\colon x \to y$ is a morphism in $\mathcal{C}$ such that $F(f)\colon F(x) \to F(y)$ is an equivalence in $\mathcal{D}$, then f itself is an equivalence.

Observation 2.1.2

A functor $F\colon \mathcal{C} \to \mathcal{D}$ is conservative if and only if its opposite functor $F^{\mathrm{op}}\colon \mathcal{C}^{\mathrm{op}} \to \mathcal{D}^{\mathrm{op}}$ is conservative.

Proposition 2.1.3

Left and right fibrations between ∞-categories are conservative.

M. Land, *Introduction to Infinity-Categories*, Compact Textbooks in Mathematics, https://doi.org/10.1007/978-3-030-61524-6_2

Proof By passing to opposite categories, it suffices to treat the case of left fibrations. Suppose that a morphism $f \colon \Delta^1 \to \mathcal{C}$ is given which becomes an equivalence in $\mathcal{D}$. Consider the diagram

$$
\begin{array}{ccc}
\Lambda^2_0 & \longrightarrow & \mathcal{C} \\
\downarrow & \nearrow\!\!\!\cdot & \downarrow p \\
\Delta^2 & \dashrightarrow & \mathcal{D}
\end{array}
$$

where the map $\Lambda^2_0 \to \mathcal{C}$ is given by f on the edge $\Delta^{\{0,1\}}$ and by the identity on $\Delta^{\{0,2\}}$. Since the image in $\mathcal{D}$ is an equivalence, there exists a dashed arrow making the diagram commute. Since $\mathcal{C} \to \mathcal{D}$ is a left fibration, the dotted arrow exists as well. This proves that f admits a left inverse in $\mathcal{C}$ which becomes a left inverse of $p(f)$ after applying p, and thus an equivalence after applying p. The same argument for this morphism proves that it itself admits a left inverse, which shows that the first constructed left inverse of f is an equivalence. Therefore, f is an equivalence as well. $\square$

Definition 2.1.4

A inner fibration $\mathcal{C} \to \mathcal{D}$ of ∞-categories is called an *isofibration* if every lifting problem

$$
\begin{array}{ccc}
\{0\} & \longrightarrow & \mathcal{C} \\
\downarrow & \nearrow & \downarrow \\
\Delta^1 & \xrightarrow{\ f\ } & \mathcal{D}
\end{array}
$$

where f represents an equivalence of $\mathcal{D}$ has a solution which represents an equivalence of $\mathcal{C}$.

Lemma 2.1.5

An inner fibration $\mathcal{C} \to \mathcal{D}$ between ∞-categories is an isofibration if and only if the induced functor $\mathrm{N}(h\mathcal{C}) \to \mathrm{N}(h\mathcal{D})$ is an isofibration.

Proof Exercise 79. $\square$

Corollary 2.1.6

A functor $\mathcal{C} \to \mathcal{D}$ between ∞-categories is an isofibration if and only if $\mathcal{C}^{\mathrm{op}} \to \mathcal{D}^{\mathrm{op}}$ is an isofibration.

Proof Exercise 80. $\qquad\qquad\qquad\qquad\qquad\qquad\qquad\qquad\qquad\qquad\qquad\qquad\qquad\qquad$ $\square$

Proposition 2.1.7

Left and right fibrations between ∞-categories are conservative isofibrations.

Proof Left and right fibrations are conservative by Proposition 2.1.3. Now let $p\colon \mathcal{C} \to \mathcal{D}$ be a left fibration and consider a lifting problem as in Definition 2.1.4. Since $\mathcal{C} \to \mathcal{D}$ is a left fibration and $\{0\} \to \Delta^1$ is left-anodyne, a lift as needed exists. By conservativity, any such lift is an equivalence. For right fibrations, use Corollary 2.1.6. $\qquad\qquad$ $\square$

Theorem 2.1.8

Let $\mathcal{C} \to \mathcal{D}$ be an inner fibration between ∞-categories and let $\phi\colon \Delta^1 \to \mathcal{C}$ be a morphism in $\mathcal{C}$. Then for $n \geq 2$, a lifting problem

$$
\begin{array}{ccc}
\Delta^{\{0,1\}} & \longrightarrow \Lambda_0^n \longrightarrow & \mathcal{C} \\
\downarrow & \nearrow & \downarrow \\
\Delta^n & \longrightarrow & \mathcal{D}
\end{array}
$$

where the top composite is ϕ can be solved if ϕ is an equivalence in $\mathcal{C}$.

Proof In order to prove the lifting property under the assumption that ϕ is an equivalence, we consider a diagram

$$
\begin{array}{ccc}
\Delta^{\{0,1\}} & \longrightarrow \Lambda_0^n \longrightarrow & \mathcal{C} \\
\downarrow & \nearrow & \downarrow \\
\Delta^n & \longrightarrow & \mathcal{D}
\end{array}
$$

where the top horizontal composite is an equivalence in $\mathcal{C}$, say ϕ. , and we wish to show the existence of the dashed arrow. For this purpose, we observe that the map $\Lambda_0^n \to \Delta^n$ is isomorphic to the join-pushout product

$$
\{0\} \star \Delta^{-2+n} \amalg_{\{0\}\star\partial\Delta^{-2+n}} \Delta^1 \star \partial\Delta^{n-2} \to \Delta^1 \star \Delta^{-2+n},
$$

where Δ^{-2+n} is the short form for $\Delta^{\{2,\dots,n\}}$. (This result is found in Exercise 71.) The map $\Delta^{\{0,1\}} \to \Lambda_0^n$ identifies with the canonical composite

$$
\Delta^1 \to \Delta^1 \star \partial\Delta^{-2+n} \to \{0\} \star \Delta^{-2+n} \amalg_{\{0\}\star\partial\Delta^{-2+n}} \Delta^1 \star \partial\Delta^{n-2}.
$$

The above diagram is, by adjunction, equivalent to the diagram

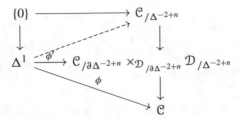

We claim that all of the three top horizontal maps in the diagram

$$\mathcal{C}_{/\Delta^{-2+n}} \longrightarrow \mathcal{C}_{/\partial\Delta^{-2+n}} \times_{\mathcal{D}_{/\partial\Delta^{-2+n}}} \mathcal{D}_{/\Delta^{-2+n}} \longrightarrow \mathcal{C}_{/\partial\Delta^{-2+n}} \longrightarrow \mathcal{C}$$

$$\mathcal{D}_{/\Delta^{-2+n}} \longrightarrow \mathcal{D}_{/\partial\Delta^{-2+n}}$$

are right fibrations and thus conservative by Proposition 2.1.3: The first map is the dual version of Theorem 1.4.23, part (1); the last map is explicitly stated in Corollary 1.4.24, part (1); and the middle map is a pullback of $\mathcal{D}_{/\Delta^{-2+n}} \to \mathcal{D}_{/\partial\Delta^{-2+n}}$ which is a right fibration by the same reasoning as the first map, hence the pullback is also a right fibration. From the assumption that ϕ is an equivalence, it follows that ϕ' is also an equivalence. Hence the dashed arrow exists due to the fact that right fibrations are isofibrations by Proposition 2.1.7.

$\square$

Remark 2.1.9

For $\mathcal{D} = \Delta^0$, we find that a lifting problem

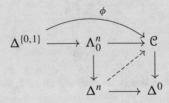

can be solved if and only if ϕ is an equivalence. The "if" case was just dealt with. Assume conversely that any such lifting problem can be solved. From the case $n = 2$, we find a left inverse α of f, witnessed by a 2-simplex τ. We can then consider a 3-horn

(continued)

2.1.9 (continued)

in $\mathcal{C}$ whose restriction to the spine is given by (f, α, f); more precisely, we consider the map $\sigma: \Lambda_0^3 \to \mathcal{C}$ with the following faces:

1. $\sigma_{|\Delta^{\{0,1,2\}}} = \tau$,
2. $\sigma_{|\Delta^{\{0,1,3\}}} = s_1(f)$,
3. $\sigma_{|\Delta^{\{0,2,3\}}} = s_0(f)$.

Then the face $\sigma_{|\Delta^{\{1,2,3\}}}$ is a witness that α is also a right inverse of f, and thus that f is an equivalence.

The general case of Theorem 2.1.8 reads as follows.

Corollary 2.1.10

An inner fibration $p: \mathcal{C} \to \mathcal{D}$ between ∞-categories is conservative if and only if for every $n \geq 2$ and every lifting problem

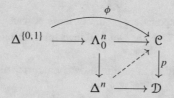

where $p(\phi)$ is an equivalence in $\mathcal{D}$, there exists a solution.

Proof Suppose that p is conservative. Then the assumption that $p(\phi)$ is an equivalence implies that ϕ itself is an equivalence, and hence any such lifting problem can be solved by Theorem 2.1.8. Conversely, suppose that any such lifting problem has a solution. In order to show that p is conservative, let us consider a morphism $\phi: \Delta^1 \to \mathcal{C}$ such that $p(\phi)$ is an equivalence in $\mathcal{D}$. Consider the map $\Lambda_0^2 \to \mathcal{C}$ whose restriction to $\Delta^{\{0,1\}}$ is ϕ and whose restriction to $\Delta^{\{0,2\}}$ is the identity. Since ϕ becomes an equivalence in $\mathcal{D}$, there exists the solid arrows in the lifting problem

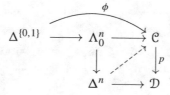

which can be solved by assumption. This provides a left inverse ψ of ϕ. As in the proof of Proposition 2.1.3, it follows that $p(\psi)$ is an equivalence. Running the same argument for ψ

instead of ϕ, we again find that ψ itself admits a left inverse and hence is an equivalence. Therefore, ϕ is an equivalence as well, and consequently, p is conservative. $\square$

Remark 2.1.11

The opposite of the inclusion $\Delta^{\{0,1\}} \to \Lambda_0^n$ is given by the map $\Delta^{\{n-1,n\}} \to \Lambda_n^n$. Since inner fibrations and conservative functors are invariant under passing to opposites, we find that the analogues statements of Theorem 2.1.8 and Corollary 2.1.10, where we replace the inclusion $\Delta^{\{0,1\}} \to \Lambda_0^n$ by the map $\Delta^{\{n-1,n\}} \to \Lambda_n^n$, hold as well.

Notice that for us, the important direction in Corollary 2.1.10 is that such lifting problems can be solved provided that p is conservative, as the next corollary shows.

Corollary 2.1.12

∞-groupoids are Kan complexes.

Proof By definition, ∞-groupoids are precisely those ∞-categories $\mathcal{C}$ where the canonical map $\mathcal{C} \to *$ is conservative. Thus the claim follows from Corollary 2.1.10 $\square$

We are now in the position to define the ∞-category of ∞-categories:

Definition 2.1.13

The ∞-category Cat_∞ of ∞-categories is the coherent nerve of the simplicial category with objects being ∞-categories and hom-simplicial sets given by the maximal ∞-groupoid inside the functor ∞-category $\mathrm{Fun}(\mathcal{C}, \mathcal{D})$. (The definition relies on the fact that the formation of the maximal ∞-groupoid is a monoidal functor from ∞-categories to Kan complexes: It is right-adjoint to the inclusion and thus preserves products.)

Definition 2.1.14

A functor $f : \mathcal{C} \to \mathcal{D}$ between ∞-categories is a *Joyal equivalence* (or categorical equivalence), if the corresponding 1-simplex in Cat_∞ is an equivalence in the sense of Definition 1.2.22.

Remark 2.1.15

Concretely, this means that there is a 2-simplex $\sigma : \Delta^2 \to \mathrm{Cat}_\infty$ such that $\sigma_{|\Delta^{\{0,1\}}} = f$ and $\sigma_{|\Delta^{\{0,2\}}} = \mathrm{id}_{\mathcal{C}}$. From Observation 1.2.66, we find that for $g = \sigma_{|\Delta^{\{1,2\}}}$ we have specified a 1-simplex in $\mathrm{Fun}(\mathcal{C}, \mathcal{D})^{\simeq}$ from gf to $\mathrm{id}_{\mathcal{C}}$. In other words, f is an

(continued)

2.1.15 (continued)

equivalence if and only if there are a functor $g\colon \mathcal{D} \to \mathcal{C}$ and natural equivalences $gf \simeq \mathrm{id}_{\mathcal{C}}$ and $fg \simeq \mathrm{id}_{\mathcal{D}}$, in the sense of the next definition.

Definition 2.1.16

Two functors $f, f'\colon \mathcal{C} \to \mathcal{D}$ are called *(naturally) equivalent* if the corresponding morphisms in Cat_∞ are equivalent in the sense of Definition 1.2.22.

Remark 2.1.17

Unwinding these definitions, we find that f is equivalent to f' precisely if there exists a natural equivalence $\tau\colon f \to f'$, i.e., if τ is an equivalence between f and f' in the ∞-category $\mathrm{Fun}(\mathcal{C}, \mathcal{D})$.

We shall continue with more applications of Joyal's extension theorem. For this, recall that J denotes the contractible groupoid with two objects 0 and 1.

Corollary 2.1.18

Equivalences in an ∞-category $\mathcal{C}$ are represented precisely by those maps $\Delta^1 \to \mathcal{C}$ which extend over the canonical map $\Delta^1 \to J$.

Proof The fact that any map $\Delta^1 \to \mathcal{C}$ which extends over J is an equivalence is discussed in Exercise 61. Conversely, an equivalence is represented by a map $\Delta^1 \to \mathcal{C}^\simeq \subseteq \mathcal{C}$. In order to show that this map extends over the extension $\Delta^1 \to J$, it suffices by Corollary 2.1.12 to observe that $\mathcal{C}^\simeq$ is a Kan complex and that the map $\Delta^1 \to J$ is anodyne. (Note that its geometric realization is a homotopy equivalence, since both are contractible.) See also Lemma 2.4.5 for a purely simplicial proof of the fact that this map is anodyne. $\square$

Corollary 2.1.19

The pullback of a conservative inner fibration $\mathcal{C} \to \mathcal{D}$ along any functor $\mathcal{D}' \to \mathcal{D}$ of ∞-categories is again a conservative inner fibration.

Proof We use the lifting criterion for conservative inner fibrations as established in Corollary 2.1.10 and consider the diagram

$$
\begin{array}{ccccccc}
\Delta^{\{0,1\}} & \longrightarrow & \Lambda_0^n & \longrightarrow & \mathcal{C}' & \longrightarrow & \mathcal{C} \\
\downarrow & & \downarrow & \nearrow & \downarrow{\scriptstyle q} & & \downarrow{\scriptstyle p} \\
& & \Delta^n & \longrightarrow & \mathcal{D}' & \longrightarrow & \mathcal{D}
\end{array}
$$

where the composite $\Delta^1 \to \mathcal{D}'$ represents an equivalence, and we want to show that the dashed arrow exists. Since p is conservative, the dotted arrow exists. Hence the dashed arrow exists as well, because the right square is a pullback. Therefore, q is conservative. □

Proposition 2.1.20
A inner fibration $p \colon \mathcal{C} \to \mathcal{D}$ between ∞-categories is an isofibration if and only if the induced functor $\mathcal{C}^\simeq \to \mathcal{D}^\simeq$ is a Kan fibration.

Proof First, let us suppose that p is an isofibration. As a first step, we show that the induced functor $p^\simeq \colon \mathcal{C}^\simeq \to \mathcal{D}^\simeq$ is also an isofibration. For this purpose, consider a lifting problem

$$
\begin{array}{ccccc}
\Lambda_j^n & \longrightarrow & \mathcal{C}^\simeq & \longrightarrow & \mathcal{C} \\
\downarrow & \nearrow & \downarrow & & \downarrow \\
\Delta^n & \longrightarrow & \mathcal{D}^\simeq & \longrightarrow & \mathcal{D}
\end{array}
$$

Since p is an inner fibration, a dotted arrow exists. We claim that the dotted arrow must already land in $\mathcal{C}^\simeq$, thus giving rise to the dashed arrow. This simply follows from the fact that its restriction to the spine lands in $\mathcal{C}^\simeq$, which implies that the whole n-simplex lies in $\mathcal{C}^\simeq$. Clearly, $p^\simeq$ satisfies the further lifting property of isofibrations: The definition says that any lifting problem

$$
\begin{array}{ccccc}
\{0\} & \longrightarrow & \mathcal{C}^\simeq & \longrightarrow & \mathcal{C} \\
\downarrow & \nearrow & \downarrow & & \downarrow \\
\Delta^1 & \longrightarrow & \mathcal{D}^\simeq & \longrightarrow & \mathcal{D}
\end{array}
$$

has a solution. (We simply spell out that certain 1-simplices are required to be equivalences.)

Next, we observe that $p^\simeq$ is clearly conservative, as is any functor from an ∞-groupoid. Therefore, we know that $p^\simeq \colon \mathcal{C}^\simeq \to \mathcal{D}^\simeq$ is a conservative inner fibration as well as an isofibration. In order to see that it is a Kan fibration, we first show that it is a left fibration. The left horn $\Delta^0 \to \Delta^1$ can be extended, because $p^\simeq$ is an isofibration. For the higher-dimensional left horns, we use the criterion for conservative inner fibrations given by

Corollary 2.1.10, which tells us that lifts exist provided that certain edges of the horn map to equivalences. But this condition is tautologically fulfilled, because $\mathcal{C}^{\simeq}$ is an ∞-groupoid. Running the same argument (using the version of Joyal lifting with the right outer horn), we also find that $p^{\simeq}$ is a right fibration, and hence a Kan fibration.

The converse direction of the proof is obvious: The map $\{0\} \to \Delta^1$ is a horn inclusion, and thus admits a lift for $\mathcal{C}^{\simeq} \to \mathcal{D}^{\simeq}$, because we assume it to be a Kan fibration. This shows that p is an isofibration.   □

Proposition 2.1.21

An inner fibration $\mathcal{C} \to \mathcal{D}$ is an isofibration if and only if it has the RLP with respect to $\Delta^0 \to J$. In particular, the isofibrations are precisely the Joyal fibrations between ∞-categories according to Definition 1.3.27.

Proof It is clear that having the lifting property with respect to $\Delta^0 \to J$ implies that the map is an isofibration. In order to show the converse direction of the proof, we observe that every diagram

$$\begin{array}{ccc} \Delta^0 & \longrightarrow & \mathcal{C} \\ \downarrow & & \downarrow \\ J & \longrightarrow & \mathcal{D} \end{array}$$

factors through the subcategories of equivalences of $\mathcal{C}$ and $\mathcal{D}$, respectively. Therefore, it suffices to show that a map $f \colon \mathcal{C} \to \mathcal{D}$ is an isofibration if and only if the induced map $\mathcal{C}^{\simeq} \to \mathcal{D}^{\simeq}$ is a Kan fibration, which was done in Proposition 2.1.20.   □

Corollary 2.1.22

Isofibrations are stable under pullback.

2.2 Pointwise Criterion for Natural Equivalences

The goal of this section is to prove that a natural transformation τ between functors $f, g \colon \mathcal{C} \to \mathcal{D}$ of ∞-categories which is pointwise an equivalence is itself an equivalence when viewed as a morphism in the ∞-category of functors $\mathrm{Fun}(\mathcal{C}, \mathcal{D})$. Unlike the case of ordinary categories (where taking the unique inverse pointwise is easily seen to assemble into an inverse natural transformation), this is not at all obvious in the situation of ∞-categories. Therefore, we need to prove the following theorem.

Theorem 2.2.1

Let $L \to K$ be a map between simplicial sets which induces a bijection $L_0 \to K_0$. Then, for every ∞-category $\mathcal{C}$, the induced functor

$$\mathrm{Fun}(K, \mathcal{C}) \longrightarrow \mathrm{Fun}(L, \mathcal{C})$$

is conservative.

Corollary 2.2.2

The canonical functor

$$\mathrm{Fun}(K, \mathcal{C}) \longrightarrow \prod_{x \in K_0} \mathcal{C}$$

is conservative. In other words, let $f\colon \Delta^1 \to \mathrm{Fun}(K, \mathcal{C})$ be a natural transformation between two functors $F, G\colon K \to \mathcal{C}$. If, for all $x \in K$, the induced morphism $\Delta^1 \to \mathcal{C}$ is an equivalence, then f is an equivalence.

In order to prove Theorem 2.2.1, we observe that it is enough to treat the case of Corollary 2.2.2, in which $L \to K$ is the map $K_0 \to K$, with $K_0 = \mathrm{sk}_0(K)$ being the discrete simplicial set on the zero simplices of K. This is readily seen by considering the diagram

$$
\begin{array}{ccc}
\mathrm{Fun}(K, \mathcal{C}) & \longrightarrow & \mathrm{Fun}(L, \mathcal{C}) \\
\downarrow & & \downarrow \\
\mathrm{Fun}(K_0, \mathcal{C}) & \overset{\cong}{\longrightarrow} & \mathrm{Fun}(L_0, \mathcal{C})
\end{array}
$$

where the lower horizontal functor is an isomorphism, and the vertical functors are conservative by assumption. It follows that the upper horizontal functor is also conservative.

In this case, we consider the skeletal filtration on K and obtain a tower of simplicial sets

$$\ldots \longrightarrow \mathrm{Fun}(\mathrm{sk}_n(K), \mathcal{C}) \longrightarrow \mathrm{Fun}(\mathrm{sk}_{n-1}(K), \mathcal{C}) \longrightarrow \ldots \longrightarrow \mathrm{Fun}(\mathrm{sk}_0(K), \mathcal{C})$$

whose inverse limit is $\mathrm{Fun}(K, \mathcal{C})$. Note that all maps appearing in the above tower are inner fibrations by Theorem 1.3.37.

We first show that the functor $\mathrm{Fun}(K, \mathcal{C}) \to \mathrm{Fun}(\mathrm{sk}_0(K), \mathcal{C})$ is conservative if each of the functors $\mathrm{Fun}(\mathrm{sk}_n(K), \mathcal{C}) \to \mathrm{Fun}(\mathrm{sk}_{n-1}(K), \mathcal{C})$ are conservative. To do so, we wish to employ Corollary 2.1.10 to show that the functor $\mathrm{Fun}(K, \mathcal{C}) \to \mathrm{Fun}(K_0, \mathcal{C})$ is conservative. So we consider a lifting problem

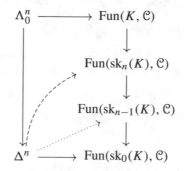

where the special edge $\Delta^{\{0,1\}} \subseteq \Lambda_0^n$ is sent to an equivalence in $\mathrm{Fun}(\mathrm{sk}_0(K), \mathcal{C})$. Inductively, we wish to show that, given the dotted arrow, we can construct the dashed arrow, which makes all diagrams commute. This is possible by Corollary 2.1.10 and the assumption that the maps induced by $\mathrm{sk}_{n-1}(K) \to \mathrm{sk}_n(K)$ are conservative.

It remains to show that for $n \geq 1$, the functors $\mathrm{Fun}(\mathrm{sk}_n(K), \mathcal{C}) \to \mathrm{Fun}(\mathrm{sk}_{n-1}(K), \mathcal{C})$ are in fact conservative. There are pullback diagrams

$$
\begin{array}{ccc}
\mathrm{Fun}(\mathrm{sk}_n(K), \mathcal{C}) & \longrightarrow & \prod \mathrm{Fun}(\Delta^n, \mathcal{C}) \\
\downarrow & & \downarrow \\
\mathrm{Fun}(\mathrm{sk}_{n-1}(K), \mathcal{C}) & \longrightarrow & \prod \mathrm{Fun}(\partial\Delta^n, \mathcal{C})
\end{array}
$$

and since products of conservative functors are conservative, all vertical maps are inner fibrations, and pullbacks of conservative inner fibrations are conservative, it suffices to show that for each $n \geq 1$, the functor

$$
\mathrm{Fun}(\Delta^n, \mathcal{C}) \longrightarrow \mathrm{Fun}(\partial\Delta^n, \mathcal{C})
$$

is conservative. First, we treat the case where $n \geq 2$. In this case, the inclusion $I^n \to \Delta^n$ factors through $\partial\Delta^n$, so we may consider the composite

$$
\mathrm{Fun}(\Delta^n, \mathcal{C}) \longrightarrow \mathrm{Fun}(\partial\Delta^n, \mathcal{C}) \longrightarrow \mathrm{Fun}(I^n, \mathcal{C}).
$$

Since the inclusion $I^n \to \Delta^n$ is inner anodyne by Proposition 1.3.22, it follows from Theorem 1.3.37 that the above composite is a trivial fibration and hence conservative by Proposition 2.1.3. (Notice that trivial fibrations are left and right fibrations, in particular.) It follows that $\mathrm{Fun}(\Delta^n, \mathcal{C}) \to \mathrm{Fun}(\partial\Delta^n, \mathcal{C})$ is also conservative, as claimed. For finishing the proof of Theorem 2.2.1, it hence suffices to prove the following proposition.

Proposition 2.2.3
Let $\mathcal{C}$ be an ∞-category. Then the canonical functor $\mathrm{Fun}(\Delta^1, \mathcal{C}) \to \mathrm{Fun}(\partial\Delta^1, \mathcal{C}) = \mathcal{C} \times \mathcal{C}$ is conservative.

Proof As in the proof of Proposition 2.1.3, it suffices to show that any lifting problem of the kind

$$
\begin{array}{ccc}
\Lambda_0^2 & \xrightarrow{(f,\mathrm{id})} & \mathrm{Fun}(\Delta^1, \mathcal{C}) \\
\downarrow & \nearrow & \downarrow \\
\Delta^2 & \xrightarrow{\sigma} & \mathrm{Fun}(\partial\Delta^1, \mathcal{C})
\end{array}
$$

where f is sent to an equivalence in $\mathrm{Fun}(\partial\Delta^1, \mathcal{C})$ is an equivalence, admits a solution. This implies that f itself admits a left inverse g, whose image in $\mathrm{Fun}(\partial\Delta^1, \mathcal{C})$ is again an equivalence. Applying the same reasoning for g, we find that g, and hence f, is invertible.

In order to solve the lifting problem, by adjunction we may equivalently solve the following lifting problem:

$$
\begin{array}{ccc}
\Lambda_0^2 \times \Delta^1 \cup \Delta^2 \times \partial\Delta^1 & \longrightarrow & \mathcal{C} \\
\downarrow & \nearrow & \\
\Delta^2 \times \Delta^1 & &
\end{array}
$$

We will do this by attaching one 2-simplex and three 3-simplices to $F_0 = \Lambda_0^2 \times \Delta^1 \cup \Delta^2 \times \partial\Delta^1$ to obtain $\Delta^2 \times \Delta^1$, and extending the map to $\mathcal{C}$ step by step: Describing $\Delta^2 \times \Delta^1$ as the nerve of the category depicted by the diagram

$$
\begin{array}{ccc}
10 & \longrightarrow & 11 & \longrightarrow & 12 \\
\uparrow & & \uparrow & & \uparrow \\
00 & \longrightarrow & 01 & \longrightarrow & 02
\end{array}
$$

we first attach the 2-simplex σ_2 given by the composable maps $01 \to 02 \to 12$. Its intersection with F_0 is the inner 2-horn, so we can extend the given map to $F_1 = F_0 \cup \sigma_2$, since $\mathcal{C}$ is an ∞-category. Next, we consider the 3-simplex $\tau_{3,1}$ given by the composable maps $00 \to 01 \to 02 \to 12$. One can readily check that its intersection with F_1 is again an inner 3-horn, so we can extend the given map to $F_2 = F_1 \cup \tau_{3,1}$. Next, we consider $\tau_{3,2}$ given by $00 \to 01 \to 11 \to 12$. Again, one finds that its intersection with F_2 is an inner horn, so the map can be extended to $F_3 = F_2 \cup \tau_{3,2}$. Finally, we consider $\tau_{3,3}$ given by $00 \to 10 \to 11 \to 12$. This time, the intersection with F_3 is given by the left outer horn Λ_0^3, but by assumption, the special edge $\Delta^{\{0,1\}} \subseteq \Lambda_0^3$ is sent to an equivalence. Hence we may extend the given map to $F_4 = F_3 \cup \tau_{3,3}$ thanks to Joyal's lifting theorem (Theorem 2.1.8). Noting that $F_4 = \Delta^2 \times \Delta^1$, the proposition is shown. $\qquad\square$

Corollary 2.2.4

Proposition 1.3.48 holds true. More precisely, given a monomorphism $L \to K$ of simplicial sets which is a bijection on 0-simplices, then the fibre of the induced map $\mathrm{Fun}(K, \mathcal{C}) \to \mathrm{Fun}(L, \mathcal{C})$ over any point $\Delta^0 \to \mathrm{Fun}(L, \mathcal{C})$ is an ∞-groupoid. In particular, for an ∞-category $\mathcal{C}$, the ∞-category $\mathrm{map}_{\mathcal{C}}(x, y)$ is an ∞-groupoid.

Proof By Theorem 2.2.1, the functor $\mathrm{Fun}(K, \mathcal{C}) \to \mathrm{Fun}(L, \mathcal{C})$ is a conservative inner fibration, therefore the pullback along $\Delta^0 \to \mathrm{Fun}(L, \mathcal{C})$ is a conservative inner fibration as well (by Corollary 2.1.19). But $X \to *$ is a conservative inner fibration if and only if X is an ∞-groupoid. $\square$

Note that the "in particular" part of Corollary 2.2.4 is a direct application of Proposition 2.2.3.

We finish this section with some stability properties of isofibrations.

Proposition 2.2.5

Let $p \colon \mathcal{C} \to \mathcal{D}$ be an inner fibration and let $i \colon K \to L$ be a monomorphism of simplicial sets. Suppose that

(1) p is an isofibration, or
(2) i induces a bijection on 0-simplices.

Then the induced functor

$$\mathcal{C}^L \to \mathcal{C}^K \times_{\mathcal{D}^K} \mathcal{D}^L$$

is an isofibration.

Proof By Theorem 1.3.37, we know that this map is an inner fibration. It thus suffices to show that any lifting problem

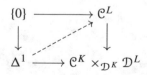

where the bottom horizontal map is an equivalence, has a solution which is again an equivalence. By Theorem 2.2.1, this is the case if for every object of L, the induced morphism in $\mathcal{C}$ is an equivalence. In particular, we see that if $K \to L$ is a bijection on 0-simplices, the right vertical map in the above lifting problem is conservative, so that any lift of an equivalence is automatically an equivalence. By adjunction, this lifting problem is thus equivalent to the lifting problem

$$\{0\} \times L \amalg_{\{0\} \times K} \Delta^1 \times K \longrightarrow \mathcal{C}$$
$$\Delta^1 \times L \longrightarrow \mathcal{D}$$

where the composite of the dashed map with the map $\Delta^1 \times \Delta^0 \to \Delta^1 \times L$ given by an arbitrary object of L is an equivalence. We claim that the set of morphisms $K \to L$ for which the conclusion holds is a saturated class. (We leave the verification of this claim as Exercise 85.) It hence suffices to show the claim of the proposition for the boundary inclusions $\partial \Delta^n \to \Delta^n$ in case (1) and the boundary inclusions with $n \geq 1$ in case (2). If $n = 0$, then we have the lifting problem

$$\{0\} \longrightarrow \mathcal{C}$$
$$\Delta^1 \longrightarrow \mathcal{D}$$

where the lower horizontal map is an equivalence. A dashed arrow representing an equivalence in $\mathcal{C}$ exists because $\mathcal{C} \to \mathcal{D}$ is an isofibration by assumption. For the remaining cases $n \geq 1$, we need to consider diagrams of the form

$$\{0\} \times \Delta^n \amalg_{\{0\} \times \partial \Delta^n} \Delta^1 \times \partial \Delta^n \longrightarrow \mathcal{C}$$
$$\Delta^1 \times \Delta^n \longrightarrow \mathcal{D}$$

and since we will not use the fact that p is an isofibration, the following argument settles all remaining cases of (1) and (2). One constructs a suitable filtration on $\Delta^1 \times \Delta^n$, starting with $\{0\} \times \Delta^n \cup \Delta^1 \times \partial \Delta^n$, by adding the missing simplices (see e.g. [Rez20, Lemma 62.2] for a concrete such filtration). One will then find that the relevant simplices are either attached along inner horn inclusions, or along special outer horns (where the edge from 0 to 1 or $n - 1$ to n is labelled with an equivalence). Using Joyal lifting, the proposition follows. □

Corollary 2.2.6
Let $K \to L$ be a monomorphism of simplicial sets which induces a bijection on 0-simplices. Then the map

$$\{0\} \times L \amalg_{\{0\} \times K} J \times K \to J \times L$$

has the LLP with respect to inner fibrations between ∞-categories.

Proof Let us consider an inner fibration $p \colon \mathcal{C} \to \mathcal{D}$ and a lifting problem

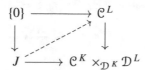

By adjunction, this lifting problem is equivalent to the lifting problem

$$
\begin{array}{ccc}
\{0\} & \longrightarrow & \mathcal{C}^L \\
\downarrow & \nearrow & \downarrow \\
J & \longrightarrow & \mathcal{C}^K \times_{\mathcal{D}^K} \mathcal{D}^L
\end{array}
$$

and by Proposition 2.2.5 the right vertical map is an isofibration. Therefore, the lifting problem can be solved by Proposition 2.1.21. □

Corollary 2.2.7

Let $f \colon \mathcal{C} \to \mathcal{D}$ be a functor between ∞-categories and $K \to L$ a monomorphism of simplicial sets. Then

$$
(\mathcal{C}^K \times_{\mathcal{D}^K} \mathcal{D}^L)^{\simeq} = (\mathcal{C}^K)^{\simeq} \times_{(\mathcal{D}^K)^{\simeq}} (\mathcal{D}^L)^{\simeq}.
$$

Proof Both simplicial sets are subsets of the pullback $\mathcal{C}^K \times_{\mathcal{D}^K} \mathcal{D}^L$, thus we easily find the inclusion "$\subseteq$". In order to show the converse, it suffices to prove that the right-hand side is in fact an ∞-groupoid, since in this case it certainly contains the smallest ∞-groupoid contained in $\mathcal{C}^K \times_{\mathcal{D}^K} \mathcal{D}^L$, which is the left-hand side. By applying Proposition 2.2.5 to the isofibration $\mathcal{D} \to \Delta^0$, we find that $\mathcal{D}^L \to \mathcal{D}^K$ is also an isofibration. Therefore, by Proposition 2.1.20, the map $(\mathcal{D}^L)^{\simeq} \to (\mathcal{D}^K)^{\simeq}$ is a Kan fibration between Kan complexes, so that any pullback along a map from a Kan complex is again a Kan complex. □

Proposition 2.2.8

Let $f \colon \mathcal{C} \to \mathcal{D}$ be a functor between ∞-categories. Then f is a Joyal equivalence if and only if, for every ∞-category $\mathcal{E}$, the induced map

$$
f^* \colon \operatorname{Fun}(\mathcal{D}, \mathcal{E}) \to \operatorname{Fun}(\mathcal{C}, \mathcal{E})
$$

is a Joyal equivalence.

Proof In order to prove the "only if" part, choose an inverse $\mathcal{D} \to \mathcal{C}$ and natural transformations (η_0, η_1) witnessing that g is inverse to f. Then the quadruple $(f^*, g^*, \eta_0^*, \eta_1^*)$ determines an equivalence between $\mathrm{Fun}(\mathcal{C}, \mathcal{E})$ and $\mathrm{Fun}(\mathcal{D}, \mathcal{E})$.

We now show that f is in fact a Joyal equivalence if, for all ∞-categories $\mathcal{E}$, the functor f^* is a Joyal equivalence. By Exercise 87, we obtain a bijection

$$f^* \colon \pi_0(\mathrm{Fun}(\mathcal{D}, \mathcal{E})^{\simeq}) \xrightarrow{\cong} \pi_0(\mathrm{Fun}(\mathcal{C}, \mathcal{E})^{\simeq})$$

Now consider the case where $\mathcal{E} = \mathcal{C}$: This bijection shows the existence of a functor $g \colon \mathcal{D} \to \mathcal{C}$ such that $f^*(g) = gf$ is equivalent to $\mathrm{id}_{\mathcal{C}}$. Taking $\mathcal{E} = \mathcal{D}$, we see that

$$f^*(fg) = fgf \simeq f,$$

and thus $fg \simeq \mathrm{id}_{\mathcal{D}}$ as needed. $\qquad\square$

Definition 2.2.9

A map $f \colon X \to Y$ between simplicial sets is called a *Joyal equivalence* (or categorical equivalence) if, for all ∞-categories $\mathcal{C}$, the induced map

$$\mathrm{Fun}(Y, \mathcal{C}) \to \mathrm{Fun}(X, \mathcal{C})$$

is a Joyal equivalence between ∞-categories.

Observation 2.2.10

This does not change the definition if X and Y are already ∞-categories by Proposition 2.2.8.

Observation 2.2.11

A Joyal equivalence between Kan complexes is precisely a homotopy equivalence.

Proposition 2.2.12

A *trivial Kan fibration* $f \colon X \to Y$ *is a Joyal equivalence.*

Proof Consider the following pullback squares of simplicial sets:

$$
\begin{array}{ccc}
\mathrm{Hom}_{/Y}(Y, X) & \longrightarrow & \mathrm{Hom}(Y, X) \\
\downarrow & & \downarrow{\scriptstyle f_*} \\
\Delta^0 & \xrightarrow{\ \mathrm{id}_Y\ } & \mathrm{Hom}(Y, Y)
\end{array}
\qquad\qquad
\begin{array}{ccc}
\mathrm{Hom}_{/Y}(X, X) & \longrightarrow & \mathrm{Hom}(X, X) \\
\downarrow & & \downarrow{\scriptstyle f_*} \\
\Delta^0 & \xrightarrow{\ f\ } & \mathrm{Hom}(X, Y)
\end{array}
$$

The right vertical maps in each square are trivial fibrations, therefore the same is true for their pullbacks. But a trivial fibration over Δ^0 has a contractible Kan complex as a source. Choose a 0-simplex $s \in \operatorname{Hom}_{/Y}(Y, X)$ such that $f_*(s) = fs = \operatorname{id}_Y$. (For this, recall that f_*, as every trivial fibration, admits a section, and that any 0-simplex $s \in \operatorname{Hom}(Y, X)$ with $fs = \operatorname{id}_Y$ lies in $\operatorname{Hom}_{/Y}(Y, X)$.) Then sf determines a 0-simplex in $\operatorname{Hom}_{/Y}(X, X)$, and therefore, there must be a 1-simplex connecting it to the identity (again because these spaces of sections are contractible). Explicitly, we can find a map

$$\Delta^1 \to \operatorname{Hom}_{/Y}(X, X)$$

whose restriction to 0 is sf and whose restriction to 1 is id_X. For an arbitrary ∞-category, we can compose this map with the canonical map

$$\operatorname{Hom}_{/Y}(X, X) \to \operatorname{Hom}(X, X) \to \operatorname{Hom}(\mathcal{C}^X, \mathcal{C}^X)$$

and see that the resulting map

$$\Delta^1 \to \operatorname{Hom}_{/Y}(X, X) \to \operatorname{Hom}(\mathcal{C}^X, \mathcal{C}^X)^{\simeq}$$

determines a natural equivalence between $(sf)^*$ and $\operatorname{id}_{\mathcal{C}^X}$. Since $(fs)^* = \operatorname{id}_{\mathcal{C}^Y}$, we have shown that f^* and s^* determine inverse equivalences of $\mathcal{C}^X$ and $\mathcal{C}^Y$, for any ∞-category Y, so that f^* is a Joyal equivalence. Therefore, f itself is a Joyal equivalence. $\qquad\square$

Corollary 2.2.13

An inner-anodyne map is a Joyal equivalence.

Proof Let $i \colon A \to B$ be an inner-anodyne map and $\mathcal{C}$ an ∞-category. We need to show that $\mathcal{C}^B \to \mathcal{C}^A$ is a Joyal equivalence. By Theorem 1.3.37, part (2), this map is a trivial fibration, thus the claim is shown by Proposition 2.2.12. $\qquad\square$

Corollary 2.2.14

Every simplicial set is Joyal-equivalent to an ∞-category.

Proof By the small object argument, Proposition 1.3.9, for any simplicial set X one can factor the map $X \to *$ into an inner-anodyne map followed by an inner fibration. This yields a map $X \to \mathcal{C}$ where $\mathcal{C}$ is an ∞-category and $X \to \mathcal{C}$ is inner-anodyne. By Corollary 2.2.13, this map is a Joyal equivalence. $\qquad\square$

Lemma 2.2.15
A Kan fibration $p\colon X \to Y$ between Kan complexes which induces a surjection on π_0 is in fact surjective on 0-simplices. Likewise, a Kan fibration between Kan complexes which induces an injection on π_0 and a surjection on π_1 has the property that any lifting problem

$$
\begin{array}{ccc}
\partial\Delta^1 & \longrightarrow & X \\
\downarrow & \nearrow & \downarrow \\
\Delta^1 & \longrightarrow & Y
\end{array}
$$

can be solved.

Proof Since any homotopy equivalence induces a bijection on simplicial path components, we find for a 0-simplex $y\colon \Delta^0 \to Y$ a commutative diagram

$$
\begin{array}{ccc}
\{0\} & \xrightarrow{\ x'\ } & X \\
\downarrow & \nearrow & \downarrow \\
\Delta^1 & \xrightarrow{\ h\ } & Y
\end{array}
$$

where $h_{|\{1\}} = y$. Since p is a Kan fibration, a dashed arrow exists. Its restriction to $\{1\}$ provides a preimage of y in X.

As for the second claim, pick two objects x, x' of X which give rise to the top horizontal map in the commutative diagram

$$
\begin{array}{ccc}
\partial\Delta^1 & \longrightarrow & X \\
\downarrow & \nearrow & \downarrow \\
\Delta^1 & \xrightarrow{\ h\ } & Y
\end{array}
$$

We wish to show that this diagram admits a dashed arrow. The assumption that p induces a surjection on π_1 implies that the same is true not only for loops at a point of x, but also at homotopy classes of paths from x to x'. (This is where we use the fact that p induces an injection on path components: Namely, the assumptions imply that there exists a path from x to x' in X which we can use to compare the set of homotopy classes of paths from x to x' to the set of homotopy classes of loops at x.) One can thus find a path $\alpha\colon \Delta^1 \to X$ which becomes equivalent to h after applying p. This means that we can find a 2-cell $\sigma\colon \Delta^2 \to Y$ such that $\sigma_{|\Delta\{0,1\}} = p(\alpha)$, $\sigma_{|\Delta\{0,2\}} = h$ and $\sigma_{|\Delta\{1,2\}} = \mathrm{id}_{x'}$. Since we can lift both $p(\alpha)$ and the identity, we obtain a lifting problem

$$
\begin{array}{ccc}
\Lambda^2_1 & \longrightarrow & X \\
\downarrow & \nearrow & \downarrow \\
\Delta^2 & \xrightarrow{\ \sigma\ } & Y
\end{array}
$$

which can be solved, because p is a Kan fibration. The resulting map solves the original lifting problem. $\square$

Remark 2.2.16

In fact, a Kan fibration which is also a weak equivalence is a trivial fibration. This is a classical fact in simplicial homotopy theory, which we can deduce from the previous lemma together with the following observation.

Lemma 2.2.17

A functor $p \colon \mathcal{C} \to \mathcal{D}$ between ∞-categories is a trivial fibration if and only if it is a Joyal equivalence and an isofibration.

Proof Trivial fibrations are Joyal equivalences by Proposition 2.2.12, and they are isofibrations because a trivial fibration is a left fibration which in turn is a conservative isofibration by Proposition 2.1.7.

In order to see the converse direction of the proof, we need to show that, for any monomorphism $K \to L$, a lifting problem

$$
\begin{array}{ccc}
K & \longrightarrow & \mathcal{C} \\
\downarrow & \nearrow & \downarrow{\scriptstyle p} \\
L & \longrightarrow & \mathcal{D}
\end{array}
$$

has a solution. This is equivalent to the lifting problem

$$
\begin{array}{ccc}
\emptyset & \longrightarrow & \mathcal{C}^L \\
\downarrow & \nearrow & \downarrow{\scriptstyle \overline{p}} \\
\Delta^0 & \longrightarrow & \mathcal{C}^K \times_{\mathcal{D}^K} \mathcal{D}^L
\end{array}
$$

which amounts to showing the surjectivity of the right map on 0-simplices. For the proof of this claim, we may shift to the underlying groupoid cores, since this shift does not change the 0-simplices.

By Proposition 2.2.5, $\overline{p}$ is an isofibration because p is an isofibration. In particular, the induced map

$$(\mathcal{C}^L)^{\simeq} \xrightarrow{\overline{p}} (\mathcal{C}^K \times_{\mathcal{D}^K} \mathcal{C}^L)^{\simeq}$$

is a Kan fibration. Furthermore, by Corollary 2.2.7 the latter map is equal to (along the canonical map) the pullback of the groupoid cores, therefore we find that the map

$$(\mathcal{C}^L)^{\simeq} \to (\mathcal{C}^K)^{\simeq} \times_{(\mathcal{D}^K)^{\simeq}} (\mathcal{D}^L)^{\simeq}$$

is a Kan fibration. We wish to show that this map is surjective on 0-simplices. By Lemma 2.2.15, it suffices to show that it induces a surjection on π_0. For this, we observe that since $\mathcal{C} \to \mathcal{D}$ is a Joyal equivalence, so are the maps

$$\mathcal{C}^K \to \mathcal{D}^K \text{ and } \mathcal{C}^L \to \mathcal{D}^L$$

by Exercise 90. Since they are isofibrations, passing to groupoid cores gives us Kan fibrations, which are in addition Joyal equivalences.

Next, we consider the diagram

$$
\begin{array}{ccc}
(\mathcal{C}^L)^{\simeq} \longrightarrow & (\mathcal{C}^K)^{\simeq} \times_{(\mathcal{D}^K)^{\simeq}} (\mathcal{D}^L)^{\simeq} & \longrightarrow (\mathcal{D}^L)^{\simeq} \\
& \downarrow & \downarrow \\
& (\mathcal{C}^K)^{\simeq} \xrightarrow{\hspace{3cm}} & (\mathcal{D}^K)^{\simeq}
\end{array}
$$

and note that all maps are Kan fibrations and that both the composite and the lower horizontal map are Joyal equivalences, and thus homotopy equivalences. We wish to show that the second top horizontal map induces a bijection on π_0, so that the first map induces a surjection on π_0. For this, we observe that the lower horizontal map has the RLP with respect to $\partial \Delta^1 \to \Delta^1$ by Lemma 2.2.15. As a pullback, the same applies to the second top horizontal map, which implies that this map is injective on π_0.

Using Lemma 2.2.15, we find that the map

$$(\mathcal{C}^L)^{\simeq} \to (\mathcal{C}^K)^{\simeq} \times_{(\mathcal{D}^K)^{\simeq}} (\mathcal{D}^L)^{\simeq}$$

induces a surjection on 0-simplices. This proves the lemma. $\square$

Corollary 2.2.18

A Kan fibration $p: X \to Y$ between Kan complexes is a trivial Kan fibration if and only if it is a homotopy equivalence. In particular, among Kan fibrations, the trivial Kan fibrations satisfy the 3-for-2 property.

In this context, a set of maps S is said to have the 3-for-2 property if for any two composable maps g and f, if two of the three maps f, g, and gf are contained in S, then so is the third.

Proof A Kan fibration which is a homotopy equivalence is an isofibration which is a Joyal equivalence. The "in particular" part follows from the 3-for-2 property for homotopy equivalences. □

Remark 2.2.19

Corollary 2.2.18 implies that given an isofibration $p \colon \mathcal{C} \to \mathcal{D}$ which is in addition a Joyal equivalence, the induced map

$$(\mathcal{C}^L)^{\simeq} \xrightarrow{\overline{p}} (\mathcal{C}^K \times_{\mathcal{D}^K} \mathcal{C}^L)^{\simeq}$$

is in fact a trivial fibration. Indeed, since $\mathcal{C} \to \mathcal{D}$ is a Joyal equivalence, the maps

$$\mathcal{C}^K \to \mathcal{D}^K \quad \text{and} \quad \mathcal{C}^L \to \mathcal{D}^L$$

are also Joyal equivalences. Hence, in the composite

$$(\mathcal{C}^L)^{\simeq} \to (\mathcal{C}^K)^{\simeq} \times_{(\mathcal{D}^K)^{\simeq}} (\mathcal{D}^L)^{\simeq} \to (\mathcal{D}^L)^{\simeq},$$

both the composite and the latter map are trivial fibrations: As for the composite, it is a Kan fibration which is a homotopy equivalence and thus a trivial fibration; as for the latter map, it is a pullback of the map $(\mathcal{C}^K)^{\simeq} \to (\mathcal{D}^K) \simeq$ which is a trivial Kan fibration by the same reasoning. By the 3-for-2 property for trivial fibrations (among Kan fibrations), it follows that the map

$$(\mathcal{C}^L)^{\simeq} \to (\mathcal{C}^K)^{\simeq} \times_{(\mathcal{D}^K)^{\simeq}} (\mathcal{D}^L)^{\simeq}$$

is also a trivial fibration.

2.3 Fully Faithful and Essentially Surjective Functors

The goal of this section is to prove that, as in ordinary category theory, functors which are essentially surjective and fully faithful are in fact invertible. The proof which we will present follows an argument which we learned from Gijs Heuts. It uses some results from classical simplicial homotopy theory.

Remark 2.3.1
Note that the proof of this fact in ordinary category theory, which we present in Exercise 36, goes through verbatim in ∞-categories once the necessary machinery is established: The main thing to prove is that an essentially surjective and fully faithful functor admits an adjoint, which will then be an inverse. Of course, for this argument to make sense we will have to speak of adjunctions, and our way to do this will require the straightening-unstraightening equivalence. We will come to adjunctions in Sect. 5.1 and to this approach in Exercise 153.

Definition 2.3.2
A functor $f\colon \mathcal{C} \to \mathcal{D}$ is called *fully faithful* if, for all objects $x, y \in \mathcal{C}$, the induced map $\mathrm{map}_{\mathcal{C}}(x, y) \to \mathrm{map}_{\mathcal{D}}(fx, fy)$ is a Joyal equivalence, i.e., a homotopy equivalence.

Definition 2.3.3
A functor $f\colon \mathcal{C} \to \mathcal{D}$ is called *essentially surjective* if the induced functor $hf\colon h\mathcal{C} \to h\mathcal{D}$ is essentially surjective. In other words, if for every object $d \in \mathcal{D}$, there exists an object $x \in \mathcal{C}$ and an equivalence $fx \simeq d$ in $\mathcal{D}$.

Lemma 2.3.4
Let $f, g\colon \mathcal{C} \to \mathcal{D}$ be functors and let $\tau\colon \Delta^1 \to \mathrm{Fun}(\mathcal{C}, \mathcal{D})$ be a natural transformation from f to g. Then the diagram

$$
\begin{array}{ccc}
\mathrm{map}_{\mathcal{C}}(x, y) & \longrightarrow & \mathrm{map}_{\mathcal{D}}(fx, fy) \\
\downarrow & & \downarrow \\
\mathrm{map}_{\mathcal{D}}(gx, gy) & \longrightarrow & \mathrm{map}_{\mathcal{D}}(fx, gy)
\end{array}
$$

commutes up to homotopy. If τ is a natural equivalence, then the lower horizontal map and the right vertical map are equivalences. In particular, if $\mathcal{D} = \mathcal{C}$ and $g = \mathrm{id}_{\mathcal{C}}$, we find that if f is equivalent to $\mathrm{id}_{\mathcal{C}}$, then the map

$$
\mathrm{map}_{\mathcal{C}}(x, y) \to \mathrm{map}_{\mathcal{D}}(fx, fy)
$$

is a homotopy equivalence.

Proof Recall that there is a functor $\mathrm{Fun}(\mathcal{C}, \mathcal{D}) \to \mathrm{Fun}(\mathcal{C}^{\Delta^1}, \mathcal{D}^{\Delta^1})$ induced by post-composition. The given transformation τ thus induces a functor $\Delta^1 \to \mathrm{Fun}(\mathcal{C}^{\Delta^1}, \mathcal{D}^{\Delta^1})$ which is (by adjunction) a functor

$$
\mathcal{C}^{\Delta^1} \to \mathcal{D}^{\Delta^1 \times \Delta^1}.
$$

Unravelling this construction, we find that this functor sends a morphism $x \to y$ to the diagram

$$
\begin{array}{ccc}
fx & \xrightarrow{\ \tau_x\ } & gx \\
\downarrow & & \downarrow \\
fy & \xrightarrow{\ \tau_y\ } & gy
\end{array}
$$

Therefore, we find a diagram

$$
\begin{array}{ccc}
\mathcal{C}^{\Delta^1} \longrightarrow \mathcal{D}^{\Delta^1 \times \Delta^1} & \longrightarrow & \mathcal{D}^{\Delta^1} \times_{\mathcal{D}} \mathcal{D}^{\Delta^1} \\
\downarrow & \searrow & \downarrow \\
\mathcal{D}^{\Delta^1} \times_{\mathcal{D}} \mathcal{D}^{\Delta^1} & \longrightarrow & \mathcal{D}^{\Delta^1}
\end{array}
$$

where the horizontal functor from $\mathcal{D}^{\Delta^1 \times \Delta^1}$ is given by restriction along the map $\Lambda_1^2 \to \Delta^1 \times \Delta^1$, which singles out one corner of the square, and the vertical functor is given by restriction along the map $\Lambda_1^2 \to \Delta^1 \times \Delta^1$, which singles out the other corner. The diagonal map is given by restriction along the inclusion $\Delta^1 \to \Delta^1 \times \Delta^1$, sending 0 to $(0,0)$ and 1 to $(1,1)$. The remaining two functors are given by composition. Both of the triangles commute up to a natural equivalence (one has to choose a section of the trivial fibration $\mathcal{D}^{\Delta^2} \to \mathcal{D}^{\Lambda_1^2}$).

Now we fix two objects x and y of $\mathcal{C}$ and consider the inclusion $\mathrm{map}_{\mathcal{C}}(x, y) \to \mathcal{C}^{\Delta^1}$. By closer inspection, we find that the diagram

$$
\begin{array}{ccc}
\mathrm{map}_{\mathcal{C}}(x, y) & \dashrightarrow & \{\tau_x\} \times \mathrm{map}_{\mathcal{D}}(gx, gy) \\
\downarrow & & \downarrow \\
\mathcal{C}^{\Delta^1} & \longrightarrow & \mathcal{D}^{\Delta^1} \times_{\mathcal{D}} \mathcal{D}^{\Delta^1}
\end{array}
$$

commutes, i.e., the lower composite factors as indicated by the dashed arrow.

The same holds for the restriction of the vertical map to $\mathrm{map}_{\mathcal{C}}(x, y)$, so that in total we obtain a diagram

$$
\begin{array}{ccc}
\mathrm{map}_{\mathcal{C}}(x, y) & \longrightarrow & \{\tau_x\} \times \mathrm{map}_{\mathcal{D}}(gx, gy) \\
\downarrow & & \downarrow \\
\mathrm{map}_{\mathcal{D}}(fx, fy) \times \{\tau_y\} & \longrightarrow & \mathrm{map}_{\mathcal{D}}(fx, gy)
\end{array}
$$

which commutes up to a natural equivalence as needed.   □

Proposition 2.3.5
A Joyal equivalence between ∞-categories is fully faithful and essentially surjective.

Proof The functor $h\mathcal{C} \to h\mathcal{D}$ is an equivalence, see Exercise 89, and thus essentially surjective. Therefore, $\mathcal{C} \to \mathcal{D}$ is essentially surjective according to Definition 2.3.3. In order to show full faithfulness, choose an inverse g of f. Then, for a pair of objects $x, y \in \mathcal{C}$, we get

$$\mathrm{map}_{\mathcal{C}}(x, y) \to \mathrm{map}_{\mathcal{D}}(fx, fy) \to \mathrm{map}_{\mathcal{C}}(gfx, gfy) \to \mathrm{map}_{\mathcal{D}}(fgfx, fgfy).$$

Since the functor gf is naturally equivalent to $\mathrm{id}_{\mathcal{C}}$ and the functor fg is naturally equivalent to $\mathrm{id}_{\mathcal{D}}$, we see that both the first two maps and the latter two maps compose into an equivalence. This implies that the middle map is itself an equivalence (having both a left and a right inverse) and thus that the first map is also an equivalence (being the right-inverse of an equivalence). □

Remark 2.3.6
Alternatively, one can argue as follows: By Exercise 65, we find that every ∞-category gives rise to a category enriched in the homotopy category of Kan complexes, because composition is well-defined up to homotopy. With this result, we find that a functor $f\colon \mathcal{C} \to \mathcal{D}$ is a Joyal equivalence if and only if the induced functor of $h(\mathrm{Kan})$-enriched categories is an equivalence. Likewise, it is fully faithful and essentially surjective if and only if the induced functor of $h(\mathrm{Kan})$-enriched categories is fully faithful. From this analysis, we deduce that if f is naturally equivalent to g, then f is fully faithful if and only if g is. In the above argument, we can apply this result to gf, which is equivalent to $\mathrm{id}_{\mathcal{C}}$ and thus must be fully faithful itself, so that the required map is in fact a homotopy equivalence.

For future reference, we note the following two lemmas.

Lemma 2.3.7
The inclusion of a full subcategory $\mathcal{C}_0 \subseteq \mathcal{C}$ is a fully faithful functor.

Proof We first observe that for every ∞-category $\mathcal{D}$, the ∞-category $\mathrm{Fun}(\mathcal{D}, \mathcal{C}_0)$ is the full subcategory of $\mathrm{Fun}(\mathcal{D}, \mathcal{C})$ on those functors which factor through $\mathcal{C}_0 \subseteq \mathcal{C}$, see Exercise 47. With this result, we can deduce that the diagram

$$\begin{array}{ccc} \mathrm{Fun}(\Delta^1, \mathcal{C}_0) & \longrightarrow & \mathrm{Fun}(\Delta^1, \mathcal{C}) \\ \downarrow & & \downarrow \\ \mathcal{C}_0 \times \mathcal{C}_0 & \longrightarrow & \mathcal{C} \times \mathcal{C} \end{array}$$

is a pullback. This implies the lemma by passing to fibres over objects (x, y) of $\mathcal{C}_0 \times \mathcal{C}_0$. $\square$

Lemma 2.3.8

Let $\mathcal{C}$ be an ∞-category and let x, y be objects of $\mathcal{C}$. Then the map

$$\mathrm{map}_{\mathcal{C}^\simeq}(x, y) \to \mathrm{map}_{\mathcal{C}}(x, y)$$

is the inclusion of those path components whose points are equivalences of $\mathcal{C}$.

Proof We will show that any map $\Delta^n \to \mathrm{map}_{\mathcal{C}}(x, y)$ lifts to $\mathrm{map}_{\mathcal{C}^\simeq}(x, y)$ if for every $i \in \Delta^n$, the corresponding morphism from x to y is an equivalence. For this, we consider a map $\Delta^n \to \mathrm{map}_{\mathcal{C}}(x, y)$ such that, for all $i \in \Delta^n$, the restriction of its adjoint map

$$\{i\} \times \Delta^1 \to \Delta^n \times \Delta^1 \to \mathcal{C}$$

represents an equivalence in $\mathcal{C}$. Then we observe that, for all $\epsilon = 0, 1$,

$$\Delta^n \times \{\epsilon\} \to \Delta^n \times \Delta^1 \to \mathcal{C}$$

is constant (at either x or y) and hence represents an equivalence of $\mathcal{C}$ as well. Since all morphisms in $\Delta^n \times \Delta^1$ are composites of morphisms of the previous form, we can deduce that the map

$$\Delta^n \times \Delta^1 \to \mathcal{C}$$

factors through $\mathcal{C}^\simeq$, which shows the claim. $\square$

Next, we aim at proving the converse of Proposition 2.3.5, namely that a functor which is fully faithful and essentially surjective is a Joyal equivalence. In order to do so, we need the following preparatory statements.

Lemma 2.3.9

If $f \colon \mathcal{C} \to \mathcal{D}$ is fully faithful and essentially surjective, then so is $f^\simeq \colon \mathcal{C}^\simeq \to \mathcal{D}^\simeq$.

Proof Essential surjectivity is clear, since $\mathcal{C}^{\simeq}$ and $\mathcal{C}$ have the same objects. Full faithfulness is also fine: Since the $\mathrm{map}_{\mathcal{C}^{\simeq}}(x, y) \subseteq \mathrm{map}_{\mathcal{C}}(x, y)$ is a collection of path components (see Lemma 2.3.8), it suffices to know the conclusion of the lemma in the case where $\mathcal{C}$ and $\mathcal{D}$ are ordinary categories (by means of the homotopy category), which is an explicit and easy check.   □

We will make use of the following fundamental property of Kan fibrations, see, e.g., [GJ09, Lemma 7.3].

Lemma 2.3.10
Let $f \colon X \to Y$ be a Kan fibration between Kan complexes. Let x be a point in X and let F be the fibre of f over the point $y = f(x)$. Then there exists a long exact sequence

$$\ldots \to \pi_{n+1}(Y, y) \xrightarrow{\partial} \pi_n(F, x) \to \pi_n(X, x) \xrightarrow{f_*} \pi_n(Y, y) \xrightarrow{\partial} \ldots \to \pi_0(X) \to \pi_0(Y),$$

natural in morphisms of fibrations, which is an exact sequence of groups for $n \geq 1$ and an exact sequence of pointed sets for $n = 0$.

Additionally, we will need Whitehead's theorem:

Proposition 2.3.11
Let $f \colon X \to Y$ be a map between Kan complexes which induces a bijection on path components. Then f is a homotopy equivalence if and only if, for all points x in X and all $n \geq 1$, the induced map

$$f_* \colon \pi_n(X, x) \to \pi_n(Y, y)$$

is a bijection.

For a proof, combine for instance [GJ09, I.11.3 & II.1.10].

Corollary 2.3.12
A fully faithful and essentially surjective functor $f \colon X \to Y$ between Kan complexes is a homotopy equivalence.

Proof Since the functor induces an equivalence of homotopy categories, we find that the map f induces a bijection $\pi_0(X) \to \pi_0(Y)$. We wish to show that for all x in X and all $n \geq 1$, the induced map

$$\pi_n(X, x) \to \pi_n(Y, y)$$

is also a bijection, where $y = f(x)$. For this, we consider the following diagram:

$$
\begin{array}{ccc}
\mathrm{map}_X(x, x) \longrightarrow P_x(X) \longrightarrow \mathrm{Fun}(\Delta^1, X) \\
\downarrow \qquad\qquad \downarrow \qquad\qquad\qquad \downarrow \\
\Delta^0 \xrightarrow{(\mathrm{id}, x)} \Delta^0 \times X \xrightarrow{(x, \mathrm{id}_X)} X \times X
\end{array}
$$

Since the rightmost vertical map is an isofibration between Kan complexes, it is a Kan fibration. Hence, the middle vertical map is a Kan fibration as well. This construction is clearly natural in X, therefore we obtain a diagram of Kan fibre sequences

$$
\begin{array}{ccc}
\mathrm{map}_X(x, x) \longrightarrow P_x(X) \longrightarrow X \\
\downarrow \qquad\qquad \downarrow \qquad\quad \downarrow \\
\mathrm{map}_Y(y, y) \longrightarrow P_y(Y) \longrightarrow Y
\end{array}
$$

where the vertical maps are induced by the map $f \colon X \to Y$. We will now show that $P_x(X)$ is contractible. For this purpose, note that it also sits in the pullback diagram

$$
\begin{array}{ccc}
P_x(X) \longrightarrow \mathrm{Fun}(\Delta^1, X) \\
\downarrow \qquad\qquad \downarrow s \\
\Delta^0 \xrightarrow{\quad x \quad} X
\end{array}
$$

where t stands for the target map, i.e., the map obtained from restriction along $\{0\} \to \Delta^1$. Since this map is anodyne and X is a Kan complex, the resulting map $\mathrm{Fun}(\Delta^1, X) \to X$ is a trivial fibration, and hence the map $P_x(X) \to \Delta^0$ is also a trivial fibration. This implies that $P_x(X)$ is indeed contractible.

We hence obtain that for all $n \geq 1$ there is a commutative diagram

$$
\begin{array}{ccc}
\pi_n(X, x) \longrightarrow \pi_{n-1}(\mathrm{map}_X(x, x), \mathrm{id}_x) \\
\downarrow \qquad\qquad\qquad \downarrow \\
\pi_n(Y, y) \longrightarrow \pi_{n-1}(\mathrm{map}_Y(y, y), \mathrm{id}_y)
\end{array}
$$

where the horizontal maps are isomorphisms by the long exact sequence of Lemma 2.3.10 and the contractability of $P_x(X)$ and $P_y(Y)$. Since f is fully faithful, the map $\mathrm{map}_X(x, x) \to \mathrm{map}_Y(y, y)$ is a homotopy equivalence, and hence induces bijections on all homotopy groups. It follows that the left vertical map is also an isomorphism. $\qquad\square$

Corollary 2.3.13
A fully faithful and essentially surjective functor between ∞-categories induces a homotopy equivalence on groupoid cores.

Proof By Lemma 2.3.9, the induced functor on groupoid cores is still essentially surjective and fully faithful, so that Corollary 2.3.12 applies.   □

For Lemma 2.3.16, we will need *Reedy's lemma*, which is the following statement. In fact, it holds in any model category. For a proof for general model categories, we refer to [Hir03, Prop. 13.1.2].

Lemma 2.3.14
Consider a pullback diagram

$$
\begin{array}{ccc}
C \times_A B & \longrightarrow & B \\
\downarrow & & \downarrow{\scriptstyle p} \\
C & \xrightarrow[f]{\simeq} & A
\end{array}
$$

where p is a fibration and f is a weak equivalence between fibrant objects. Then the map $C \times_A B \to B$ is also a weak equivalence.

Remark 2.3.15
For simplicial sets, let us assume that we have long exact sequences in homotopy groups for Kan fibrations at our disposal. In this case, we easily find from a diagram chase, using the long exact sequence for the vertical fibrations, that the induced map

$$
\pi_i(C \times_A B, x) \to \pi_i(B, x')
$$

is a bijection for every basepoint x of $C \times_A B$ and for every $i \geq 1$. In order to prove Reedy's lemma for the case that interests us, it hence remains to see that the map $C \times_A B \to B$ induces a bijection on path components. For this, the arguments are very similar to the ones used in the proof of Lemma 2.2.15.

Let us prove surjectivity of this map first: Pick a point b in B representing a class $[b] \in \pi_0(B)$ and consider the point $p(b)$ in A. Since $C \to A$ induces a bijection on path components, we can find a point c in C and a path $\Delta^1 \to A$ connecting $f(c)$ to $p(b)$. As in Lemma 2.2.15, this yields a lifting problem

(continued)

2.3.15 (continued)

which can be solved due to p being a Kan fibration. We thus find that there exists a b' in B such that $p(b') = f(c)$ and such that $[b'] = [b]$ in $\pi_0(B)$. Therefore, the pair (c, b') determines an element of $\pi_0(C \times_A B)$ which is sent to $[b]$ in $\pi_0(B)$. This shows that the map $\pi_0(C \times_A B) \to \pi_0(B)$ is indeed surjective.

In order to show injectivity, consider two points (c, b) and (c', b') of $C \times_A B$ and assume that there is a path $\alpha \colon \Delta^1 \to B$ connecting b and b' in B. Then $p(\alpha) \colon \Delta^1 \to A$ connects $p(b) = f(c)$ with $p(b') = f(c')$. Since the map f is a homotopy equivalence, there is a path $\beta \colon \Delta^1 \to C$ such that $f_*(\beta)$ is equivalent to $p_*(\alpha)$. More precisely, we find a 2-cell $\sigma \colon \Delta^2 \to A$ such that

(1) $\sigma_{|\Delta^{\{0,1\}}} = f_*(\beta)$,
(2) $\sigma_{|\Delta^{\{1,2\}}} = \mathrm{id}_{f(c')}$, and
(3) $\sigma_{|\Delta^{\{0,2\}}} = p_*(\alpha)$.

Since we can lift both $p_*(\alpha)$ and $\mathrm{id}_{f(c')}$ along p, similarly as in Lemma 2.2.15, we find a diagram

which admits a dashed arrow as indicated, since p is a Kan fibration. It follows that there exists a path $\gamma \colon \Delta^1 \to B$ connecting b and b' such that $p_*(\gamma) = f_*(\beta)$. Therefore, γ and β combine to a map $\Delta^1 \to C \times_A B$ connecting (c, b) and (c', b'). This shows that the map in question is injective.

Lemma 2.3.16

Consider a diagram of Kan complexes

$$
\begin{array}{ccccc}
C & \longrightarrow\!\!\!\!\!\to & A & \longleftarrow & B \\
\downarrow{\scriptstyle\simeq} & & \downarrow{\scriptstyle\simeq} & & \downarrow{\scriptstyle\simeq} \\
C' & \longrightarrow\!\!\!\!\!\to & A' & \longleftarrow & B'
\end{array}
$$

(continued)

Lemma 2.3.16 (continued)
where the vertical maps are weak equivalences and the left horizontal maps are Kan fibrations. Then the induced map on pullbacks

$$C \times_A B \to C' \times_{A'} B'$$

is again a weak equivalence.

Proof We first reduce the case in question to the situation where the maps $B \to A$ and $B' \to A'$ are also fibrations: By the small object argument, we find a commutative diagram

$$
\begin{array}{ccccc}
A & \twoheadleftarrow & D & \xleftarrow{\simeq} & B \\
\downarrow & & \downarrow & & \downarrow \\
A' & \twoheadleftarrow & D' & \xleftarrow{\simeq} & B'
\end{array}
$$

by functorially factoring the map $B \to A$ into a weak equivalence followed by a fibration (in our case an anodyne map followed by a Kan fibration). It follows that both D and D' are fibrant, and we obtain a commutative diagram

$$
\begin{array}{ccc}
C \times_A B & \longrightarrow & C' \times_{A'} B' \\
\simeq \downarrow & & \downarrow \simeq \\
C \times_A D & \longrightarrow & C' \times_{A'} D'
\end{array}
$$

for which we wish to show that the top horizontal map is an equivalence. We claim that both vertical maps are equivalences: E.g., the left vertical map sits inside a pullback diagram

$$
\begin{array}{ccccc}
C \times_A B & \longrightarrow & C \times_A D & \longrightarrow & C \\
\downarrow & & \downarrow & & \downarrow \\
B & \xrightarrow{\simeq} & D & \longrightarrow & A
\end{array}
$$

so that Reedy's lemma implies that the top horizontal map is an equivalence. (The map $C \times_A D \to D$ is a fibration, since it is pulled back from $C \to A$, which is a fibration by assumption.) The argument for the right vertical map is analogous.

We may thus assume that in the statement of the lemma, all horizontal maps are in fact fibrations. Note that the map in question factors into the composite

$$C \times_A B \to (C' \times_{A'} A) \times_A B = C' \times_{A'} B \to C' \times_{A'} B'.$$

Now we can use Reedy's lemma three times:

(1) The map $C \to C' \times_{A'} A$ is an equivalence: By Reedy's lemma, the map $C' \times_{A'} A \to C'$ is a weak equivalence, since it sits in the pullback

$$
\begin{array}{ccc}
C' \times_{A'} A & \longrightarrow & C' \\
\downarrow & & \downarrow \\
A & \xrightarrow{\;\simeq\;} & A'
\end{array}
$$

Hence in the composite

$$
C \to C' \times_{A'} A \to C',
$$

both the second map and the composite are equivalences. Then the 3-for-2 property for equivalences confirms the claim.

(2) The map $C \times_A B \to (C' \times_{A'} A) \times_A B$ is an equivalence: It sits in the pullback square

$$
\begin{array}{ccccc}
C \times_A B & \longrightarrow & (C' \times_{A'} A) \times_A B & \longrightarrow & B \\
\downarrow & & \downarrow & & \downarrow \\
C & \longrightarrow & C' \times_{A'} A & \longrightarrow & A
\end{array}
$$

where the middle vertical map is a fibration, since it is pulled back from the map $B \to A$, which is now a fibration by assumption. The lower horizontal map is an equivalence by the previous step, therefore the claim follows again by Reedy's lemma.

(3) The map $C' \times_{A'} B \to C' \times_{A'} B'$ is an equivalence: It sits in a pullback square

$$
\begin{array}{ccccc}
C' \times_{A'} B & \longrightarrow & C' \times_{A'} B' & \longrightarrow & C' \\
\downarrow & & \downarrow & & \downarrow \\
B & \longrightarrow & B' & \longrightarrow & A'
\end{array}
$$

where the right vertical map is a fibration, since it is pulled back from $C' \to A'$, which is also a fibration by assumption. Now the map $B \to B'$ is an equivalence, so we conclude again using Reedy's lemma. □

We will need a similar invariance statement for inverse limits. A more general version of the following result can be found in [Hir03, Theorem 19.9.1].

Lemma 2.3.17

Consider a natural transformation between functors $\mathbb{N}^{\mathrm{op}} \to \mathrm{sSet}$

$$
\begin{array}{ccccccccc}
\cdots & \longrightarrow & X_3 & \longrightarrow & X_2 & \longrightarrow & X_1 & \longrightarrow & X_0 \\
\downarrow{\simeq} & & \downarrow{\simeq} & & \downarrow{\simeq} & & \downarrow{\simeq} & & \downarrow{\simeq} \\
\cdots & \longrightarrow & Y_3 & \longrightarrow & Y_2 & \longrightarrow & Y_1 & \longrightarrow & Y_0
\end{array}
$$

and assume that all horizontal maps are fibrations, that all vertical maps are weak equivalences and that all objects are Kan complexes. Then the induced map

$$
\lim_i X_i \to \lim_i Y_i
$$

is an equivalence.

Remark 2.3.18

Again, one can prove this result by using long exact sequences in homotopy groups. Namely, it turns out that there is an exact sequence

$$
0 \longrightarrow \lim_i{}^1 \pi_{k+1}(X_i) \longrightarrow \pi_k(\lim_i X_i) \longrightarrow \lim_i \pi_k(X_i) \longrightarrow 0
$$

so applying (carefully!) a diagram chase argument shows that in our situation, the induced map on inverse limits induces a bijection on all homotopy groups. For such a proof, see, e.g., [Hir15].

Lemma 2.3.19

Consider a commutative diagram of Kan complexes

$$
\begin{array}{ccc}
Y' & \xrightarrow{\ f\ } & Y \\
\downarrow{p'} & & \downarrow{p} \\
X' & \xrightarrow{\ f'\ } & X
\end{array}
$$

where the map p is a fibration. Suppose that f' is a homotopy equivalence, and that for each 0-simplex x' of X' the induced map $p'^{-1}(x') \to p^{-1}(x)$ between vertical fibres is an equivalence as well. Then the map f is a homotopy equivalence.

Proof By Reedy's lemma, we may assume without loss of generality that f' is the identity: We can simply replace the map p by the canonical map $X' \times_X Y \to X'$ and leave the fibres unchanged, while knowing that the map $X' \times_X Y \to Y$ is an equivalence.

If long exact sequences in homotopy groups are available, it takes an easy diagram chase to see that for every point y' in Y' and every $i \geq 1$, the map

$$\pi_i(Y', y') \to \pi_i(Y, y)$$

is a bijection, where $y = f(y')$. It hence remains to show that the map f induces a bijection on path components.

In order to show injectivity, assume that two points x, y in Y' are given whose images under f in Y are connected by a path $\alpha \colon \Delta^1 \to Y$. In other words, we have $\alpha(0) = f(x)$ and $\alpha(1) = f(y)$. It follows that $p\alpha \colon \Delta^1 \to X$ is a path between $p(f(x)) = p'(x)$ and $p(f(y)) = p'(y)$. Consider the lifting problem

$$
\begin{array}{ccc}
\{0\} & \xrightarrow{\ x\ } & Y' \\
\downarrow & \overset{\beta}{\nearrow} & \downarrow p \\
\Delta^1 & \xrightarrow{\ p\alpha\ } & X
\end{array}
$$

which can be solved, since p' is a fibration. We hence have $p'\beta = p\alpha$, and since $p' = pf$, we obtain $pf\beta = p\alpha$. Furthermore, we have $f\beta(0) = f(x) = \alpha(0)$. We hence obtain a lifting problem

$$
\begin{array}{ccc}
\Lambda_0^2 & \xrightarrow{(f\beta, \alpha)} & Y \\
\downarrow & \overset{\tau}{\nearrow} & \downarrow p \\
\Delta^2 & \xrightarrow{\ \sigma\ } & X
\end{array}
$$

where σ is a degeneration of the path $p\alpha$. This lifting problem can be solved, since p is a Kan fibration. By restricting the dashed arrow τ to $\Delta^{\{1,2\}}$, we obtain a path from $f(y)$ to $f(\beta(1))$ which is sent by p to the constant path at $p'(y)$. In other words, $\tau_{|\Delta^{\{1,2\}}}$ is a path in $p^{-1}(p'(y))$. Since f restricts to a homotopy equivalence on this fibre, there is also a path between y and $\beta(1)$ inside $p'^{-1}(p'(y))$. Since $\beta(1)$ is connected (via β) with x, we find that x and y are connected by a path in Y'.

In order to show surjectivity, consider a point y of Y and let $x = p(y)$. By definition, y lies in the fibre F_x of p over x. This shows that F_x is not empty. By assumption, the map f' restricts to a homotopy equivalence $F_x \simeq F'_x$, where the latter denotes the fibre of p' over x. Pick a point y in F'_x which corresponds to $[y]$ under the bijection $\pi_0(F'_x) \to \pi_0(F_x)$. Then the map $\pi_0(Y') \to \pi_0(Y)$ sends $[y']$ to $[y]$, i.e., the map in question is indeed surjective. $\square$

We are now in the position to prove the characterization of Joyal equivalences as the essentially surjective and fully faithful functors.

Theorem 2.3.20
A fully faithful and essentially surjective functor $f : \mathcal{C} \to \mathcal{D}$ between ∞-categories is a Joyal equivalence.

Proof We will prove the theorem by showing that for any simplicial set X, the canonical functor

$$f_* : (\mathcal{C}^X)^{\simeq} \to (\mathcal{D}^X)^{\simeq}$$

is a homotopy equivalence. Once this is shown, one can consider $X = \mathcal{D}$ and, by inverting the homotopy equivalence, obtain a diagram

$$
\begin{array}{ccc}
\Delta^0 & \xrightarrow{\ g\ } & (\mathcal{C}^{\mathcal{D}})^{\simeq} \\
\downarrow & & \downarrow{\scriptstyle f_*} \\
\Delta^1 & \xrightarrow{\ h\ } & (\mathcal{D}^{\mathcal{D}})^{\simeq}
\end{array}
$$

where h is a path from $\mathrm{id}_{\mathcal{D}}$ to fg, i.e., h provides a natural equivalence between fg and $\mathrm{id}_{\mathcal{D}}$. In order to see that gf is also naturally equivalent to $\mathrm{id}_{\mathcal{C}}$, we consider the homotopy equivalence

$$f_* : (\mathcal{C}^{\mathcal{C}})^{\simeq} \to (\mathcal{D}^{\mathcal{C}})^{\simeq}$$

and observe that $\mathrm{id}_{\mathcal{C}}$ is sent to f and that gf is sent to fgf. But since fg is connected to $\mathrm{id}_{\mathcal{D}}$, we find that fgf is also connected to f through a natural equivalence. Since the above map is a homotopy equivalence, this implies that there must also be a path between $\mathrm{id}_{\mathcal{C}}$ and fg, so that any such path provides a natural equivalence $fg \simeq \mathrm{id}_{\mathcal{C}}$, and thus that f and g are mutually inverse functors. Hence, f is a Joyal equivalence.

We will now prove the remaining claim. For this, we first consider the case where $X = \Delta^0$. This is equivalent to the statement that f induces a homotopy equivalence of groupoid cores, which we settled in Corollary 2.3.12. Next, we treat the case $X = \Delta^1$. We recall that the source-target map $\mathcal{C}^{\Delta^1} \to \mathcal{C} \times \mathcal{C}$ is an isofibration, hence the resulting map

$$(\mathcal{C}^{\Delta^1})^{\simeq} \to \mathcal{C}^{\simeq} \times \mathcal{C}^{\simeq}$$

is a Kan fibration which fits into the commutative square

$$
\begin{array}{ccc}
(\mathcal{C}^{\Delta^1})^{\simeq} & \longrightarrow & \mathcal{C}^{\simeq} \times \mathcal{C}^{\simeq} \\
\downarrow & & \downarrow{\scriptstyle \simeq} \\
(\mathcal{D}^{\Delta^1})^{\simeq} & \longrightarrow & \mathcal{D}^{\simeq} \times \mathcal{D}^{\simeq}
\end{array}
$$

where the right vertical map is a homotopy equivalence by the previous step (and the observation that products of Joyal equivalences are again Joyal equivalences). According to Lemma 2.3.19, in order to show that the left vertical map is a homotopy equivalence, it suffices to show that the induced map on fibres over a point $(x, y) \in \mathcal{C}^{\simeq} \times \mathcal{C}^{\simeq}$ is a homotopy equivalence as well. For this, we observe that the diagram

$$
\begin{array}{ccc}
(\mathcal{C}^{\Delta^1})^{\simeq} & \longrightarrow & \mathcal{C}^{\Delta^1} \\
\downarrow & & \downarrow \\
\mathcal{C}^{\simeq} \times \mathcal{C}^{\simeq} & \longrightarrow & \mathcal{C} \times \mathcal{C}
\end{array}
$$

is a pullback, since the right vertical functor is conservative by Theorem 2.2.1 so that we can allude to Exercise 78. Now the fibre of the respective horizontal maps in the above square over the point (x, y), respectively over the point (px, py), is given by the corresponding mapping space, which establishes the remaining claim.

Next, we deal with the case $X = I^n$, i.e., the n-dimensional spine. We will prove that the map

$$
(\mathcal{C}^{I^n})^{\simeq} \to (\mathcal{D}^{I^n})^{\simeq}
$$

is a homotopy equivalence by induction on n. The case $n = 1$ was done in the previous step. Now we claim that there is a pullback diagram as follows:

$$
\begin{array}{ccc}
(\mathcal{C}^{I^n})^{\simeq} & \longrightarrow & (\mathcal{C}^{I^{n-1}})^{\simeq} \\
\downarrow & & \downarrow \\
(\mathcal{C}^{\Delta^1})^{\simeq} & \longrightarrow & \mathcal{C}^{\simeq}
\end{array}
$$

Since $I^n = I^{n-1} \amalg \Delta^1$, the diagram is a pullback before applying groupoid cores, and the two maps with target $\mathcal{C}$ are isofibrations (because $\mathcal{C}$ is an ∞-category and the map $\mathcal{C} \to \Delta^0$ is an isofibration). As in the proof of Corollary 2.2.7, it hence suffices to observe that the pullback of groupoid cores is itself an ∞-groupoid. But this is the case, because the right vertical map is a Kan fibration (since it is an isofibration before applying the groupoid core). The map $\mathcal{C} \to \mathcal{D}$ induces a map from this square to the corresponding square where $\mathcal{C}$ is replaced by $\mathcal{D}$ throughout. On all spots except the top-left spot, this map is a homotopy equivalence by the inductive assumption. Therefore, we can conclude the claim by Lemma 2.3.16.

Next, we deal with the case $X = \Delta^n$. For this, we consider the diagram

$$
\begin{array}{ccc}
(\mathcal{C}^{\Delta^n})^{\simeq} & \longrightarrow & (\mathcal{D}^{\Delta^n})^{\simeq} \\
\downarrow & & \downarrow \\
(\mathcal{C}^{I^n})^{\simeq} & \longrightarrow & (\mathcal{D}^{I^n})^{\simeq}
\end{array}
$$

induced by the functor $\mathcal{C} \to \mathcal{D}$ and the inclusion $I^n \to \Delta^n$. By the previous step, the lower horizontal map is a homotopy equivalence, and by Theorem 1.3.37 the vertical maps are trivial fibrations before applying the groupoid core, and thus remain trivial fibrations after applying the groupoid core. (The square obtained by restricting to groupoid cores is a pullback, since trivial fibrations are conservative.) Since trivial fibrations are homotopy equivalences, we conclude the claim by the 3-for-2 property for homotopy equivalences.

Next we deal with an arbitrary but finite-dimensional simplicial set X. We prove the statement by induction over the dimension. For 0-dimensional X, it follows again from the fact that products of Joyal equivalences are Joyal equivalences. Let us prove the inductive step and assume that X is an n-dimensional simplicial set. Consider its skeletal pushout

$$
\begin{array}{ccc}
\coprod \partial \Delta^n & \longrightarrow & \mathrm{sk}_{n-1}(X) \\
\downarrow & & \downarrow \\
\coprod \Delta^n & \longrightarrow & X
\end{array}
$$

which induces a pullback square

$$
\begin{array}{ccc}
(\mathcal{C}^X)^{\simeq} & \longrightarrow & (\mathcal{C}^{\mathrm{sk}_{n-1}(X)})^{\simeq} \\
\downarrow & & \downarrow \\
\prod(\mathcal{C}^{\Delta^n})^{\simeq} & \longrightarrow & \prod(\mathcal{C}^{\partial \Delta^n})^{\simeq}
\end{array}
$$

where the lower horizontal map is a product of Kan fibrations, and hence itself a Kan fibration. We conclude this case by Lemma 2.3.16.

In order to prove the general case, we now write an arbitrary simplicial set X as the $\mathbb{N}$-indexed colimit over its skeleta. We then obtain an isomorphism

$$
(\mathcal{C}^X)^{\simeq} \cong \lim_n (\mathcal{C}^{\mathrm{sk}_n(X)})^{\simeq},
$$

and all transition maps in the diagram describing the inverse limit are Kan fibrations (since they are restrictions along monomorphisms). We conclude the claim by using Lemma 2.3.17.   □

2.4   Localizations

In this section, we want to study a further construction of ∞-categories which will play a role later as well, namely by universally inverting a chosen set of morphisms into a given ∞-category. Such a construction is called a Dwyer–Kan localization and appears in many contexts of mathematics. Universally inverting morphisms in the context of ordinary category theory has been known to be fruitful for a long time; getting a concrete handle on the localization, however, is typically hard. The same holds true in the context of ∞-categories, and we will discuss particularly nice Dwyer–Kan localizations called Bousfield localizations in the later Sect. 5.1.

Definition 2.4.1

Let $\mathcal{C}$ be an ∞-category and let $S \subseteq \mathcal{C}_1$ be a subset of the morphisms of $\mathcal{C}$. For an auxiliary ∞-category $\mathcal{D}$, we let $\text{Fun}^S(\mathcal{C}, \mathcal{D}) \subseteq \text{Fun}(\mathcal{C}, \mathcal{D})$ be the full subcategory consisting of those functors $f: \mathcal{C} \to \mathcal{D}$ such that $f(S) \subseteq \mathcal{D}^\simeq$, i.e., where f maps the morphisms of S to equivalences in $\mathcal{D}$. If S consists of all morphisms, we will write $\text{Fun}^\simeq(\mathcal{C}, \mathcal{D})$ for $\text{Fun}^{\mathcal{C}_1}(\mathcal{C}, \mathcal{D})$.

Definition 2.4.2

Let $\mathcal{C}$ be an ∞-category and let $S \subseteq \mathcal{C}_1$ be a subset of the morphisms of $\mathcal{C}$. A functor $\mathcal{C} \to \mathcal{C}[S^{-1}]$ is called a *Dwyer–Kan localization* of $\mathcal{C}$ along S, if for every auxiliary ∞-category $\mathcal{D}$, the restriction functor

$$\text{Fun}(\mathcal{C}[S^{-1}], \mathcal{D}) \longrightarrow \text{Fun}(\mathcal{C}, \mathcal{D})$$

is fully faithful and its essential image consists of those functors that send S to equivalences.

Remark 2.4.3

By Theorem 2.3.20, this definition is equivalent to saying that the restriction functor factors through a Joyal equivalence $\text{Fun}(\mathcal{C}[S^{-1}], \mathcal{D}) \to \text{Fun}^S(\mathcal{C}, \mathcal{D})$.

Lemma 2.4.4

If a localization exists, then it is uniquely determined up to Joyal equivalence.

Proof Let $i: \mathcal{C} \to X$ and $j: \mathcal{C} \to Y$ be localizations of $\mathcal{C}$ along S. By the universal property, we obtain a diagram

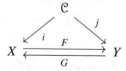

where F is a functor such that $Fi \simeq j$ and G is a functor such that $Gj \simeq i$. We want to show that $FG \simeq \text{id}_Y$ and that $GF \simeq \text{id}_X$. By the universal property, it suffices again to show that these equations hold after precomposition with j and i, respectively. There, we find that

$$FGj \simeq Fi \simeq j,$$

and likewise that

$$GFi \simeq Gj \simeq i.$$

□

In order to prove that localizations exist, we first need the following lemma.

Lemma 2.4.5
The map $\Delta^1 \to J$ is a localization at the unique morphism from 0 to 1.

Proof Let $\mathcal{D}$ be an ∞-category. We already know that the restriction map factors as follows:

$$\mathrm{Fun}(J, \mathcal{D}) \to \mathrm{Fun}^{\simeq}(\Delta^1, \mathcal{D}) \subseteq \mathrm{Fun}(\Delta^1, \mathcal{D})$$

Now we need to show that the first map is a Joyal equivalence. We will show that it is in fact a trivial fibration and then conclude the lemma using Proposition 2.2.12. For this, we consider a filtration $F_k(J)$ of J with $F_1(J) = \Delta^1$. In order to define $F_k(J)$ for $k > 1$, we consider the non-degenerate k-simplex $v_k \colon \Delta^k \to J$ given by the string of composable morphisms

$$0 \to 1 \to 0 \to \ldots,$$

and we let $F_k(J)$ be the smallest sub-simplicial set of J which contains this k-simplex. Note that $v_1(\Delta^1) \subseteq J$ is the canonical inclusion. In addition, for each $k \geq 2$, we have that $v_k(\Delta^{\{0,1\}}) = v_1(\Delta^1) \subseteq J$. We claim that there is a pushout diagram as follows:

$$
\begin{array}{ccc}
\Lambda_0^k & \longrightarrow & F_{k-1}(J) \\
\downarrow & & \downarrow \\
\Delta^k & \longrightarrow & F_k(J)
\end{array}
$$

In order to prove this, we observe that clearly the image of $v_k \cup F_{k-1}(J)$ equals $F_k(J)$. It then suffices to see that their intersection is given by Λ_0^k. For this, we consider the composite $\Delta^{[k]\setminus\{i\}} \to \Delta^k \to F_k(J)$. For $i = 0$, it is given by the sequence of $k - 1$ composable maps

$$1 \to 0 \to 1 \to \ldots,$$

which is not contained in $F_{k-1}(J)$. However, if $i \neq 0$, then it is given by a sequence starting with 0 of length $k - 1$ and is hence contained in $F_{k-1}(J)$ by definition.

We want to show that the map

$$\mathrm{Fun}(J, \mathcal{D}) \to \mathrm{Fun}^{\simeq}(\Delta^1, \mathcal{D})$$

is a Joyal equivalence. This map factors as follows:

$$\mathrm{Fun}(J, \mathcal{D}) \to \mathrm{Fun}^{\Delta^1}(F_k(J), \mathcal{D}) \to \mathrm{Fun}^{\Delta^1}(F_{k-1}(J), \mathcal{D}) \to \mathrm{Fun}^{\simeq}(\Delta^1, \mathcal{D})$$

We will show that the map in the middle is a trivial fibration for all $k \geq 2$. Then it follows that the map

$$\mathrm{Fun}(J, \mathcal{D}) \cong \lim_k \mathrm{Fun}^{\Delta^1}(F_k(J), \mathcal{D}) \to \mathrm{Fun}^{\simeq}(\Delta^1, \mathcal{D})$$

is also a trivial fibration and hence a Joyal equivalence. We thus need to show that for every commutative diagram

$$
\begin{array}{ccccc}
\partial\Delta^n & \longrightarrow & \mathrm{Fun}^{\Delta^1}(F_k(J), \mathcal{D}) & \longrightarrow & \mathrm{Fun}^{\Delta^1}(\Delta^k, \mathcal{D}) \\
\downarrow & & \downarrow & & \downarrow \\
\Delta^0 \xrightarrow{\;x\;} \Delta^n & \longrightarrow & \mathrm{Fun}^{\Delta^1}(F_{k-1}(J), \mathcal{D}) & \longrightarrow & \mathrm{Fun}^{\Delta^1}(\Lambda_0^k, \mathcal{D})
\end{array}
$$

there exists a dashed arrow which makes everything commute. We claim that it suffices to find a dotted arrow: It is clear that the right square is a pullback if we drop the superscript Δ^1, so that the pullback consists of all functors $F_k(J) \to \mathcal{D}$ whose restriction to Δ^k sends Δ^1 to an equivalence. This shows that the right square is a pullback.

By adjunction, this lifting problem corresponds to the lifting problem

$$
\begin{array}{ccccccc}
\Delta^1 & \longrightarrow & \Lambda_0^k & \longrightarrow & \mathrm{Fun}(\Delta^n, \mathcal{D}) & \xrightarrow{\;\mathrm{ev}_x\;} & \mathcal{D} \\
& & \downarrow & & \downarrow & & \\
& & \Delta^k & \longrightarrow & \mathrm{Fun}(\partial\Delta^n, \mathcal{D}) & &
\end{array}
$$

and we observe that the top horizontal composite is an equivalence for every object x of Δ^n. Therefore, the dashed arrow exists by Joyal's extension theorem, because the right vertical map is an inner fibration and the functor $\mathrm{Fun}(\Delta^n, \mathcal{D}) \to \prod_x \mathcal{D}$ is conservative by Theorem 2.2.1. $\qquad\square$

Lemma 2.4.6
For every ∞-category $\mathcal{C}$, there exists a localization along all morphisms of $\mathcal{C}$.

Proof We first construct an anodyne map $f : \mathcal{C} \to X$ to a Kan complex (an ∞-groupoid) X, by letting Y be the pushout

$$
\begin{array}{ccc}
\coprod_{\alpha \in \mathcal{C}_1} \Delta^1 & \longrightarrow & \mathcal{C} \\
\downarrow & & \downarrow f \\
\coprod_{\alpha \in \mathcal{C}_1} J & \longrightarrow & Y
\end{array}
$$

Since $\Delta^1 \to J$ is anodyne (this follows from the proof of Lemma 2.4.5), it follows that f is indeed anodyne. Then we take an inner-anodyne map $g : Y \to X$ with X an ∞-category: By the small object argument, we can factor the map $Y \to *$ through an inner-anodyne map followed by an inner fibration. Since g is inner-anodyne, the composite gf is anodyne. We claim that X is in fact an ∞-groupoid. For this, we need to show that its homotopy category is a groupoid.

Since the map $Y \to X$ is inner-anodyne, it is a Joyal equivalence, and thus induces an equivalence on homotopy categories. Furthermore, taking homotopy categories is left-adjoint to the nerve functor, and hence preserves pushouts, so that there is a pushout of categories

$$
\begin{array}{ccc}
\coprod_{\alpha \in \mathcal{C}_1}[1] & \longrightarrow & h\mathcal{C} \\
\downarrow & & \downarrow \\
\coprod_{\alpha \in \mathcal{C}_1} J & \longrightarrow & hY
\end{array}
$$

Therefore, hY is obtained from $h\mathcal{C}$ by inverting all morphisms in $\mathcal{C}$ (Exercise 96). In particular, it is a groupoid, and thus hX is a groupoid as well.

Finally, we claim that, for every ∞-category $\mathcal{D}$, the restriction functor $\mathrm{Fun}(X, \mathcal{D}) \to \mathrm{Fun}(\mathcal{C}, \mathcal{D})$ factors through a trivial fibration

$$
\mathrm{Fun}(X, \mathcal{D}) \to \mathrm{Fun}^{\simeq}(\mathcal{C}, \mathcal{D}).
$$

Since trivial fibrations are Joyal equivalences by Proposition 2.2.12, the map $\mathcal{C} \to X$ is a localization.

It is clear that $\mathrm{Fun}(X, \mathcal{D}) \to \mathrm{Fun}(\mathcal{C}, \mathcal{D})$ factors through $\mathrm{Fun}^{\simeq}(\mathcal{C}, \mathcal{D})$, since the fact that X is an ∞-groupoid implies that every morphism in $\mathcal{C}$ maps to an equivalence in X. The map of interest now factors as

$$
\mathrm{Fun}(X, \mathcal{D}) \to \mathrm{Fun}(Y, \mathcal{D}) \to \mathrm{Fun}^{\simeq}(\mathcal{C}, \mathcal{D}) \subseteq \mathrm{Fun}(\mathcal{C}, \mathcal{D}).
$$

The first map is a trivial fibration by Theorem 1.3.37, part (2), since $Y \to X$ is an inner fibration. The second map sits inside a diagram

$$
\begin{array}{ccc}
\mathrm{Fun}(Y,\mathcal{D}) & \longrightarrow \mathrm{Fun}^{\simeq}(\mathcal{C},\mathcal{D}) & \longrightarrow \mathrm{Fun}(\mathcal{C},\mathcal{D}) \\
\downarrow & \downarrow & \downarrow \\
\prod_{\alpha \in \mathcal{C}_1} \mathrm{Fun}(J,\mathcal{C}) \xrightarrow{\simeq} \prod_{\alpha \in \mathcal{C}_1} \mathrm{Fun}^{\simeq}(\Delta^1,\mathcal{C}) & \longrightarrow & \prod_{\alpha \in \mathcal{C}_1} \mathrm{Fun}(\Delta^1,\mathcal{D})
\end{array}
$$

where the big square is a pullback by definition of Y, and the right square is a pullback by closer inspection. It follows that the left square is a pullback as well. By Lemma 2.4.5, the left lower horizontal map is a trivial fibration, thus the same applies to the upper horizontal map. This finishes the proof of the lemma. □

Remark 2.4.7

We will see later that the association of sending $\mathcal{C}$ to the localization along all morphisms is a left adjoint to the inclusion of ∞-groupoids into ∞-categories (as ∞-functors between ∞-categories), see Proposition 5.1.12.

Proposition 2.4.8

For every $S \subseteq \mathcal{C}_1$, there exists a localization of $\mathcal{C}$ along S.

Proof For every subset $S \subseteq \mathcal{C}_1$, there is a smallest subcategory $\mathcal{C}_S$ of $\mathcal{C}$ which contains S: This is clear for categories, and the statement for our case follows by pulling back the corresponding subcategory of the homotopy category of $\mathcal{C}$. It follows easily that a localization of $\mathcal{C}$ along $\mathcal{C}_S$ is a localization of $\mathcal{C}$ along S, see also Exercise 93.

We thus take a localization of $\mathcal{C}_S$ along all morphisms, more precisely an inner-anodyne map $\mathcal{C}_S \to X$ to an ∞-groupoid X as in Lemma 2.4.6. Then we consider the pushout

$$
\begin{array}{ccc}
\mathcal{C}_S & \longrightarrow & \mathcal{C} \\
\downarrow & & \downarrow \\
X & \longrightarrow & W
\end{array}
$$

and an inner-anodyne map $g \colon W \to \mathcal{D}$ with $\mathcal{D}$ an ∞-category. Then, for an auxiliary ∞-category $\mathcal{E}$, we consider the diagram

$$
\begin{array}{ccc}
\mathrm{Fun}(\mathcal{D},\mathcal{E}) \xrightarrow{g^*} \mathrm{Fun}(W,\mathcal{E}) & \longrightarrow & \mathrm{Fun}^S(\mathcal{C},\mathcal{E}) \\
\downarrow & & \downarrow \\
\mathrm{Fun}(X,\mathcal{E}) & \longrightarrow & \mathrm{Fun}^S(\mathcal{C}_S,\mathcal{E})
\end{array}
$$

and we claim that the right-hand square is a pullback diagram: For this, it suffices to observe that every functor $W \to \mathcal{E}$ sends the morphisms in the image of S to equivalences, which follows from the fact that they are sent to equivalences in X. Moreover, the lower horizontal map is a trivial fibration by the previous step, thus the upper horizontal map is also a trivial fibration. The map g^* is a trivial fibration, since $W \to \mathcal{D}$ is inner-anodyne. Therefore, the upper composite is a trivial fibration, and thus a Joyal equivalence. This shows that the map $\mathcal{C} \to \mathcal{D}$ is a localization along S.   □

Apart from the fact that the procedure of "universally inverting" morphisms produces many interesting examples of ∞-categories (even if the category that we start out with is an ordinary category), we will use it in Lemma 2.4.12 to prove a certain factorization property of functors between ∞-categories.

Example Consider the 1-category Cat_∞^1 of ∞-categories, i.e., the full subcategory of sSet whose objects are the ∞-categories. Recall that the ∞-category Cat_∞ of ∞-categories is given by the homotopy-coherent nerve $\mathrm{N}(\mathrm{Cat}_\infty^1)$ of this category, with its canonical Kan enrichment given by $\mathrm{Fun}(\mathcal{C}, \mathcal{D})^\simeq$. The identity of Cat_∞^1 canonically refines to a functor between simplicial categories, with constant simplicial enrichment on the domain and the canonical simplicial enrichment on the target. In other words, we obtain a canonical functor of ∞-categories

$$\mathrm{Cat}_\infty^1 \longrightarrow \mathrm{Cat}_\infty.$$

This functor sends Joyal equivalences to equivalences: By definition, a Joyal equivalence is a map of ∞-categories which becomes an equivalence in the ∞-category Cat_∞, see Definition 2.1.14. It follows that this functor induces a functor

$$\mathrm{Cat}_\infty^1[\mathrm{Joy}^{-1}] \longrightarrow \mathrm{Cat}_\infty$$

where Joy denotes the set of Joyal equivalences.

Example Likewise, there is a canonical functor $\mathrm{Kan} \longrightarrow \widehat{\mathrm{Spc}}$ which sends homotopy equivalences to equivalences in $\widehat{\mathrm{Spc}}$. Hence, there is an induced functor

$$\mathrm{Kan}[\mathrm{he}^{-1}] \longrightarrow \widehat{\mathrm{Spc}},$$

where he denotes the (large) set of homotopy equivalences between Kan complexes.

Lemma 2.4.9

The inclusions $\mathrm{Kan} \to \mathrm{sSet}$ *and* $\mathrm{Cat}_\infty^1 \to \mathrm{sSet}$ *induce equivalences*

$$\mathrm{Kan}[\mathrm{he}^{-1}] \simeq \mathrm{sSet}[\mathrm{we}^{-1}] \text{ and } \mathrm{Cat}_\infty^1[\mathrm{Joy}^{-1}] \simeq \mathrm{sSet}[\mathrm{Joy}^{-1}].$$

Proof The small object argument provides functors $F \colon \mathrm{sSet} \to \mathrm{Cat}_\infty^1$ and $G \colon \mathrm{sSet} \to \mathrm{Kan}$ by functorially factoring the map $X \to *$ into an inner-anodyne map followed by an inner fibration, respectively by an anodyne map followed by a Kan fibration. We claim that these functors send Joyal equivalences to Joyal equivalences, respectively weak homotopy equivalences to weak homotopy equivalences. In order to see this, suppose that $X \to Y$ is such an equivalence. Then we have a commutative diagram

$$
\begin{array}{ccc}
X & \longrightarrow & \mathcal{C} \\
\downarrow & & \downarrow \\
Y & \longrightarrow & \mathcal{D}
\end{array}
$$

where the horizontal maps are inner-anodyne and hence Joyal equivalences. By the 3-for-2 property for Joyal equivalences, we find that $\mathcal{C} \to \mathcal{D}$ is a Joyal equivalence if and only $X \to Y$ is a Joyal equivalence. The argument for weak homotopy equivalences is the same. Hence this functor induces a functor

$$\mathrm{sSet}[\mathrm{Joy}^{-1}] \longrightarrow \mathrm{Cat}_\infty^1[\mathrm{Joy}^{-1}],$$

which we claim to be an inverse to the canonical functor

$$\mathrm{Cat}_\infty^1[\mathrm{Joy}^{-1}] \longrightarrow \mathrm{sSet}[\mathrm{Joy}^{-1}]$$

as induced by the inclusion.

For this, we observe that the map $X \to \mathcal{C} = F(X)$ determines a natural transformation from the identity of sSet to the composite $iF \colon \mathrm{sSet} \to \mathrm{sSet}$. More precisely, it determines a map

$$\Delta^1 \longrightarrow \mathrm{Fun}(\mathrm{sSet}, \mathrm{sSet}) \longrightarrow \mathrm{Fun}(\mathrm{sSet}, \mathrm{sSet}[\mathrm{Joy}^{-1}])$$

which we claim to land inside the full subcategory consisting of those functors that send Joyal equivalences to equivalences: This is because F (and clearly i as well) send Joyal equivalences to Joyal equivalences, as we had just argued. Hence, by the universal property of localizations, we obtain a map

$$\Delta^1 \longrightarrow \mathrm{Fun}(\mathrm{sSet}[\mathrm{Joy}^{-1}], \mathrm{sSet}[\mathrm{Joy}^{-1}])$$

whose restriction to 0 and 1 is given by the identity and a functor whose restriction to sSet is induced by the composite Fi. For a fixed object $X \in$ sSet, the resulting morphism is given by $X \to F(X)$, which we argued to be a Joyal equivalence. In particular, the 1-simplex given above is a natural equivalence between the identity of sSet[Joy^{-1}] and the functor induced by Fi.

It remains to show that the functor induced by the composite iF is an equivalence as well. For this, we use analogous arguments to find that the map $X \to FX$ induces a transformation id $\to iF$ which is pointwise a Joyal equivalence. This shows the corollary for Joyal equivalences, and the argument for weak homotopy equivalences is the same. □

The following theorem is very important, but beyond the scope of this book. A similar result holds in general for simplicial model categories, see [Lur17, Theorem 1.3.4.20]. We will discuss some steps of the proof of this result later, see Corollary 3.2.26.

Theorem 2.4.10
The canonical functors sSet[we^{-1}] $\to$ Spc *and* sSet[Joy^{-1}] $\to$ Cat$_\infty$ *are equivalences of ∞-categories.*

We finish this section with a useful factorization construction.

Definition 2.4.11
Let $f : \mathcal{C} \to \mathcal{D}$ be a functor between ∞-categories. We define the *path-fibration* $P(f)$ of f by the pullback

$$
\begin{array}{ccc}
P(f) & \longrightarrow & \mathrm{Fun}(J, \mathcal{D}) \\
\downarrow & & \downarrow{\scriptstyle s} \\
\mathcal{C} & \xrightarrow{\ f\ } & \mathcal{D}
\end{array}
$$

of simplicial sets, where s is the source map.

Consider the map $\mathcal{C} \to \mathrm{Fun}(J, \mathcal{D})$, which is adjoint to the map $\mathcal{C} \times J \to \mathcal{C} \to \mathcal{D}$. It sends an object x to the identity morphism of $f(x)$. Clearly, the composition of this functor with s is given by f, therefore we obtain an induced map $c : \mathcal{C} \to P(f)$ which is a section of the canonical map $P(f) \to \mathcal{C}$. In particular, c is a monomorphism. Furthermore, the composite of $P(f) \to \mathrm{Fun}(J, \mathcal{D})$ with the target map $\mathrm{Fun}(J, \mathcal{D}) \to \mathcal{D}$ yields a composite

$$
\mathcal{C} \xrightarrow{c} P(f) \xrightarrow{t} \mathcal{D}.
$$

This composite is given by f, so we gained a factorization of f through $P(f)$.

Lemma 2.4.12

Let $f\colon \mathcal{C} \to \mathcal{D}$ be a functor between ∞-categories and let $\mathcal{C} \xrightarrow{c} P(f) \xrightarrow{t} \mathcal{D}$ be the factorization which we just constructed. Then c is a Joyal equivalence and t is an isofibration. In particular, any map $f\colon \mathcal{C} \to \mathcal{D}$ between ∞-categories can be factored into a Joyal equivalence which is a monomorphism, followed by an isofibration.

Proof We claim that the source map $s\colon \mathrm{Fun}(J,\mathcal{D}) \to \mathcal{D}$ is a trivial fibration. By Lemma 2.2.17, it suffices to show that it is a Joyal equivalence and an isofibration. There are many arguments that $\Delta^0 \to J$ is a Joyal equivalence (e.g., both are ordinary categories and the functor is clearly an equivalence of ordinary categories), so it follows from Proposition 2.2.8 that the induced map $\mathrm{Fun}(J,\mathcal{D}) \to \mathrm{Fun}(\Delta^0,\mathcal{D}) \cong \mathcal{D}$ is a Joyal equivalence as well. By Proposition 2.2.5, the restriction along a monomorphism which maps into an ∞-category is an isofibration, so we conclude that $\mathrm{Fun}(J,\mathcal{D}) \to \mathcal{D}$ is a trivial fibration as claimed. Thus, as a pullback of this map, the functor $P(f) \to \mathcal{C}$ is also a trivial fibration. Since the composite

$$\mathcal{C} \to P(f) \to \mathcal{C}$$

is the identity (by construction), it follows from the 3-for-2 property for Joyal equivalences that the map $\mathcal{C} \to P(f)$ is a Joyal equivalence as claimed.

In order to see that $P(f) \to \mathcal{D}$ is an isofibration, we observe that the square

$$
\begin{array}{ccc}
P(f) & \longrightarrow & \mathrm{Fun}(J,\mathcal{D}) \\
\downarrow & & \downarrow \\
\mathcal{C} \times \mathcal{D} & \xrightarrow{\ f \times \mathrm{id}\ } & \mathcal{D} \times \mathcal{D}
\end{array}
$$

is also a pullback and that, again by Proposition 2.2.5, the right vertical map is an isofibration. It follows that $P(f) \to \mathcal{C} \times \mathcal{D}$ is an isofibration as well, see Corollary 2.1.22. Since the projection $\mathcal{C} \times \mathcal{D} \to \mathcal{D}$ is also an isofibration, the lemma is proven. $\square$

2.5 Fat Joins, Fat Slices and Mapping Spaces

In this section, we will construct an alternative join and show that it is Joyal-equivalent to the construction of Definition 1.4.8, again following Rezk's notes [Rez20]. This result is then used to compare different models of mapping spaces in an ∞-category. We will also derive some more useful lifting properties for isofibrations. Fat joins and fat slices will also turn out to be particularly useful for discussing limits and colimits in ∞-categories. We will finish the section with a comparison between the mapping spaces in the coherent nerve of a simplicial category and the hom-simplicial sets present in the simplicial category, although without proof.

Definition 2.5.1

Let X and Y be simplicial sets. We define a new simplicial set $X \diamond Y$ to be the pushout

$$
\begin{array}{ccc}
X \times Y \times \partial \Delta^1 & \longrightarrow & X \amalg Y \\
\downarrow & & \downarrow \\
X \times Y \times \Delta^1 & \longrightarrow & X \diamond Y
\end{array}
$$

where the top horizontal arrow sends the triple $(x, y, 0)$ to x and the triple $(x, y, 1)$ to y. This simplicial set is called *fat join*.

Lemma 2.5.2
For fixed Y, the association $X \mapsto X \diamond Y$ extends to a colimit-preserving functor sSet $\to$ sSet$_{Y/}$. *The same holds for the association $X \mapsto Y \diamond X$.*

Proof The first statement is an easy calculation: We need to check that for any morphism $X \to X'$, the diagram

$$
\begin{array}{ccc}
Y & \longrightarrow & X \diamond Y \\
 & \searrow & \downarrow \\
 & & X' \diamond Y
\end{array}
$$

commutes. But this follows simply from the fact that the diagram

$$
\begin{array}{ccc}
Y & \longrightarrow & X \amalg Y \\
 & \searrow & \downarrow \\
 & & X' \amalg Y
\end{array}
$$

commutes. In order to see that this functor commutes with colimits, it again suffices to show that it commutes with coproducts and coequalizers. So let X, X' and Y be simplicial sets. We need to show that the diagram

$$
\begin{array}{ccc}
Y & \longrightarrow & X \diamond Y \\
\downarrow & & \downarrow \\
X' \diamond Y & \longrightarrow & (X \amalg X') \diamond Y
\end{array}
$$

is a pushout in simplicial sets. But this follows immediately from the definition. Next, we recall that coequalizers in $\mathrm{sSet}_{Y/}$ are calculated via the forgetful functor $\mathrm{sSet}_{Y/} \to \mathrm{sSet}$, see Lemma 1.4.14. So let

$$X \rightrightarrows X' \longrightarrow C$$

be a coequalizer diagram. We need to show that applying $- \diamond Y$ yields again a coequalizer diagram. Then the functors $- \times Y \times \partial\Delta^1$ and $- \times Y \times \Delta^1$ are left-adjoint and hence preserve this coequalizer diagram. The diagram

$$X \amalg Y \rightrightarrows X' \amalg Y \longrightarrow C \amalg Y$$

is indeed a coequalizer: Given a map $X' \amalg Y \to T$ whose precomposition with the two given maps leads to the same map $X \amalg Y \to T$, we find that there is a unique map from C to T and from Y to T as needed. Now since two colimits always commute, we may take the pushout over the coequalizers and obtain the coequalizer of the pushout. This proves the lemma. □

Observation 2.5.3
There exists a canonical map of simplicial sets $X \diamond Y \to \Delta^1$ induced by the commutative diagram

$$
\begin{array}{ccc}
X \times Y \times \partial\Delta^1 & \longrightarrow & X \amalg Y \\
\downarrow & & \downarrow \\
X \times Y \times \Delta^1 & \xrightarrow{\ \mathrm{pr}\ } & \Delta^1
\end{array}
$$

where the right vertical map is given by $X \to \Delta^{\{0\}}$, combined with $Y \to \Delta^{\{1\}}$.

Lemma 2.5.4
Let X and Y be simplicial sets. Then there exists a canonical map $X \diamond Y \to X \star Y$ which commutes with the projections to Δ^1 and the inclusions of X and Y. This map is functorial in X and Y.

Proof Recall from Lemma 1.4.10 that for every map $p\colon K \to \Delta^1$ of simplicial sets, there exists a canonical factorization into $K \to K_0 \star K_1 \to \Delta^1$, where $K_i = p^{-1}(\{i\})$. By applying this result to the map $X \diamond Y \to \Delta^1$ which we have just constructed, we find that $(X \diamond Y)_0$ is given by the pushout

$$
\begin{array}{ccc}
X \times Y \times \{0\} & \longrightarrow & X \\
\downarrow & & \downarrow \\
X \times Y \times \{0\} & \longrightarrow & X
\end{array}
$$

and likewise that $(X \diamond Y)_1$ is given by Y. $\square$

We will now need Lemma 2.3.14 (Reedy's lemma) in the following context.

Proposition 2.5.5
Consider a pullback diagram of ∞-categories

$$
\begin{array}{ccc}
\mathcal{C} & \xrightarrow{\ f\ } & \mathcal{D} \\
{\scriptstyle p'}\downarrow & & \downarrow{\scriptstyle p} \\
\mathcal{C}' & \xrightarrow{\ f'\ } & \mathcal{D}'
\end{array}
$$

where the map p is an isofibration and the map f' is a Joyal equivalence. Then the map f is also a Joyal equivalence.

Proof We will show that for every ∞-category $\mathcal{E}$, the induced map

$$
\pi_0(\mathrm{Fun}(\mathcal{E}, \mathcal{C})^{\simeq}) \xrightarrow{\ f_*\ } \pi_0(\mathrm{Fun}(\mathcal{E}, \mathcal{D})^{\simeq})
$$

is a bijection. Once this is shown, we can choose $\mathcal{E} = \mathcal{D}$ and find a functor $g \colon \mathcal{D} \to \mathcal{C}$ such that $f_*(g) = fg \simeq \mathrm{id}_{\mathcal{D}}$. It is then easy to see that g is an inverse of f.

In order to show that f_* above is indeed a bijection, observe that the diagram

$$
\begin{array}{ccc}
\mathrm{Fun}(\mathcal{E}, \mathcal{C})^{\simeq} & \longrightarrow & \mathrm{Fun}(\mathcal{E}, \mathcal{D})^{\simeq} \\
\downarrow & & \downarrow \\
\mathrm{Fun}(\mathcal{E}, \mathcal{C}')^{\simeq} & \longrightarrow & \mathrm{Fun}(\mathcal{E}, \mathcal{D}')^{\simeq}
\end{array}
$$

is a pullback diagram of Kan complexes, where the right vertical map is a Kan fibration and the lower horizontal map is a homotopy equivalence. In Lemma 2.3.14, we have shown that this implies that the top horizontal arrow is also a homotopy equivalence, and hence in particular induces a bijection on path components. $\square$

Remark 2.5.6
In fact, in the proof of Lemma 2.3.14, we have explicitly shown that in the situation of a pullback diagram of Kan complexes

$$
\begin{array}{ccc}
C \times_A B & \longrightarrow & B \\
\downarrow & & \downarrow \\
C & \longrightarrow & A
\end{array}
$$

(continued)

2.5.6 (continued)

such that the map $B \to A$ is a Kan fibration and the map $C \to A$ is a homotopy equivalence, the induced map $\pi_0(C \times_A B) \to \pi_0(B)$ is bijective, which is all that we have used in the above argument.

Lemma 2.5.7

Consider a diagram of ∞-categories

$$
\begin{array}{ccc}
\mathcal{C} & \longrightarrow\!\!\!\!\!\!\rightarrow \mathcal{D} & \longleftarrow \mathcal{E} \\
\downarrow{\scriptstyle\simeq} & \downarrow{\scriptstyle\simeq} & \downarrow{\scriptstyle\simeq} \\
\mathcal{C}' & \longrightarrow\!\!\!\!\!\!\rightarrow \mathcal{D}' & \longleftarrow \mathcal{E}'
\end{array}
$$

where the vertical maps are Joyal equivalences and the left horizontal maps are isofibrations. Then the induced map on pullbacks

$$
\mathcal{C} \times_{\mathcal{D}} \mathcal{E} \to \mathcal{C}' \times_{\mathcal{D}'} \mathcal{E}'
$$

is again a Joyal equivalence.

Proof Copying the proof of Lemma 2.3.16, we again first explain how to reduce the proof to the case where the two maps $\mathcal{E} \to \mathcal{D}$ and $\mathcal{E}' \to \mathcal{D}'$ are also isofibrations. Here, we simply factor the maps functorially, as in Lemma 2.4.12, into a monomorphism which is a Joyal equivalence followed by an isofibration, and we obtain a diagram

$$
\begin{array}{ccccc}
\mathcal{C} & \longrightarrow\!\!\!\!\!\!\rightarrow \mathcal{D} & \longleftarrow\!\!\!\!\!\!\leftarrow \mathcal{P} & \longleftarrow_{\simeq} \mathcal{E} \\
\downarrow{\scriptstyle\simeq} & \downarrow{\scriptstyle\simeq} & \downarrow{\scriptstyle\simeq} & \downarrow{\scriptstyle\simeq} \\
\mathcal{C}' & \longrightarrow\!\!\!\!\!\!\rightarrow \mathcal{D}' & \longleftarrow\!\!\!\!\!\!\leftarrow \mathcal{P}' & \longleftarrow_{\simeq} \mathcal{E}'
\end{array}
$$

where all vertical maps are Joyal equivalences. Note that $\mathcal{P}$ and $\mathcal{P}'$ are also ∞-categories. We then obtain a commutative diagram

$$
\begin{array}{ccc}
\mathcal{C} \times_{\mathcal{D}} \mathcal{E} & \longrightarrow & \mathcal{C}' \times_{\mathcal{D}'} \mathcal{E}' \\
\downarrow & & \downarrow \\
\mathcal{C} \times_{\mathcal{D}} \mathcal{P} & \longrightarrow & \mathcal{C}' \times_{\mathcal{D}'} \mathcal{P}'
\end{array}
$$

and we claim that its vertical maps are Joyal equivalences: For instance, the left vertical map sits inside a diagram

$$
\begin{array}{ccccc}
\mathcal{C} \times_{\mathcal{D}} \mathcal{E} & \longrightarrow & \mathcal{C} \times_{\mathcal{D}} \mathcal{P} & \longrightarrow & \mathcal{C} \\
\downarrow & & \downarrow & & \downarrow \\
\mathcal{E} & \xrightarrow{\;\simeq\;} & \mathcal{P} & \longrightarrow & \mathcal{D}
\end{array}
$$

where both squares are pullbacks. We may thus apply Proposition 2.5.5 to the left square. (The argument for the right vertical map above is the same.)

By now, we have reduced the proof to the case where all horizontal maps in the diagram of the statement of the lemma are isofibrations.

A threefold application of Proposition 2.5.5 completes the proof as in Lemma 2.3.16, by considering the following factorization of the map in question:

$$
\mathcal{C} \times_{\mathcal{D}} \mathcal{E} \to (\mathcal{C}' \times_{\mathcal{D}'} \mathcal{D}) \times_{\mathcal{D}} \mathcal{E} \cong \mathcal{C}' \times_{\mathcal{D}'} \mathcal{E} \to \mathcal{C}' \times_{\mathcal{D}'} \mathcal{E}'.
$$

$\square$

Corollary 2.5.8
Suppose to be given a diagram of simplicial sets

$$
\begin{array}{ccccc}
X & \longleftarrow & Y & \longrightarrow & Z \\
\downarrow & & \downarrow & & \downarrow \\
X' & \longleftarrow & Y' & \longrightarrow & Z'
\end{array}
$$

where all vertical maps are Joyal equivalences and the left horizontal maps are monomorphisms. Then the induced map

$$
X \amalg_Y Z \to X' \amalg_{Y'} Z'
$$

is again a Joyal equivalence.

Proof Let $\mathcal{C}$ be an ∞-category. By definition, we need to show that the map

$$
\mathrm{Fun}(X' \amalg_{Y'} Z', \mathcal{C}) \to \mathrm{Fun}(X \amalg_Y Z, \mathcal{C})
$$

is a Joyal equivalence. But this map is isomorphic to the map

$$
\mathrm{Fun}(X', \mathcal{C}) \times_{\mathrm{Fun}(Y', \mathcal{C})} \mathrm{Fun}(Z', \mathcal{C}) \to \mathrm{Fun}(X, \mathcal{C}) \times_{\mathrm{Fun}(Y, \mathcal{C})} \mathrm{Fun}(Z, \mathcal{C})
$$

which is a Joyal equivalence by Lemma 2.5.7 and the assumptions. $\square$

Corollary 2.5.9
For all monomorphisms $A \to B$, the pushout product map

$$J \times A \amalg_{\{0\} \times A} \{0\} \times B \longrightarrow J \times B$$

is a Joyal equivalence.

Proof We consider the following diagram:

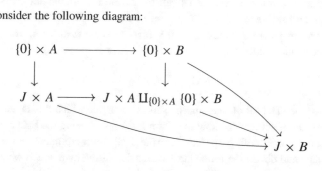

The horizontal arrows are monomorphisms, and both the left vertical arrow and the right bended arrow are Joyal equivalences: This is because $\{0\} \to J$ is a Joyal equivalence and Joyal equivalences are closed under finite products, as shown in Exercise 101. By Corollary 2.5.8, the right vertical map is also a Joyal equivalence, therefore the lemma follows from the 3-for-2 property for Joyal equivalences. $\square$

Definition 2.5.10
A map of simplicial sets is called J-anodyne if it belongs to the saturated set generated by inner-anodyne maps and maps of the form $J \times A \amalg_{\{0\} \times A} \{0\} \times B \longrightarrow J \times B$ for a monomorphism $A \to B$.

Corollary 2.5.11
Every J-anodyne map is a Joyal equivalence.

Proof We claim that monomorphisms which are in addition Joyal equivalences form a saturated set. We leave the proof of this claim as Exercise 102. It then suffices to note that inner-anodyne maps are Joyal equivalences (by Corollary 2.2.13) and that maps of the form $J \times A \amalg_{\{0\} \times A} \{0\} \times B \longrightarrow J \times B$ are also Joyal equivalences (by Corollary 2.5.9). $\square$

Next, we can use Corollary 2.5.11 to provide a smaller generating set for J-anodyne maps, all of whose domains are finite simplicial sets. This will help us to apply the small object argument as presented in this book, since it needs this technical assumption.

Proposition 2.5.12
The set of J-anodyne maps is the smallest saturated set containing inner-anodyne maps and the map $\{0\} \to J$.

Proof For notational convenience, let us call the smallest saturated set containing inner-anodyne maps and the map $\{0\} \to J$ the set of *super-J-anodyne* maps. It is clear that super-J-anodyne maps are J-anodyne, so it suffices to show that, for all monomorphisms $A \to B$, the pushout product map

$$\{0\} \times B \amalg_{\{0\} \times A} J \times A \to J \times B$$

is super-J-anodyne. The set of monomorphisms $A \to B$ for which this is the case is itself saturated, therefore it suffices to show the claim for $A \to B$ being the boundary inclusions $\partial \Delta^n \to \Delta^n$ for $n \geq 0$. For this purpose, we use the small object argument for the set of super-J-anodyne maps and obtain the following factorization of the map which we are interested in:

$$\{0\} \times \Delta^n \amalg_{\{0\} \times \partial \Delta^n} J \times \partial \Delta^n \xrightarrow{i} \mathcal{C} \xrightarrow{p} J \times \Delta^n$$

where i is super-J-anodyne and the second map satisfies the RLP with respect to inner horn inclusions and $\{0\} \to J$. Since $J \times \Delta^n$ is an ∞-category, we can deduce that the map $\mathcal{C} \to J \times \Delta^n$ is an isofibration between ∞-categories. Since J-anodyne maps are Joyal equivalences by Corollary 2.5.11, and since super-J-anodyne maps are J-anodyne, we find that the map $\mathcal{C} \to J \times \Delta^n$ is in fact a trivial fibration and thus admits a solution $s : J \times \Delta^n \to \mathcal{C}$ to the lifting problem

$$
\begin{array}{ccc}
\{0\} \times \Delta^n \amalg_{\{0\} \times \partial \Delta^n} J \times \partial \Delta^n & \xrightarrow{\quad i \quad} & \mathcal{C} \\
\downarrow & \nearrow{\scriptstyle s} & \downarrow{\scriptstyle p} \\
J \times \Delta^n & =\!=\!=\!=\!= & J \times \Delta^n
\end{array}
$$

Considering the diagram

$$
\begin{array}{ccc}
\{0\} \times \Delta^n \amalg_{\{0\} \times \partial \Delta^n} J \times \partial \Delta^n & \longrightarrow & J \times \Delta^n \\
\| & & \downarrow{\scriptstyle s} \\
\{0\} \times \Delta^n \amalg_{\{0\} \times \partial \Delta^n} J \times \partial \Delta^n & \xrightarrow{\quad i \quad} & \mathcal{C} \\
\| & & \downarrow{\scriptstyle p} \\
\{0\} \times \Delta^n \amalg_{\{0\} \times \partial \Delta^n} J \times \partial \Delta^n & \longrightarrow & J \times \Delta^n
\end{array}
$$

we find that the map that we are interested in is a retract of the super-J-anodyne map i, and hence it is itself super-J-anodyne.   □

If the target of a monomorphism is an ∞-category, then we can prove the following stronger version of Corollary 2.5.11.

Proposition 2.5.13

Let $i \colon A \to \mathcal{B}$ be a monomorphism with $\mathcal{B}$ an ∞-category. Then i is J-anodyne if and only if i is a Joyal equivalence.

Proof By Corollary 2.5.11, it suffices to show the "if" part. For this, we apply the small object argument and factor the map $A \to \mathcal{B}$ into a composition

$$A \to \mathcal{B}' \to \mathcal{B}$$

where the map $A \to \mathcal{B}'$ is J-anodyne and the map $\mathcal{B}' \to \mathcal{B}$ satisfies the RLP with respect to J-anodyne maps. Since $\mathcal{B}$ is an ∞-category, the same applies to $\mathcal{B}'$ and the map $\mathcal{B}' \to \mathcal{B}$ is an isofibration. By Corollary 2.5.11 and the 3-for-2 property for Joyal fibrations, the map $\mathcal{B}' \to \mathcal{B}$ is an isofibration and a Joyal equivalence, and thus a trivial fibration by Lemma 2.2.17. Choosing a solution $s \colon \mathcal{B} \to \mathcal{B}'$ of the lifting problem

$$
\begin{array}{ccc}
A & \longrightarrow & \mathcal{B}' \\
\downarrow & \overset{s}{\nearrow} & \downarrow \\
\mathcal{B} & = & \mathcal{B}
\end{array}
$$

we find, as in the proof of Proposition 2.5.12, that the map $A \to \mathcal{B}$ is a retract of the J-anodyne map $A \to \mathcal{B}'$, and thus is itself J-anodyne.   □

At a later stage of the book (when we show that the association $x \mapsto \mathrm{map}_{\mathcal{C}}(x, -)$ is itself functorial in x, and in the analogue of Theorem 1.4.23 for the fat slices), we will need the following theorem. Thanks to Hoang Kim Nguyen for the explanation of the needed reduction steps!

Theorem 2.5.14

Let $p \colon \mathcal{C} \to \mathcal{D}$ be an isofibration between ∞-categories and let $A \to B$ be a monomorphism which is in addition a Joyal equivalence. Then any lifting problem

$$
\begin{array}{ccc}
A & \longrightarrow & \mathcal{C} \\
\downarrow & \overset{}{\nearrow} & \downarrow p \\
B & \longrightarrow & \mathcal{D}
\end{array}
$$

admits a solution as indicated by the dashed arrow.

Proof Using the small object argument, we can factor the map $B \to \mathcal{D}$ through an J-anodyne map $B \to \mathcal{B}$ followed by a map $\mathcal{B} \to \mathcal{D}$ satisfying the RLP with respect to J-anodyne maps. It follows that $\mathcal{B}$ is an ∞-category. Since J-anodyne maps are Joyal equivalences, we find that the composite $A \to \mathcal{B}$ is a monomorphism and a Joyal equivalence. By Proposition 2.5.13, this map is J-anodyne. Since isofibrations between ∞-categories have the RLP with respect to J-anodyne maps, we can find a dashed arrow in the diagram

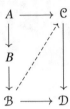

which also solves the original lifting problem.   □

Let us finish this intermezzo on J-anodyne maps with a nice fact about inner-anodyne maps. The following lemma is taken from Stevenson [Ste18a, Lemma 2.19].

Lemma 2.5.15
Let $\mathcal{C}$ be an ∞-category and let $i : A \to \mathcal{C}$ be a monomorphism which is a bijection on 0-simplices and a Joyal equivalence. Then i is inner-anodyne.

Proof By the small object argument, we may factor this map into a composite $A \xrightarrow{j} \mathcal{B} \xrightarrow{p} \mathcal{C}$, with j an inner-anodyne map and p an inner fibration. Since $\mathcal{C}$ is an ∞-category, p satisfies the assumptions of Exercise 104 and is therefore a trivial fibration. As in the proof of Proposition 2.5.13, this shows that i is a retract of j and thus inner-anodyne as well.   □

Corollary 2.5.16
Let $K \to L$ be a monomorphism which is a bijection on 0-simplices. Then the map

$$\{0\} \times L \amalg_{\{0\} \times K} J \times K \to J \times L$$

is inner-anodyne.

Proof It suffices to prove the claim for the maps $\partial \Delta^n \to \Delta^n$ with $n \geq 1$. By construction, this map is a bijection on 0-simplices, a Joyal equivalence by Corollary 2.5.9, and has an ∞-

category as target, because Δ^n and J are ∞-categories. Applying Lemma 2.5.15 concludes the corollary.

$\square$

Now we come back to important properties of fat joins and its relation to the ordinary join from Sect. 1.4.

Corollary 2.5.17

Let $X \to X'$ be a Joyal equivalence between simplicial sets and let Y be a simplicial set. Then the map $X \diamond Y \to X' \diamond Y$ is a Joyal equivalence. Likewise, the map $Y \diamond X \to Y \diamond X'$ is a Joyal equivalence.

Proof Since $X \diamond Y$ is the pushout

$$
\begin{array}{ccc}
X \times Y \times \partial\Delta^1 & \longrightarrow & X \sqcup Y \\
\downarrow & & \downarrow \\
X \times Y \times \Delta^1 & \longrightarrow & X \diamond Y
\end{array}
$$

where the left vertical map is a monomorphism, it suffices by Corollary 2.5.8 to show that the maps induced by $X \to X'$ on the other three corners are Joyal equivalences. But this follows from Exercise 101.

$\square$

Definition 2.5.18

A map of simplicial sets $f\colon X \to Y$ is said to admit a pre-inverse if maps $g\colon Y \to X$, $\tau\colon \Delta^1 \to \mathrm{Hom}(X, X)$ and $\tau'\colon \Delta^1 \to \mathrm{Hom}(Y, Y)$ exist such that

(1) $\tau_\varepsilon = \mathrm{id}_X$ and $\tau_{1+\varepsilon} = gf$, where $\varepsilon \in \{0, 1\} \cong \mathbb{Z}/2$,
(2) $\tau'_\varepsilon = \mathrm{id}_Y$ and $\tau'_{1+\varepsilon} = fg$, where again $\varepsilon \in \{0, 1\} \cong \mathbb{Z}/2$,
(3) for all objects x of X, the morphism $\tau(x)\colon \Delta^1 \to X$ represents a degenerate edge of X, and for all objects y of Y, $\tau'(y)\colon \Delta^1 \to Y$ represents a degenerate edge of Y.

Proposition 2.5.19

Let X and Y be simplicial sets. Then the canonical map $X \diamond Y \to X \star Y$ of Lemma 2.5.4 is a Joyal equivalence.

Proof As noted in Lemmas 2.5.2 and 1.4.15, both functors $- \diamond Y$ and $- \star Y$ commute with filtered colimits. Also, when viewed as taking values in $\mathrm{sSet}_{Y/}$, they both commute with coproducts. We may therefore reduce the general situation to the case where X has only finitely many non-degenerate simplices and is connected. In this case, we can write X as a pushout $X' \amalg_{\partial \Delta^n} \Delta^n$. Since pushouts are connected colimits, we also obtain isomorphisms

$$X \diamond Y \cong X' \diamond Y \amalg_{\partial \Delta^n \diamond Y} \Delta^n \diamond Y$$

and likewise for $\star$ instead of $\diamond$. It hence suffices to show the claim for X', $\partial \Delta^n$ and Δ^n. By the same reasoning, the statement for Δ^n (for all n) implies the statement for $\partial \Delta^n$ and X' by induction. Hence it remains to show that

$$\Delta^n \diamond Y \to \Delta^n \star Y$$

is a Joyal equivalence. Now the inclusion $I^n \to \Delta^n$ is inner-anodyne by Proposition 1.3.22, and thus in the diagram

$$
\begin{array}{ccc}
I^n \diamond Y & \longrightarrow & I^n \star Y \\
\downarrow & & \downarrow \\
\Delta^n \diamond Y & \longrightarrow & \Delta^n \star Y
\end{array}
$$

both vertical maps are Joyal equivalences: For the left vertical map, this is Corollary 2.5.17, and for the right vertical map, it is the fact that $- \star Y$ preserves inner-anodyne maps, see Lemma 1.4.22, part (1). (In fact, it even sends right-anodyne maps to inner-anodyne maps.) Therefore, it suffices to prove the statement for I^n. Since $I^n \cong I^{n-1} \amalg_{\Delta^0} \Delta^1$, it finally suffices to treat the case where X is either Δ^0 or Δ^1. Now we observe that Δ^0 is a retract of Δ^1, therefore the map

$$\Delta^0 \diamond Y \to \Delta^0 \star Y$$

is a retract of the map

$$\Delta^1 \diamond Y \to \Delta^1 \star Y.$$

Since retracts of Joyal equivalences are Joyal equivalences (see Exercise 102), it suffices to show that the latter is a Joyal equivalence for all Y. Performing the same reductions to Y, it finally suffices to show that the map

$$\Delta^1 \diamond \Delta^1 \to \Delta^1 \star \Delta^1 \cong \Delta^3$$

is a Joyal equivalence. In order to do so, we will (almost) construct a *pre-inverse* for this map. In fact, we will construct a zig-zag of such pre-inverses connecting the identity to a to-be-constructed inverse. In order to construct this map, we first observe from the definitions that there is a canonical quotient map $(\Delta^1)^{\times 3} \to \Delta^1 \diamond \Delta^1$. We claim that the composite

$$\mathrm{can} \colon (\Delta^1)^{\times 3} \to \Delta^1 \diamond \Delta^1 \to \Delta^3$$

is given by the formula

$$(a, b, c) = \begin{cases} a & \text{if } c = 0, \\ b+2 & \text{if } c = 1. \end{cases}$$

This is just an explicit check of the definitions. Then we consider the 3-simplex σ of $(\Delta^1)^{\times 3}$ represented by

$$(000) \to (100) \to (101) \to (111)$$

and its image in $\Delta^1 \diamond \Delta^1$. First, we observe that the composite

$$\Delta^3 \to \Delta^1 \diamond \Delta^1 \to \Delta^3$$

is the identity.

Then we must construct a map $\Delta^1 \to \mathrm{Hom}(\Delta^1 \diamond \Delta^1, \Delta^1 \diamond \Delta^1)$ which exhibits the map σ as pre-inverse of the canonical map $\Delta^1 \diamond \Delta^1 \to \Delta^3$. We will do this in three steps: First, we will construct two maps $\Delta^1 \to \mathrm{Hom}((\Delta^1)^{\times 3}, (\Delta^1)^{\times 3})$ connecting the identity to an auxiliary map Φ and Φ to $\sigma \circ \mathrm{can}$; secondly, we will show that they descend to maps between $\Delta^1 \diamond \Delta^1$; and finally, we will show that both resulting maps have the property that for fixed object, the induced edge of $\Delta^1 \diamond \Delta^1$ is a degenerate edge.

In the following picture, the left cube represents the identity of $(\Delta^1)^3$, the middle cube represents the map Φ, and the right cube represents the composite $\sigma \circ \mathrm{can}$. Note that there are evident maps from the middle cube to both outer cubes.

Now one needs to check that these maps descend, after post-composition with the projection, to the quotient $\Delta^1 \diamond \Delta^1$. For this, we first observe that all cubes restrict to endomorphisms of $\Delta^1 \times \Delta^1 \times \partial \Delta^1$. More concretely, this means that if we only look at the front layer and the back layer (i.e., we neglect the diagonal maps) and project to the third coordinate, only identity morphisms remain. Next, we find that there exist compatible endomorphisms of $\Delta^1 \amalg \Delta^1$, since the identity of $\Delta^1 \amalg \Delta^1$ is compatible for both Φ and $\sigma \circ \mathrm{can}$. This shows that Φ and $\sigma \circ \mathrm{can}$ descend to the pushout $\Delta^1 \diamond \Delta^1$.

Alternatively, one sees that $\Delta^1 \diamond \Delta^1$ is the quotient of $(\Delta^1)^{\times 3}$, where the sub-simplicial set $\Delta^1 \times \Delta^1 \times \{0\}$ is collapsed (via the first projection) to Δ^1 and the sub-simplicial set $\Delta^1 \times \Delta^1 \times \{1\}$ is collapsed (via the second projection) to Δ^1. We thus have to check if Φ and $\sigma \circ \mathrm{can}$ followed by this projection are suitably invariant, i.e., if they satisfy the conditions that $F(x, y, 0)$ is independent of y and $F(x, y, 1)$ is independent of x. But this is an explicit check.

Next, we observe that any morphism in $\Delta^1 \times \Delta^1 \times \partial\Delta^1$ which is mapped to an identity (a degenerate edge) of $\Delta^1 \amalg \Delta^1$ is also mapped to a degenerate edge in $\Delta^1 \diamond \Delta^1$. This shows that all maps from the middle cube to the left and right cube are degenerate edges in $\Delta^1 \diamond \Delta^1$, which implies that the horizontal maps between the cubes also descend to 1-simplices of $\mathrm{Hom}(\Delta^1 \diamond \Delta^1, \Delta^1 \diamond \Delta^1)$, and furthermore that, for any object x of $\Delta^1 \diamond \Delta^1$, these maps are degenerate. This finally implies that the map σ is a pre-inverse to the map can.   □

Corollary 2.5.20

Let $X \to X'$ be a Joyal equivalence and Y be a simplicial set. Then the map $X \star Y \to X' \star Y$ is again a Joyal equivalence.

Proof In the commutative diagram

$$
\begin{array}{ccc}
X \diamond Y & \longrightarrow & X' \diamond Y \\
\downarrow & & \downarrow \\
X \star Y & \longrightarrow & X' \star Y
\end{array}
$$

the top horizontal map is a Joyal equivalence by Corollary 2.5.17, and the vertical maps are Joyal equivalences by Proposition 2.5.19.   □

Definition 2.5.21

Let $p \colon Y \to W$ be an object of $\mathrm{sSet}_{Y/}$. We define the *fat slice* of p to be the simplicial set $W^{p/}$ defined by

$$
(W^{p/})_n = \mathrm{Hom}_{\mathrm{sSet}_{Y/}}(Y \diamond \Delta^n, W),
$$

and the simplicial set $W^{/p}$ to be given by

$$
(W^{/p})_n = \mathrm{Hom}_{\mathrm{sSet}_{Y/}}(\Delta^n \diamond Y, W).
$$

Lemma 2.5.22

The functor $\mathrm{sSet}_{Y/} \to \mathrm{sSet}$ given by sending $p \colon Y \to W$ to $W^{/p}$ is right-adjoint to the functor $- \diamond Y$. Likewise, the functor $p \mapsto W^{p/}$ is right-adjoint to the functor $Y \diamond -$.

Proof By definition, the adjunction bijection holds for representables, and hence for all simplicial sets, since the functors $- \diamond Y$ and $Y \diamond -$ preserve colimits by Lemma 2.5.2. $\square$

Lemma 2.5.23

Let Y be a simplicial set and $p \colon Y \to W$ a map of simplicial sets. Then there are canonical maps $W_{/p} \to W^{/p}$ and $W_{p/} \to W^{p/}$.

Proof On n-simplices, we have to provide a map

$$\mathrm{Hom}_{\mathrm{sSet}_{Y/}}(Y \star \Delta^n, W) \to \mathrm{Hom}_{\mathrm{sSet}_{/Y}}(Y \diamond \Delta^n, W).$$

For this, it suffices to recall that there is a map $Y \diamond \Delta^n \to Y \star \Delta^n$ in $\mathrm{sSet}_{Y/}$ which is natural with respect to maps in the simplex category Δ. Likewise for the other slice. $\square$

As a consequence of Theorem 2.5.14, we can establish an analog of Theorem 1.4.23 for fat slices.

Lemma 2.5.24

Let

$$S \xrightarrow{i} T \xrightarrow{f} \mathcal{C} \xrightarrow{p} \mathcal{D}$$

be maps of simplicial sets such that i is a monomorphism and p is an isofibration between ∞-categories. Then the functor

$$\mathcal{C}^{f/} \longrightarrow \mathcal{D}^{pf/} \times_{\mathcal{D}^{pfi/}} \mathcal{C}^{fi/}$$

is a left fibration. Likewise, the functor

$$\mathcal{C}^{/f} \longrightarrow \mathcal{D}^{/pf} \times_{\mathcal{D}^{/pfi}} \mathcal{C}^{/fi}$$

is a right fibration.

Proof Let $A \to B$ be a left-anodyne map and consider a lifting problem

$$
\begin{array}{ccc}
A & \longrightarrow & \mathcal{C}^{f/} \\
\downarrow & \nearrow & \downarrow \\
B & \longrightarrow & \mathcal{D}^{pf/} \times_{\mathcal{D}^{pfi/}} \mathcal{C}^{fi/}
\end{array}
$$

By adjunction, this is equivalent to the lifting problem

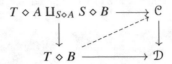

where the left vertical map is a monomorphism. We wish to show that it is a Joyal equivalence, so that we can allude to Theorem 2.5.14 to conclude the lemma. For this purpose, we claim that in the diagram

$$T \diamond A \amalg_{S \diamond A} S \diamond B \longrightarrow T \star A \amalg_{S \star A} S \star B$$
$$\downarrow \qquad\qquad\qquad\qquad \downarrow$$
$$T \diamond B \longrightarrow T \star B$$

both horizontal maps are Joyal equivalences: For the lower horizontal map, this is precisely Proposition 2.5.19; and for the upper horizontal map, we use the fact that these are pushouts along monomorphisms, so that a 3-fold application of Proposition 2.5.19 together with Corollary 2.5.8 confirms the claim. Next, we recall from Lemma 1.4.22 that the right vertical map is inner-anodyne and hence a Joyal equivalence. Again, the argument for the other slice is the same.   □

Remark 2.5.25

In fact, the conclusion of Lemma 2.5.24 holds more generally if p is replaced by an inner fibration between arbitrary simplicial sets, see [Ste18b, Theorem 1.2].

Corollary 2.5.26

Let $p\colon Y \to \mathcal{C}$ be a diagram with $\mathcal{C}$ an ∞-category. Then the functor $\mathcal{C}^{p/} \to \mathcal{C}$ is a left fibration and the functor $\mathcal{C}^{/p} \to \mathcal{C}$ is a right fibration. In particular, both $\mathcal{C}^{p/}$ and $\mathcal{C}^{/p}$ are ∞-categories.

Proof This is Lemma 2.5.24 for the special case $\emptyset \to Y \to \mathcal{C} \to \Delta^0$.   □

Proposition 2.5.27

Let $p\colon Y \to \mathcal{C}$ be a diagram. Then the canonical functor

$$\mathcal{C}_{p/} \longrightarrow \mathcal{C}^{p/}$$

is a Joyal equivalence. The same is true for $\mathcal{C}_{/p} \to \mathcal{C}^{/p}$.

Proof We only sketch a proof. The idea is to show that both $X \star -$ and $X \diamond -$ are left Quillen functors and weakly equivalent by Proposition 2.5.19, so that the induced functors on homotopy categories are isomorphic. Then it follows that the respective right adjoints are also isomorphic in the homotopy category, which is what we need. For a functor, to be left Quillen means that it preserves monomorphisms as well as monomorphisms which are also Joyal equivalences. For monomorphisms, this is clear; and for Joyal equivalences, it follows from Corollaries 2.5.20 and 2.5.17. Therefore, the respective right adjoints preserve trivial fibrations and fibrations between ∞-categories. It follows (this is where the proof remains a sketch) that the two slice functors descend to right adjoints of the two join functors on the homotopy category associated to the Joyal model structure. From this, we can deduce that the slice functors, being right-adjoint to the two isomorphic join functors, are also isomorphic. □

With the help of ordinary slices, one can also define right and left mapping spaces in an ∞-category.

Definition 2.5.28

Let $\mathcal{C}$ be an ∞-category and let x and y be objects of $\mathcal{C}$. We define the *right mapping space* by the pullback

$$
\begin{array}{ccc}
\mathrm{map}_{\mathcal{C}}^{R}(x, y) & \longrightarrow & \mathcal{C}_{/y} \\
\downarrow & & \downarrow \\
\Delta^{0} & \xrightarrow{\;\;x\;\;} & \mathcal{C}
\end{array}
$$

and the *left mapping space* by

$$
\begin{array}{ccc}
\mathrm{map}_{\mathcal{C}}^{L}(x, y) & \longrightarrow & \mathcal{C}_{x/} \\
\downarrow & & \downarrow \\
\Delta^{0} & \xrightarrow{\;\;y\;\;} & \mathcal{C}
\end{array}
$$

Remark 2.5.29

The map $\mathrm{map}_{\mathcal{C}}^{R}(x, y) \to \Delta^{0}$ is a right fibration and the map $\mathrm{map}_{\mathcal{C}}^{L}(x, y) \to \Delta^{0}$ is a left fibration. By Exercise 84, both are in fact Kan fibrations, so that the left and right mapping spaces are Kan complexes.

Next, we want to compare these mapping spaces to the mapping space which we have already defined in Definition 1.3.47. The following lemma will take care of this.

Lemma 2.5.30
The following two diagrams are pullback diagrams:

$$
\begin{array}{ccc}
\mathcal{C}^{/y} & \longrightarrow & \mathrm{Fun}(\Delta^1, \mathcal{C}) \\
\downarrow & & \downarrow \\
\mathcal{C} & \xrightarrow{\ (\mathrm{id}, y)\ } & \mathcal{C} \times \mathcal{C}
\end{array}
\qquad
\begin{array}{ccc}
\mathcal{C}^{x/} & \longrightarrow & \mathrm{Fun}(\Delta^1, \mathcal{C}) \\
\downarrow & & \downarrow \\
\mathcal{C} & \xrightarrow{\ (x, \mathrm{id})\ } & \mathcal{C} \times \mathcal{C}
\end{array}
$$

Proof We only present the proof for the left square. (The argument for the right square is similar.) W e have to show that the diagram is a pullback on all n-simplices, so we consider the following diagram:

$$
\begin{array}{ccccc}
\mathrm{Hom}_{/y}(\Delta^n \diamond \Delta^0, \mathcal{C}) & \longrightarrow & \mathrm{Hom}(\Delta^n \diamond \Delta^0, \mathcal{C}) & \longrightarrow & \mathrm{Hom}(\Delta^n \times \Delta^0 \times \Delta^1, \mathcal{C}) \\
\downarrow & & \downarrow & & \downarrow \\
\mathrm{Hom}(\Delta^n, \mathcal{C}) & \xrightarrow{(\mathrm{id}, y)} & \mathrm{Hom}(\Delta^n, \mathcal{C}) \times \mathrm{Hom}(\Delta^0, \mathcal{C}) & \longrightarrow & \mathrm{Hom}(\Delta^n \times \Delta^0 \times \partial\Delta^1, \mathcal{C}) \\
\downarrow & & \downarrow & & \\
\Delta^0 & \xrightarrow{\ y\ } & \mathrm{Hom}(\Delta^0, \mathcal{C}) & &
\end{array}
$$

In order to show that the big top square is a pullback, notice first that the rightmost square is a pullback by the very definition of the fat join $\Delta^n \diamond \Delta^0$. The lower left square is a pullback by inspection. The combination of the left two squares is also a pullback by definition. It follows that the top left square is a pullback. Hence, combining the two top squares, we obtain a pullback as needed. □

Passing to fibres, we obtain the following corollary.

Corollary 2.5.31
The following diagrams are pullbacks:

$$
\begin{array}{ccc}
\mathrm{map}_{\mathcal{C}}(x, y) & \longrightarrow & \mathcal{C}^{/y} \\
\downarrow & & \downarrow \\
\Delta^0 & \xrightarrow{\ x\ } & \mathcal{C}
\end{array}
\qquad
\begin{array}{ccc}
\mathrm{map}_{\mathcal{C}}(x, y) & \longrightarrow & \mathcal{C}^{x/} \\
\downarrow & & \downarrow \\
\Delta^0 & \xrightarrow{\ y\ } & \mathcal{C}
\end{array}
$$

We can then use the following lemma to compare the various definitions of mapping spaces.

Lemma 2.5.32
Consider a diagram

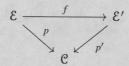

where p and p' are isofibrations and where f is a Joyal equivalence. Then, for all objects x in $\mathcal{C}$, the induced map on fibres $\mathcal{E}_x \to \mathcal{E}'_x$ is also a Joyal equivalence.

Proof This follows immediately from Lemma 2.5.7.   □

Remark 2.5.33
The converse of this statement is not correct (i.e., one cannot deduce that f is a Joyal equivalence if it is so fibrewise, but see Exercise 134). However, this statement is true for left and right fibrations (in fact, more generally for cocartesian and cartesian fibrations, as we will show in Theorem 3.1.27).

Corollary 2.5.34
Let $\mathcal{C}$ be an ∞-category and let x and y be objects of $\mathcal{C}$. Then the maps

$$\mathrm{map}^R_{\mathcal{C}}(x, y) \to \mathrm{map}_{\mathcal{C}}(x, y) \leftarrow \mathrm{map}^L_{\mathcal{C}}(x, y)$$

are homotopy equivalences.

Proof Apply Lemma 2.5.32 to the diagrams

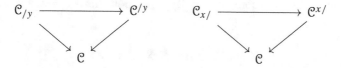

and use Corollary 2.5.26 and Proposition 2.5.27.   □

Finally, we wish to compare the mapping spaces of the coherent nerve of a Kan-enriched category with the mapping-Kan-complexes that prevail in the simplicial category. The proofs are beyond the scope of this book, see, e.g., [Lur09, Section 2.2.2]. A direct combinatorial argument for the below theorem was recently given by Hebestreit and Krause [HK20].

Theorem 2.5.35
Let $\mathcal{C}$ be a Kan-enriched category and let x and y be objects of $\mathcal{C}$. Then there is a canonical map

$$\mathrm{Hom}_{\mathcal{C}}(x, y) \longrightarrow \mathrm{map}_{\mathrm{N}(\mathcal{C})}(x, y)$$

which is a homotopy equivalence. The homotopy class of this map is natural in x and y.

Corollary 2.5.36
Let $F \colon \mathcal{C} \to \mathcal{C}'$ be a weak equivalence of Kan-enriched categories. Then the induced functor $\mathrm{N}(F) \colon \mathrm{N}(\mathcal{C}) \to \mathrm{N}(\mathcal{C}')$ is a Joyal equivalence of ∞-categories.

Proof The functor $\mathrm{N}(F)$ is essentially surjective if and only if F is weakly essentially surjective in the sense of Definition 1.2.49. Furthermore, Theorem 2.5.35 shows that F is weakly fully faithful if and only if $\mathrm{N}(F)$ is fully faithful. Hence we conclude the proof by Theorem 2.3.20. $\qquad\qquad\qquad\qquad\qquad\qquad\qquad\qquad\qquad\qquad\qquad\qquad\qquad\qquad\qquad\square$

Recall that we defined the simplicial category CW whose objects are CW-complexes and whose simplicial set of maps is given by $\mathcal{S}(\mathrm{map}(X, Y))$. Its coherent nerve was denoted by Spc. Furthermore, we defined the simplicial category Kan whose objects are Kan complexes and whose simplicial set of maps is given by the internal hom-set $\underline{\mathrm{Hom}}(A, B)$. Now we claim that there is a functor CW $\to$ Kan constructed as follows: On objects, it sends X to $\mathcal{S}(X)$; and on morphisms, we have to provide a simplicial map

$$\mathcal{S}(\mathrm{map}(X, Y)) \longrightarrow \underline{\mathrm{Hom}}(\mathcal{S}(X), \mathcal{S}(Y))$$

which is compatible with composition. By adjunction, this is equivalent to providing a map

$$\mathcal{S}(\mathrm{map}(X, Y)) \times \mathcal{S}(X) \longrightarrow \mathcal{S}(Y).$$

Next, recall that $\mathcal{S}$ is a right adjoint and hence preserves products. It hence suffices to provide a simplicial map

$$\mathcal{S}(\mathrm{map}(X, Y) \times X) \longrightarrow \mathcal{S}(Y)$$

where we can use the ordinary composition map $\mathrm{map}(X, Y) \times X \to Y$ and apply the functor S.

Corollary 2.5.37

The previously described functor $\mathrm{Spc} = \mathrm{N(CW)} \to \mathrm{N(Kan)} = \widehat{\mathrm{Spc}}$ *is a Joyal equivalence.*

Proof We show that the functor $\mathrm{CW} \to \mathrm{Kan}$ is a weak equivalence of simplicial categories. It is weakly essentially surjective because every Kan complex X is a homotopy equivalence to $S(|X|)$ and thus, up to equivalence, in the image of the functor $\mathrm{CW} \to \mathrm{Kan}$. In order to show that the functor is fully faithful, we have to show that the map

$$S(\mathrm{map}(X, Y)) \longrightarrow \underline{\mathrm{Hom}}(S(X), S(Y))$$

is a homotopy equivalence. This is the case if and only if the composite

$$\mathrm{map}(X, Y) \longrightarrow |S(\mathrm{map}(X, Y))| \longrightarrow |\underline{\mathrm{Hom}}(S(X), S(Y))|$$

is a homotopy equivalence. But for this, we can use the fact that, for any two Kan complexes A and B, the canonical map

$$|\underline{\mathrm{Hom}}(A, B)| \longrightarrow \mathrm{map}(|A|, |B|)$$

is a homotopy equivalence. $\qquad\square$

Corollary 2.5.38

The canonical functor $\mathrm{Spc} \to \mathrm{Cat}_\infty$ *is fully faithful.*

Proof The functor is given by applying the coherent nerve to the functor of simplicial categories $\mathrm{Kan} \to \mathrm{Cat}^1_\infty$. Therefore, we only need to show that, for any two Kan complexes X and Y, the canonical map

$$\underline{\mathrm{Hom}}(X, Y) \to \mathrm{Fun}(X, Y)^{\simeq}$$

is a homotopy equivalence. In fact, it is an isomorphism of simplicial sets, so the proposition follows. $\qquad\square$

(Co)Cartesian Fibrations and the Construction of Functors 3

3.1 (Co)Cartesian Fibrations

In this section, we define and discuss the notion of (co)cartesian fibrations and in particular of p-(co)cartesian morphisms of X, if $p\colon X \to Y$ is an inner fibration between simplicial sets. (Co)cartesian fibrations are natural generalizations of (left) right fibrations, and they will be a key player in the straightening-unstraightening equivalence (which we discuss to some extend only in this book). We will introduce the notion of a morphism of (co)cartesian fibrations (over a fixed base-∞-category) and show that among such, equivalences can be detected fibrewise, i.e., that such a morphism is already an equivalence if its induced functor on each fibres is an equivalence. In classical category theory, (co)cartesian fibrations are known as Grothendieck-(op)-fibrations, where they arise, under the Grothendieck-construction, as associated to (pseudo-)functors with values in the 2-category of categories.

Definition 3.1.1

A morphism $f\colon \Delta^1 \to X$ is called *p-cartesian* if for $n \geq 2$ any lifting problem

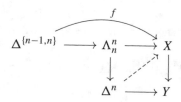

---

admits a solution. Dually, it is called *p-cocartesian* if any lifting problem

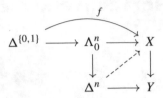

admits a solution. One calls such an f a p-(co)cartesian lift of $p(f)$.

The definition of (co)cartesian morphisms can be rephrased in terms of slices as follows.

Lemma 3.1.2
Let $p\colon X \to Y$ be an inner fibration and let $f\colon x \to y$ be a morphism in Y. Then f is p-cartesian if and only if the functor

$$X_{/f} \longrightarrow X_{/y} \times_{Y_{/p(y)}} Y_{/pf}$$

is a trivial fibration. Dually, f is p-cocartesian if and only if the functor

$$X_{f/} \longrightarrow X_{x/} \times_{Y_{p(x)/}} Y_{pf/}$$

is a trivial fibration.

Proof Let $\partial \Delta^n \to \Delta^n$ for $n \geq 0$ and consider a lifting problem

$$
\begin{array}{ccc}
\partial \Delta^n & \longrightarrow & X_{/f} \\
\downarrow & \nearrow & \downarrow \\
\Delta^n & \longrightarrow & X_{/y} \times_{Y_{/p(y)}} Y_{/pf}
\end{array}
$$

which we wish to solve. By adjunction, this corresponds to the lifting problem

$$
\begin{array}{ccc}
\partial \Delta^n \star \Delta^1 \amalg_{\partial \Delta^n \star \Delta^0} \Delta^n \star \Delta^0 & \longrightarrow & X \\
\downarrow & \nearrow & \downarrow \\
\Delta^n \star \Delta^1 & \longrightarrow & Y
\end{array}
$$

where the restriction to Δ^1 is given by f. We recall that the left vertical map is isomorphic to $\Lambda^{n+2}_{n+2} \to \Delta^{n+2}$ and that the inclusion of Δ^1 into Λ^{n+2}_{n+2} is the edge $\Delta^{\{n+1,n+2\}}$. Therefore, the diagram can be solved for all $n \geq 0$. The cocartesian case is similar. $\square$

Remark 3.1.3

Suppose that $p\colon X \to Y$ is an inner fibration between ∞-categories. Then the map $X_{/p} \to X^{/p}$ is a Joyal equivalence for any diagram $p\colon W \to X$ by Proposition 2.5.27. Furthermore, the map

$$X^{/f} \longrightarrow X^{/y} \times_{Y/p(y)} Y^{/pf}$$

is also a right fibration by Lemma 2.5.24. (We have shown this only for p being an isofibration, but see the remark following Lemma 2.5.24.) Hence, f is cartesian if and only if this map is a trivial fibration. Likewise, f is cocartesian if and only if the map

$$X^{f/} \longrightarrow X^{x/} \times_{Y p(x)/} Y^{pf/}$$

is a trivial fibration.

Lemma 3.1.4

Let $p\colon \mathcal{E} \to \mathcal{D}$ and $q\colon \mathcal{D} \to \mathcal{C}$ be inner fibrations between ∞-categories, and let $f\colon \Delta^1 \to \mathcal{E}$ be a morphism in $\mathcal{E}$ such that $p(f)$ is q-(co)cartesian. Then f is p-(co)cartesian if and only if f is qp-(co)cartesian.

Proof Consider the diagram

$$\mathcal{E}_{/f} \longrightarrow \mathcal{E}_{/y} \times_{\mathcal{D}_{/p(y)}} \mathcal{D}_{/pf} \xrightarrow{\;\simeq\;} \mathcal{E}_{/y} \times_{\mathcal{D}_{/p(y)}} \mathcal{D}_{/p(y)} \times_{\mathcal{C}_{/qp(y)}} \mathcal{C}_{/qpf}$$
$$\downarrow \cong$$
$$\mathcal{E}_{/y} \times_{\mathcal{C}_{/qp(y)}} \mathcal{C}_{/qpf}$$

where the second horizontal map is a trivial fibration by the assumption that pf is q-cartesian. Both the first horizontal map and the diagonal map are right fibrations and thus trivial fibrations if and only if they are Joyal equivalences. From this, the lemma follows. $\square$

Remark 3.1.5
The previous lemma holds more generally without the assumption that the involved simplicial sets are ∞-categories, see [Lur09, Prop. 2.4.1.3]: Since the composition of trivial fibrations is again a trivial fibration, we immediately see that if f is p-cartesian then it is also qp-cartesian. In order to see the converse, we can show that the map $\mathcal{E}_{/f} \to \mathcal{E}_{/y} \times_{\mathcal{D}_{/p(y)}} \mathcal{D}_{/pf}$ (which is a Joyal equivalence by the 3-for-2 property) in fact has contractible fibres and then allude to the general fact that a right fibration which has contractible fibres is a trivial fibration (or a Joyal equivalence and thus a trivial fibration), see [Lur09, Lemma 2.1.3.4] or Theorem 3.1.27. The fibre which we are interested in is the fibre of the map between fibres of the other two maps. Since these fibres are contractible by assumption, we are done.

Lemma 3.1.6
Let $p: \mathcal{E} \to \mathcal{C}$ be an inner fibration between ∞-categories, and let $f: \Delta^1 \to \mathcal{E}$ be a morphism. Then f is an equivalence if and only if it is p-(co)cartesian and $p(f)$ is an equivalence.

Proof Suppose that f is an equivalence. Then the same is true for $p(f)$, and the Joyal lifting theorem (Theorem 2.1.8) implies that f is both cartesian and cocartesian. Assume now conversely that $p(f)$ is an equivalence so that it is q-cartesian, where $q: \mathcal{C} \to \Delta^0$ is the projection by Exercise 115. We find that f is p-cartesian if and only if f is qp-cartesian by Lemma 3.1.4, which in turn is the case if and only if f is an equivalence by another application of Exercise 115.   □

Lemma 3.1.7
Let $p: \mathcal{E} \to \mathcal{C}$ be an inner fibration and let $\sigma: \Delta^2 \to \mathcal{E}$, which we will depict as the diagram

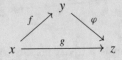

Suppose that φ is p-cartesian. Then g is p-cartesian if and only if f is p-cartesian.

Proof We observe that the inclusions $\Delta^{\{0,1\}} \to \Delta^2$ and $\Delta^{\{0,2\}} \to \Delta^2$ are left-anodyne, so it follows that in the two commutative squares

$$
\begin{array}{ccc}
\mathcal{E}_{/\sigma} & \longrightarrow & \mathcal{E}_{/z} \times_{\mathcal{C}_{/p(z)}} \mathcal{C}_{/p\sigma} \\
\downarrow & & \downarrow \\
\mathcal{E}_{/g} & \longrightarrow & \mathcal{E}_{/z} \times_{\mathcal{C}_{/p(z)}} \mathcal{C}_{/pg}
\end{array}
\qquad
\begin{array}{ccc}
\mathcal{E}_{/\sigma} & \longrightarrow & \mathcal{E}_{/y} \times_{\mathcal{C}_{/p(y)}} \mathcal{C}_{/p\sigma} \\
\downarrow & & \downarrow \\
\mathcal{E}_{/f} & \longrightarrow & \mathcal{E}_{/y} \times_{\mathcal{C}_{/p(y)}} \mathcal{C}_{/pf}
\end{array}
$$

all vertical maps are trivial fibrations. We wish to show the the left lower horizontal map is a trivial fibration if and only if the right lower horizontal map is. Since both maps are isofibrations in any case, it suffices to show that one map is a Joyal equivalence if and only if the other map is. By means of the commutative squares, it hence suffices to show that the left top horizontal map is a Joyal equivalence if and only if the right top horizontal map is. For this, we consider the diagram

$$
\mathcal{E}_{/y} \times_{\mathcal{C}_{/p(y)}} \mathcal{C}_{/p\sigma} \xleftarrow{\ \simeq\ } \mathcal{E}_{/\varphi} \times_{\mathcal{C}_{/p\varphi}} \mathcal{C}_{/p\sigma} \xrightarrow{\ \simeq\ } \mathcal{E}_{/z} \times_{\mathcal{C}_{/p(z)}} \mathcal{C}_{/p\sigma}
$$

with $\mathcal{E}_{/\sigma}$ mapping to both sides.

Here, the right horizontal map is an equivalence because φ is p-cartesian, so that the map $\mathcal{E}_{/\varphi} \to \mathcal{E}_{/z} \times_{\mathcal{C}_{/p(z)}} \mathcal{C}_{/p\varphi}$ is an equivalence. The left horizontal map is an equivalence because it is induced by the restriction $\{1\} \to \Delta^{\{1,2\}}$ which is left-anodyne. Therefore, we find that the left bended map is an equivalence if and only if the right bended map is an equivalence, which was left to show. $\square$

Remark 3.1.8

The conclusion of Lemma 3.1.7 does not require that the base $\mathcal{C}$ is an ∞-category. In fact, the reduction is not too complicated once one has established some properties of the more general notion of locally cocartesian edges, see [Lur09, 2.4.2] and the discussion following Definition 3.2.5.

The following notion will be convenient to use.

Definition 3.1.9

A commutative diagram of ∞-categories

$$
\begin{array}{ccc}
\mathcal{E} & \longrightarrow & \mathcal{E}' \\
\downarrow & & \downarrow{\scriptstyle p} \\
\mathcal{C} & \longrightarrow & \mathcal{C}'
\end{array}
$$

is called *homotopy-cartesian* if p is an isofibration and the induced map $\mathcal{E} \to \mathcal{C} \times_{\mathcal{C}'} \mathcal{E}'$ is a Joyal equivalence. For $\mathcal{C} = \Delta^0$, we will also use the term *homotopy fibre sequence* for the composition $\mathcal{E} \to \mathcal{E}' \to \mathcal{C}'$.

Remark 3.1.10
More generally, one can neglect the condition of p being an isofibration by factoring p into a Joyal equivalence followed by an isofibration $\mathcal{E}' \xrightarrow{\simeq} \mathcal{E}'' \xrightarrow{q} \mathcal{C}'$ and then requesting that the induced map $\mathcal{E} \to \mathcal{C} \times_{\mathcal{C}'} \mathcal{E}''$ is a Joyal equivalence.

Lemma 3.1.11
Let $p \colon \mathcal{C} \to \mathcal{D}$ be an isofibration between ∞-categories and let x and y be objects of $\mathcal{C}$. Then the induced map

$$\mathrm{map}_{\mathcal{C}}(x, y) \to \mathrm{map}_{\mathcal{D}}(p(x), p(y))$$

is a Kan fibration. The same holds true for map^L and map^R, even if p is only an inner fibration.

Proof We consider the diagram

$$
\begin{array}{ccccc}
\mathrm{map}_{\mathcal{C}}(x, y) & \longrightarrow & \mathrm{map}_{\mathcal{D}}(p(x), p(y)) & \longrightarrow & \Delta^0 \\
\downarrow & & \downarrow & & \downarrow{\scriptstyle x} \\
\mathcal{C}/y & \longrightarrow & \mathcal{D}/p(y) \times_{\mathcal{D}} \mathcal{C} & \longrightarrow & \mathcal{C} \\
& & \downarrow & & \downarrow{\scriptstyle p} \\
& & \mathcal{D}/p(y) & \longrightarrow & \mathcal{D}
\end{array}
$$

respectively the ones with the ordinary slice $\mathcal{C}_{/y}$ instead of $\mathcal{C}/y$ and its variant using $\mathcal{C}_{x/}$ instead of $\mathcal{C}_{/y}$. In this diagram, all squares are pullbacks: the lower one by definition, the right large one by definition as well, and hence also the small one in the top right corner; the combined large horizontal one by definition, and hence also the left small square. The lower horizontal map is a right fibration (in the case of $\mathcal{C}_{x/}$, it is a left fibration), hence so is the upper horizontal map. Since $\mathrm{map}_{\mathcal{D}}(p(x), p(y))$ is a Kan complex, it is in fact a Kan fibration as claimed.   □

Lemma 3.1.12

Let $\mathcal{C} \to \mathcal{D}$ be an isofibration between ∞-categories, let z be an object of $\mathcal{D}$, and let x and y objects of $\mathcal{C}$ with $p(x) = p(y) = z$. Then the diagram

$$
\begin{array}{ccc}
\mathrm{map}_{\mathcal{C}_z}(x, y) & \longrightarrow & \mathrm{map}_{\mathcal{C}}(x, y) \\
\downarrow & & \downarrow \\
\Delta^0 & \xrightarrow{\ \mathrm{id}_z\ } & \mathrm{map}_{\mathcal{D}}(z, z)
\end{array}
$$

is a pullback.

Proof In order to see the remaining claim, we consider the diagram

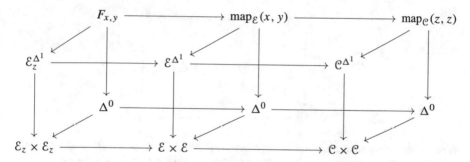

where all horizontal composites are given by the fibre inclusion over the point corresponding to id_z. Since the middle vertical square and the right vertical square are pullbacks, the leftmost vertical square is also a pullback. □

Corollary 3.1.13

Let $\mathcal{C} \to \mathcal{D}$ be an isofibration between ∞-categories, let z be an object of $\mathcal{D}$, and let x and y be objects of $\mathcal{C}_z$. Then the diagram

$$
\begin{array}{ccc}
\mathrm{map}^?_{\mathcal{C}_z}(x, y) & \longrightarrow & \mathrm{map}^?_{\mathcal{E}}(x, y) \\
\downarrow & & \downarrow \\
\Delta^0 & \xrightarrow{\ \mathrm{id}\ } & \mathrm{map}^?_{\mathcal{C}}(p(x), p(y))
\end{array}
$$

is homotopy-cartesian for $? = R, L$ or void.

Proof For the ordinary mapping spaces, this follows immediately since the diagram is a pullback and the right vertical map is a Kan fibration. In order to see the claim for the right and left mapping spaces, we use Lemma 3.1.11 and Corollary 2.5.34.   □

Lemma 3.1.14

Let $p \colon \mathcal{E} \to \mathcal{C}$ be an inner fibration between ∞-categories and let $f \colon \Delta^1 \to \mathcal{E}$ be a p-cartesian morphism from x to y. Then, for all objects z of $\mathcal{E}$, the induced map

$$\mathcal{E}^{/f} \times_{\mathcal{E}} \{z\} \longrightarrow \left(\mathcal{E}^{/y} \times_{\mathcal{C}/p(y)} \mathcal{C}^{/pf} \right) \times_{\mathcal{E}} \{z\}$$

is a trivial fibration as well.

Proof We have a commutative diagram

$$
\begin{array}{ccc}
\mathcal{E}^{/f} \times_{\mathcal{E}} \{z\} & \longrightarrow & \mathcal{E}^{/f} \\
\downarrow & & \downarrow \\
\left(\mathcal{E}^{/y} \times_{\mathcal{C}/p(y)} \mathcal{C}^{/pf} \right) \times_{\mathcal{E}} \{z\} & \longrightarrow & \mathcal{E}^{/y} \times_{\mathcal{C}/p(y)} \mathcal{C}^{/pf} \\
\downarrow & & \downarrow \\
\Delta^0 & \longrightarrow & \mathcal{E}
\end{array}
$$

where both the big square and the lower square are pullbacks. Hence, the upper square is also a pullback. Since in this square the right vertical map is a trivial fibration by the assumption that f is p-cartesian, the claim follows.   □

Remark 3.1.15

Replacing the fat slice by the ordinary slice, the same statement holds true for inner fibrations between arbitrary simplicial sets.

Corollary 3.1.16

Let $p \colon \mathcal{E} \to \mathcal{C}$ be an inner fibration between ∞-categories, let $f \colon \Delta^1 \to \mathcal{E}$ be a p-cartesian morphism of $\mathcal{E}$ from x to y, and let z be an object of $\mathcal{E}$. Then the diagram

$$
\begin{array}{ccc}
\mathrm{map}_{\mathcal{E}}(z, x) & \longrightarrow & \mathrm{map}_{\mathcal{E}}(z, y) \\
\downarrow & & \downarrow \\
\mathrm{map}_{\mathcal{C}}(p(z), p(x)) & \longrightarrow & \mathrm{map}_{\mathcal{C}}(p(z), p(y))
\end{array}
$$

(continued)

3.1.16 (continued)
is a homotopy-cartesian diagram of ∞-groupoids. The horizontal maps are induced by post-composition with f and pf, respectively.

Proof We consider the diagram

$$
\begin{array}{ccccc}
\mathrm{map}_{\mathcal{E}}(z, x) & \xleftarrow{\ \simeq\ } & \mathcal{E}^{/f} \times_{\mathcal{E}} \{z\} & \longrightarrow & \mathrm{map}_{\mathcal{E}}(z, y) \\
\downarrow & & \downarrow & & \downarrow \\
\mathrm{map}_{\mathcal{C}}(p(z), p(x)) & \xleftarrow{\ \simeq\ } & \mathcal{C}^{/pf} \times_{\mathcal{C}} \{p(z)\} & \longrightarrow & \mathrm{map}_{\mathcal{C}}(p(z), p(y))
\end{array}
$$

and observe that the left horizontal maps are trivial fibrations, since the inclusion $\{0\} \to \Delta^1$ is left-anodyne. It thus suffices to show that the right square is homotopy-cartesian, and since the right vertical map is a Kan fibration by Lemma 3.1.11, it suffices to recall that the map

$$
\mathcal{E}^{/f} \to \mathcal{E}^{/y} \times_{\mathcal{C}^{/p(y)}} \mathcal{C}^{/pf}
$$

is a trivial fibration, so that the same remains true after applying $- \times_{\mathcal{E}} \{z\}$. $\qquad\square$

Remark 3.1.17
The statement of the corollary is not quite correct, since the square which we construct does not a priori commute (only up to homotopy). But since the right vertical map is a Kan fibration, it can always be replaced by a commutative diagram without changing the homotopy types of the participants, and a concrete way of doing this is to consider the right square in the diagram appearing in the proof.

Remark 3.1.18
Given an inner fibration $p \colon \mathcal{E} \to \mathcal{C}$ and a morphism $f \colon \Delta^1 \to \mathcal{E}$ such that, for *every* object z of $\mathcal{E}$, the diagram

$$
\begin{array}{ccc}
\mathrm{map}_{\mathcal{E}}(z, x) & \longrightarrow & \mathrm{map}_{\mathcal{E}}(z, y) \\
\downarrow & & \downarrow \\
\mathrm{map}_{\mathcal{C}}(p(z), p(x)) & \longrightarrow & \mathrm{map}_{\mathcal{C}}(p(z), p(y))
\end{array}
$$

(continued)

3.1.18 (continued)

is homotopy-cartesian, then f is p-cartesian. This can be seen by observing that the map

$$\mathcal{E}^{/f} \to \mathcal{E}^{/y} \times_{\mathcal{C}^{/(y)}} \mathcal{C}^{/pf}$$

is a trivial fibration if and only if it induces an equivalence on each fibre over points in $\mathcal{E}$. As before, this statement holds in general for right fibrations via a combinatorial argument (without $\mathcal{E}$ and $\mathcal{C}$ having to be ∞-categories). In the case where $\mathcal{E}$ and $\mathcal{C}$ are ∞-categories, we will soon prove this result for so-called cartesian fibrations (see Theorem 3.1.27), and we will subsequently deduce the version for right fibrations from it.

Corollary 3.1.19

Let $p\colon \mathcal{E} \to \mathcal{C}$ be an inner fibration and let x and y be objects of $\mathcal{E}$. Let $f\colon x' \to y$ be a p-cartesian morphism with $p(x') = p(x)$. Then the diagram

$$\begin{array}{ccc} \mathrm{map}_{\mathcal{E}_{p(x)}}(x, x') & \longrightarrow & \mathrm{map}_{\mathcal{E}}(x, y) \\ \downarrow & & \downarrow \\ \Delta^0 & \xrightarrow{\;pf\;} & \mathrm{map}_{\mathcal{C}}(p(x), p(y)) \end{array}$$

is homotopy-cartesian.

Proof By Corollary 3.1.16 we have a homotopy-cartesian diagram

$$\begin{array}{ccc} \mathrm{map}_{\mathcal{E}}(x, x') & \longrightarrow & \mathrm{map}_{\mathcal{E}}(x, y) \\ \downarrow & & \downarrow \\ \mathrm{map}_{\mathcal{C}}(p(x), p(x')) & \longrightarrow & \mathrm{map}_{\mathcal{C}}(p(x), p(y)) \end{array}$$

so that the induced map of vertical fibres is an equivalence over the point $\Delta^0 \to \mathrm{map}_{\mathcal{C}}(p(x), p(x'))$ corresponding to $\mathrm{id}_{p(x)}$. By Corollary 3.1.13, the left vertical fibre is given by $\mathrm{map}_{\mathcal{E}_{p(x)}}(x, x')$, so the claim is shown. □

Therefore, inner fibrations $p\colon \mathcal{E} \to \mathcal{C}$ which have a sufficient supply of cartesian morphisms are such that the mapping spaces in $\mathcal{E}$ are controlled by those of $\mathcal{C}$ and all fibres.

Definition 3.1.20

An inner fibration $p\colon X \to Y$ is called a *cartesian fibration* if every lifting problem

$$
\begin{array}{ccc}
\{1\} & \longrightarrow & X \\
\downarrow & \nearrow & \downarrow \\
\Delta^1 & \longrightarrow & Y
\end{array}
$$

has a solution which is a p-cartesian morphism in X. Dually, p is called a *cocartesian fibration* if every lifting problem

$$
\begin{array}{ccc}
\{0\} & \longrightarrow & X \\
\downarrow & \nearrow & \downarrow \\
\Delta^1 & \longrightarrow & Y
\end{array}
$$

admits a solution which is a p-cocartesian morphism in X.

Informally, an inner fibration is a (co)cartesian fibration if it admits a p-(co)cartesian lift of any morphism in $\mathcal{C}$ (with a specified source or target).

Example Right fibrations are cartesian fibrations, and left fibrations are cocartesian fibrations.

Lemma 3.1.21

Let $p\colon \mathcal{E} \to \mathcal{C}$ be a (co)cartesian fibration between ∞-categories. Then p is an isofibration.

Proof This follows from Lemma 3.1.6, which states that a p-cartesian lift of an equivalence is an equivalence. □

Lemma 3.1.22

A cartesian fibration $\mathcal{E} \to \mathcal{C}$ is a right fibration if and only if every morphism in $\mathcal{E}$ is p-cartesian. Dually, a cocartesian fibration is a left fibration if and only if every morphism in $\mathcal{E}$ is p-cocartesian.

Proof By definition, in a cartesian fibration one can lift the right outer 1-horn. If furthermore every morphism in $\mathcal{E}$ is p-cartesian, this simply says that one can also lift all right outer horns of dimension greater or equal to 2. Conversely, a right fibration admits some lift of the diagram in the definition, and by definition of a right fibration, every morphism in p-cartesian. The argument for cocartesian fibrations is similar. □

Proposition 3.1.23

Let $p\colon \mathcal{E} \to \mathcal{C}$ be a cartesian fibration between ∞-categories. Then p is a right fibration if and only if, for all objects x of $\mathcal{C}$, the fibres $\mathcal{E}_x = \mathcal{E} \times_{\mathcal{C}} \{x\}$ are ∞-groupoids.

Proof Right fibrations are cartesian fibrations whose fibres are ∞-groupoids. Conversely, assume that $p\colon \mathcal{E} \to \mathcal{C}$ is a cartesian fibration whose fibres are ∞-groupoids. We will show that every morphism is p-cartesian and then allude to Lemma 3.1.22. For this purpose, let $f\colon \Delta^1 \to \mathcal{E}$ be a morphism from x to y and choose a p-cartesian lift φ of $p(f)$ with target y. We consider the diagram

$$
\begin{array}{ccc}
\Lambda_2^2 & \xrightarrow{(f,\varphi)} & \mathcal{E} \\
\downarrow & \nearrow & \downarrow \\
\Delta^2 & \xrightarrow{\ \sigma\ } & \mathcal{C}
\end{array}
$$

where the map σ is given by the diagram

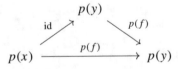

Since φ is p-cartesian, there exists a dashed arrow in this diagram. The resulting 2-simplex τ is given by the diagram

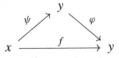

where ψ is a morphism in the fibre $\mathcal{E}_x$ over x and hence invertible by the assumption that all fibres are ∞-groupoids. Therefore, by applying Lemma 3.1.7, we find that f is cartesian because φ is. □

Remark 3.1.24

The conclusion of Proposition 3.1.23 also holds if $\mathcal{E}$ and $\mathcal{C}$ are not assumed to be ∞-categories. In fact, one can deduce the more general statement from Proposition 3.1.23, and we recommend to do so as an exercise.

Corollary 3.1.25

A cartesian fibration is conservative if and only if it is a right fibration.

Proof Right fibrations are conservative by Proposition 2.1.7. Conversely, given a conservative cartesian fibration $p\colon \mathcal{E} \to \mathcal{C}$, the fibre $\mathcal{E}_x$ over each object x of $\mathcal{C}$ is an ∞-groupoid: Each morphism in the fibre is sent to the identity of x by p. We may thus apply Proposition 3.1.23. $\qquad\qquad\square$

Definition 3.1.26

Let $p\colon X \to Y$ and $p'\colon X' \to Y$ be (co)cartesian fibrations. We say that a map $f\colon X \to X'$ is a morphism of (co)cartesian fibrations if $p'f = p$ and if f sends p-cartesian morphisms to p'-cartesian morphisms.

Example Suppose that $p\colon X \to Y$ is a cartesian fibration and that $p'\colon X' \to Y$ is a right fibration. Then any map $f\colon X \to X'$ with $p'f = p$ is a morphism of cartesian fibrations, because all morphisms in X' are p'-cartesian.

Theorem 3.1.27

Let $f\colon \mathcal{E} \to \mathcal{E}'$ be a morphism of (co)cartesian fibrations $p\colon \mathcal{E} \to \mathcal{C}$ and $p'\colon \mathcal{E}' \to \mathcal{C}$ between ∞-categories. Then f is a Joyal equivalence if and only if, for all objects z of $\mathcal{C}$, the induced map on fibres $\mathcal{E}_z \to \mathcal{E}'_z$ is a Joyal equivalence.

Proof The "only if" direction holds more generally for maps between isofibrations, see Lemma 2.5.32. Let us hence assume that all induced maps $\mathcal{E}_x \to \mathcal{E}'_x$ are Joyal equivalences. We wish to show that f is a Joyal equivalence. For this, we will first show that f is fully faithful and essentially surjective, and then conclude the theorem from Theorem 2.3.20.

In order to see that f is essentially surjective, we consider an object y' of $\mathcal{E}'$ and let $x = p'(y')$. Since the map $\mathcal{E}_x \to \mathcal{E}'_x$ is a Joyal equivalence, it is in particular essentially surjective. Hence there exists an object y in $\mathcal{E}_x$ and an equivalence between $f(y)$ and y' in $\mathcal{E}'_x$. It follows that f is essentially surjective.

In order to see that f is fully faithful, we consider two objects x and y in $\mathcal{E}$ and need to show that the map

$$\mathrm{map}_{\mathcal{E}}(x, y) \longrightarrow \mathrm{map}_{\mathcal{E}'}(f(x), f(y))$$

is a homotopy equivalence.

For this purpose, we choose a p-cartesian lift $\hat{\alpha} \colon x' \to y$ of α (which implies $p(x') = p(x)$) and note that $f(\hat{\alpha})$ is a p'-cartesian lift of $f(\alpha)$ by assumption. Furthermore, by Corollary 3.1.19 we have a diagram of homotopy fibre sequences

$$
\begin{array}{ccc}
\mathrm{map}_{\mathcal{E}_{p(x)}}(x, x') & \longrightarrow & \mathrm{map}_{\mathcal{E}'_{p(x)}}(f(x), f(x')) \\
\downarrow & & \downarrow \\
\mathrm{map}_{\mathcal{E}}(x, y) & \longrightarrow & \mathrm{map}_{\mathcal{E}'}(f(x), f(y)) \\
\downarrow & & \downarrow \\
\mathrm{map}_{\mathcal{C}}(p(x), p(y)) & =\!=\!=\!= & \mathrm{map}_{\mathcal{C}}(p(x), p(y))
\end{array}
$$

where the horizontal map on the base and the fibre are an equivalence by the assumption that f restricts to a fully faithful functor on the fibres. Therefore, the middle horizontal map is also an equivalence by Lemma 2.3.19.   □

3.2   Marked Simplicial Sets and Marked Anodyne Maps

So far, we have seen that a left/right fibration is a special kind of (co)cartesian fibration. Since left fibrations are determined by a right lifting property (with respect to left-anodyne maps), one can ask whether (co)cartesian fibrations are also characterized by a lifting property. The answer is that this not true on the nose, but that it is true in the context of marked simplicial sets, i.e., simplicial sets where one remembers a set of 1-simplices (the marked edges) as part of the datum.

Marked simplicial sets provide another model for the ∞-category of ∞-categories. Although this presentation has some technical advantages over the Joyal model structure, this point will not be of great importance to us. We will, however, relate marked simplicial sets to the previous construction of Dwyer–Kan localizations: Namely, there is a close relation between Dwyer–Kan localizations and marked equivalences, and we will see that one can view the canonical map to a Dwyer–Kan localization as a fibrant replacement in the marked model structure on marked simplicial sets.

Definition 3.2.1

A *marked simplicial set* is a pair (X, M) where M is a subset of the 1-simplices of X which contains all degenerate 1-simplices. The elements of M are called *marked edges* in X. Note that there is a corresponding category sSet$_+$ of marked simplicial sets, where morphisms are required to send marked edges to marked edges.

Example Let X be a simplicial set. Then we denote by $X^\flat$ the marked simplicial set where an edge is marked if and only if it is degenerate. We denote by $X^\sharp$ the marked simplicial set in which all morphisms are marked. This produces functors $(-)^\flat, (-)^\sharp \colon$ sSet $\to$ sSet$_+$. There are also two functors sSet$_+$ $\to$ sSet: The one forgets the marking, and the other takes the smallest sub-simplicial set spanned by the marked 1-simplices.

Example Let $p \colon X \to S$ be a map of simplicial sets. We denote by $X^\natural$ the marked simplicial set where an edge is marked if and only if it is p-cocartesian. Thus, if $p \colon X \to S$ is a cocartesian fibration, then the map $X^\natural \to S^\sharp$ is a map of marked simplicial sets.

Definition 3.2.2

The smallest saturated set containing the following maps of marked simplicial sets is called *marked left-anodyne*:

(1) for all $0 < i < n$, the maps $(\Lambda_i^n)^\flat \to (\Delta^n)^\flat$;
(2) for every $n > 0$, the map $(\Lambda_0^n)^{s\flat} \to (\Delta^n)^{s\flat}$, where the superscript $s\flat$ denotes all degenerate edges and the special edge $\Delta^{\{0,1\}}$ to be marked;
(3) the map $(\Lambda_1^2)^\sharp \amalg_{(\Lambda_1^2)^\flat} (\Delta^2)^\flat \to (\Delta^2)^\sharp$; and
(4) for every ∞-groupoid X, the map $X^\flat \to X^\sharp$.

Remark 3.2.3

A different (but equivalent) generating set of the marked left-anodyne maps is given by the following maps:

(1$'$) for all $0 < i < n$, the maps $(\Lambda_i^n)^\flat \to (\Delta^n)^\flat$;
(2$'$) the maps $(\Delta^1)^\sharp \times (\Delta^1)^\flat \cup \{0\} \times (\Delta^1)^\sharp \to (\Delta^1)^\sharp \times (\Delta^1)^\sharp$;
(3$'$) the maps $(\Delta^1)^\sharp \times (\partial \Delta^n)^\flat \cup \{0\} \times (\Delta^n)^\flat \to (\Delta^1)^\sharp \times (\Delta^n)^\flat$;
(4$'$) the map $J^\flat \to J^\sharp$.

Remark 3.2.4

We could equally well take the set generated by the maps (1), (2), (3'), and (4), see [Lur09, Proposition 3.1.1.5] for details. This will be used in Lemma 3.2.12. Likewise, we could use the set generated by (1), (2), (3), and (4'). (In fact, this is very easy to see, and we recommend it as an exercise).

In order to prove that (co)cartesian fibrations are determined by a lifting property in marked simplicial sets, it will be convenient to talk about locally cocartesian edges.

Definition 3.2.5

Let $p\colon X \to S$ be an inner fibration of simplicial sets. An edge $f\colon x \to y$ is called *locally cocartesian*, if it is q-cocartesian with q being the pullback

$$
\begin{array}{ccc}
X' & \longrightarrow & X \\
\downarrow{\scriptstyle q} & & \downarrow{\scriptstyle p} \\
\Delta^1 & \xrightarrow{\ pf\ } & S
\end{array}
$$

Lemma 3.2.6

Let $p\colon X \to S$ be an inner fibration and $f\colon x \to y$ an edge in X.

(1) If f is p-cocartesian, then it is locally cocartesian.
(2) If f is locally cocartesian and if there exists a p-cocartesian lift $g\colon x \to y'$ of $p(f)$, then f is p-cocartesian.

Proof In fact, the statement (1) can be strengthened into saying that if f is p'-cocartesian for some pullback of p, then f is locally cocartesian, see Exercise 118. We will omit the proof here. In order to show (2), we consider a lifting problem

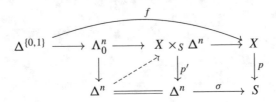

Writing $X' = X \times_S \Delta^n$, we observe that f and g are canonically edges in X', such that f is still locally cocartesian, and g is p'-cocartesian. In order to see that g gives rise to an edge of X' indeed, we only need to observe that the diagram

$$
\begin{array}{ccc}
\Delta^{\{0,1\}} & \xrightarrow{\ g\ } & X \\
\downarrow & & \downarrow{p} \\
\Delta^n & \xrightarrow{\ \sigma\ } & S
\end{array}
$$

commutes. We may thus assume that S is an ∞-category. In this case, we consider the horn obtained from g and f, and a lifting problem

$$
\begin{array}{ccccc}
\Lambda_0^2 & \xrightarrow{(g,f)} & X \times_S \Delta^1 & \longrightarrow & X \\
\downarrow & \nearrow^{\tau'} & \downarrow{q} & & \downarrow{p} \\
\Delta^2 & \longrightarrow & \Delta^1 & \xrightarrow{\ pf\ } & S
\end{array}
$$

Since g is p-cocartesian, a dashed arrow rendering the diagram commutative exists. We claim that $h = \tau'_{|\Delta\{1,2\}}$ is an equivalence in the fibre ∞-category $X_{p(y)}$. From Lemma 3.1.7, we can deduce that h is q-cocartesian and lies over the identity. It is hence an equivalence in the fibre $X_{p(y)}$ by Lemma 3.1.6. Again by Lemma 3.1.6, we deduce that h is also p cocartesian. (Notice that we have assumed S and hence X to be ∞-categories.) □

The following lemma provides another useful observation.

Lemma 3.2.7

Let $p \colon X \to S$ be an inner fibration of simplicial sets. Consider a 2-simplex of X depicted as

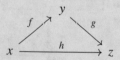

Suppose that f is p-cocartesian. Then if g (respectively h) is p-cocartesian, then h (respectively g) is locally cocartesian.

Proof Exercise 119. □

Remark 3.2.8

In [Rez19], you can find for more information about locally (co)cartesian edges and criteria for when locally (co)cartesian edges are in fact (co)cartesian.

Proposition 3.2.9

A map of marked simplicial sets $p \colon X \to S$ has the right lifting property with respect to all marked left-anodyne maps if and only if the following points hold:

(1) p is an inner fibration.

(2) An edge of X is marked if and only if it is p-cocartesian and its image is marked in S.

(3) Any lifting problem of marked simplicial sets

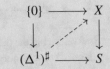

can be solved.

Proof Let us first assume that p satisfies the RLP with respect to marked left-anodyne maps and conclude that p satisfies the conditions (1) to (3) of the statement. The properties (1) and (3) follow easily. In order to see (2), we first show that any marked edge $f \colon (\Delta^1)^\sharp \to X$ is p-cocartesian. For this, consider a lifting problem

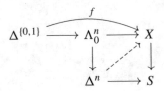

We want to show that this lifting problem admits a solution: Since the above composite is marked, it gives rise to a diagram of marked simplicial sets

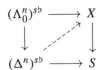

which can be solved since the left vertical map is marked anodyne.

Conversely, let us show next that any p-cocartesian edge $f : x \to y$ is marked provided that pf is marked. For this purpose, consider the diagram

$$
\begin{array}{ccc}
\{0\} & \xrightarrow{\ x\ } & X \\
\downarrow & \overset{g}{\nearrow} & \downarrow \\
(\Delta^1)^\sharp & \xrightarrow{\ pf\ } & S
\end{array}
$$

which can be solved by the assumption. The resulting morphism g is marked and thus p-cocartesian by the previous argument. Consider then the diagram

$$
\begin{array}{ccc}
(\Lambda_0^2)^{sb} & \xrightarrow{(g,f)} & X \\
\downarrow & \overset{\tau}{\nearrow} & \downarrow \\
(\Delta^2)^{sb} & \longrightarrow & S
\end{array}
$$

which can again be solved by the assumption. Here, the lower horizontal map is a degenerate 2-simplex on the morphism $pf = pg$. We denote $\tau_{|\Delta^{\{1,2\}}}$ by h. Considering the pullback diagram

$$
\begin{array}{ccc}
X \times_Y \Delta^2 & \longrightarrow & X \\
\downarrow{\scriptstyle q} & & \downarrow \\
\Delta^2 & \xrightarrow{\ p\tau\ } & S
\end{array}
$$

and the canonical map $\Delta^2 \to X \times_Y \Delta^2$, we may apply Lemma 3.1.7 to see that h is q-cocartesian, since f and g are also q-cocartesian. As $qh = \mathrm{id}_{p(y)}$, we deduce from Lemma 3.1.6 that h is an equivalence in the ∞-category $X_{p(y)}$. We consider the diagram

$$
\begin{array}{ccc}
J^\flat & \xrightarrow{\ h\ } & X_{p(y)} \longrightarrow X \\
\downarrow & \overset{\mathrm{id}_{p(y)}}{\nearrow} & \downarrow \\
J^\sharp & \longrightarrow & S
\end{array}
$$

which can again be solved, since J is an ∞-groupoid. It follows that h is marked. Next, we observe that the RLP with respect to the map $(\Lambda_1^2)^\sharp \amalg_{(\Lambda_1^2)^\flat} (\Delta^2)^\flat \to (\Delta^2)^\sharp$ implies that a composition of marked morphisms is marked. Since f is a composition of g and h, we find that f is marked as needed.

We now prove that any map $p \colon X \to S$ of marked simplicial sets satisfying the properties (1) to (3) of the statement satisfies the RLP with respect to marked left-anodyne maps. The lifting property with respect to the maps (1) and (2) of Definition 3.2.2 is immediate. In order to see that p satisfies the RLP with respect to the map $(\Lambda_1^2)^\sharp \amalg_{(\Lambda_1^2)^\flat} (\Delta^2)^\flat \to (\Delta^2)^\sharp$, we have

to show that a composite of marked edges is again marked. More precisely, assume that a
2-simplex $\sigma \colon \Delta^2 \to X$ is given, depicted as the diagram

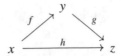

and assume that both f and g are p-cocartesian and that $p(h)$ is marked. First, we deduce
from Lemma 3.2.7 that h is locally cocartesian. By an application of the properties (2) and
(3) of the statement, we also find a p-cocartesian edge $h' \colon x \to z'$ over $p(h)$. Lemma 3.2.6
then shows that h is in fact p-cocartesian, and hence marked.

Finally, we need to argue that p has the RLP with respect to $J^\flat \to J^\sharp$, see Remark 3.2.4.
Unravelling the definitions, it will suffice to prove the following: Given an edge $f \colon x \to y$
whose classifying map takes part in a commutative diagram

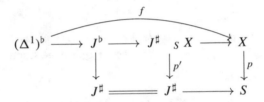

then f is a marked edge in X, or by property (2) of the statement, f is p-cocartesian. Since
property (3) of the statement supplies a p-cocartesian $g \colon x \to y'$ lift of $p(f)$, it suffices by
Lemma 3.2.6 to show that f is locally cocartesian. By construction, f is an equivalence in
the ∞-category $J^\sharp \times_S X$, and hence p'-cocartesian. In particular, it is locally cocartesian,
see Exercise 118.   □

Corollary 3.2.10
*A map of marked simplicial sets $p \colon (X, S) \to Y^\sharp$ has the right lifting property with
respect to marked left-anodyne maps if and only if S equals all p-cocartesian edges
and p is a cocartesian fibration.*

The following consequence will be of use to us later on.

Corollary 3.2.11
The map $(\Lambda_0^2)^\sharp \amalg_{(\Lambda_0^2)^\flat} (\Delta^2)^\flat \to (\Delta^2)^\sharp$ is marked anodyne.

Proof We observe that the left-hand side is the simplicial set Δ^2, where the edges $\Delta^{\{0,1\}}$ and
$\Delta^{\{0,2\}}$ are marked. We need to show that this map satisfies the LLP with respect to maps

$p\colon X \to Y$ satisfying the properties (1)–(3) of Proposition 3.2.9. Since the argument is similar to the one appearing in the proof of Proposition 3.2.9 for Λ_1^2 instead of Λ_0^2, we leave the details as an exercise. □

Lemma 3.2.12

The pushout product of a marked left-anodyne map with any monomorphism is again marked left-anodyne.

Proof We refer to [Lur09, Prop. 3.1.2.3.] for a full proof of this result. It is in spirit very similar to the arguments which we gave when showing that the pushout product of a left-/right-/inner-anodyne map with a monomorphism is again left/right/inner-anodyne: The first things to observe are:

(a) The set of monomorphisms such that the conclusion holds is saturated.
(b) The set of marked anodyne maps for which the conclusion holds is saturated.
(c) The monomorphisms in marked simplicial sets are generated by the maps $(\partial \Delta^n)^\flat \to (\Delta^n)^\flat$ and the map $(\Delta^1)^\flat \to (\Delta^1)^\sharp$.

We will only prove (c) and leave the other two items as exercises. As for (c), it is clear that the boundary inclusions as described generate monomorphisms of the form $K^\flat \to L^\flat$. For every marked simplicial set (K, M) and every monomorphism of simplicial sets $K \to L$, the following is a pushout of marked simplicial sets:

$$
\begin{array}{ccc}
K^\flat & \longrightarrow & (K, M) \\
\downarrow & & \downarrow \\
L^\flat & \longrightarrow & (L, M)
\end{array}
$$

Therefore, the right vertical map is generated by the boundary inclusions. In order to show that a general monomorphism of marked simplicial sets is generated by the maps appearing in statement (c), it suffices to see that $(L, M) \to (L, M')$ is generated by the these maps for $M \subseteq M'$. For this, we observe that there is a pushout

$$
\begin{array}{ccc}
\coprod\limits_{m' \in M' \setminus M} (\Delta^1)^\flat & \longrightarrow & (L, M) \\
\downarrow & & \downarrow \\
\coprod\limits_{m' \in M' \setminus M} (\Delta^1)^\sharp & \longrightarrow & (L, M')
\end{array}
$$

so that the claim is proven.

It hence suffices to show that the pushout product of a map of the kind (1)–(4) of Definition 3.2.2 with a map of the kind as appears in (c) is marked anodyne. There are eight cases to consider:

(1) We consider the pushout product of $(\Lambda^n_i)^\flat \subseteq (\Delta^n)^\flat$ with $(\partial\Delta^n)^\flat \subseteq (\Delta^n)^\flat$. Since $(-)^\flat$ preserves colimits, the pushout product map is given by applying $(-)^\flat$ to the pushout product of the underlying maps of simplicial sets. This is again inner-anodyne, so that it becomes marked anodyne upon applying $(-)^\flat$.

(2) The pushout product of $(\Lambda^n_i)^\flat \subseteq (\Delta^n)^\flat$ with $(\Delta^1)^\flat \to (\Delta^1)^\sharp$ is an isomorphism and thus marked anodyne. For this, note that $n \geq 2$ so that the map $\Lambda^n_i \to \Delta^n$ is an isomorphism on vertices.

(3) The pushout product of $K^\flat \to K^\sharp$ with $(\partial\Delta^n)^\flat \to (\Delta^n)^\flat$ is an isomorphism for $n > 0$, and it equals the map $K^\flat \to K^\sharp$ for $n = 0$; in either case, it is marked anodyne.

(4) The pushout product of $K^\flat \to K^\sharp$ with $(\Delta^1)^\flat \to (\Delta^1)^\sharp$ is the map $(K \times \Delta^1, M) \to (K \times \Delta^1)^\sharp$, where M is given by the pairs (a, b) and either a or b is degenerate. Since every 1-simplex (a, b) in $(K \times \Delta^1)$ is a composite of (a, id) and (id, b), and since identities are degenerate, the map we are interested in is marked anodyne (adding a composite of marked edges is marked anodyne).

(5) The pushout product of $(\Lambda^2_1)^\sharp \amalg_{(\Lambda^2_1)^\flat} (\Delta^2)^\flat \to (\Delta^2)^\sharp$ with $(\partial\Delta^n)^\flat \to (\Delta^n)^\flat$ is an isomorphism for $n \geq 1$, and the given map for $n = 0$, and hence it is marked anodyne in any case.

(6) The pushout product of $(\Lambda^2_1)^\sharp \amalg_{(\Lambda^2_1)^\flat} (\Delta^2)^\flat \to (\Delta^2)^\sharp$ with $(\Delta^1)^\flat \to (\Delta^1)^\sharp$ is almost an isomorphism: The only edge which is not marked in the domain (in the target, all edges are marked) is the edge $(0 \to 2, 0 \to 1)$. This, however, is a composite of marked edges, so that the needed map is again marked anodyne.

The remaining cases are handled most easily by using the alternative generating set of marked left-anodyne maps, i.e., by working with the set (3′) instead of (3). The key point is that the saturated set generated by the maps

$$(\Delta^1)^\sharp \times (\partial\Delta^n)^\flat \cup \{0\} \times (\Delta^n)^\flat \to (\Delta^1)^\sharp \times (\Delta^n)^\flat$$

is the same as that for the maps

$$(\Delta^1)^\sharp \times A^\flat \cup (\Delta^1)^\flat \times B^\flat \to (\Delta^1)^\sharp \times B^\flat$$

for monomorphisms $A \to B$.

(7) The pushout product of $(\Delta^1)^\sharp \times (\partial\Delta^n)^\flat \cup \{0\} \times (\Delta^n)^\flat \to (\Delta^1)^\sharp \times (\Delta^n)^\flat$ with $(\partial\Delta^n)^\flat \to (\Delta^n)^\flat$ is of the latter kind. This follows from the associativity of pushout products: It is given by the pushout product of $\{0\} \to (\Delta^1)^\sharp$ with the pushout product of $(\partial\Delta^n)^\flat \to (\Delta^n)^\flat$ with itself, which is clearly of the form $A^\flat \to B^\flat$ for a monomorphism $A \to B$.

(8) The pushout product of $(\Delta^1)^\sharp \times (\partial\Delta^n)^\flat \cup \{0\} \times (\Delta^n)^\flat \to (\Delta^1)^\sharp \times (\Delta^n)^\flat$ with $(\Delta^1)^\flat \to (\Delta^1)^\sharp$ is, as before, an isomorphism if $n > 0$. For $n = 0$, the first pushout product is simply $\{0\} \to (\Delta^1)^\sharp$, so we obtain the map $(\Delta^1 \times \Delta^1, M) \to (\Delta^1 \times \Delta^1)^\sharp$, where M consists of the degenerate edges and the edges $\{0\} \times \Delta^1$, and $\Delta^1 \times \{\varepsilon\}$ for $\varepsilon = 0, 1$.

Using property (3) and Corollary 3.2.11, we find that this is a composition of marked left-anodyne maps.

□

From this result, we can deduce the following proposition.

Proposition 3.2.13

Let $p \colon \mathcal{E} \to \mathcal{C}$ be a cocartesian fibration and let K be a simplicial set. Then $p_* \colon \mathcal{E}^K \to \mathcal{C}^K$ is again a cocartesian fibration, and an edge is p_*-cocartesian if and only if its image in $\mathcal{E}$ under the restriction along any object of K is p-cocartesian.

Proof As a special case of Lemma 3.2.12, we find that for any marked left-anodyne map $A \to B$, the map $A \times K^\flat \to B \times K^\flat$ is also marked left-anodyne. Using Proposition 3.2.9, one can solve any lifting problem

$$
\begin{array}{ccc}
A \times K^\flat & \longrightarrow & \mathcal{E}^\natural \\
\downarrow & \nearrow & \downarrow \\
B \times K^\flat & \longrightarrow & \mathcal{C}^\sharp
\end{array}
$$

Since $(-)^\flat$ is left-adjoint to the forgetful functor, this means that $(\mathcal{E}^K, S) \to (\mathcal{C}^K)^\sharp$ has the right lifting property with respect to marked anodyne maps, where S consists of those edges whose restriction to any object of K become p-cocartesian. By Corollary 3.2.10, S consists precisely of the p_* cocartesian edges and p_* is a cocartesian fibration. □

Definition 3.2.14

Let $p \colon \mathcal{E} \to \mathcal{C}$ be a cocartesian fibration and K a marked simplicial set. We denote by $\mathrm{Fun}^{\mathrm{mcc}}(K, \mathcal{E})$ the full subcategory of $\mathrm{Fun}(K, \mathcal{E})$ on functors which send all morphisms of K to p-cocartesian morphisms in $\mathcal{E}$. If K is equipped with a map $f \colon K \to \mathcal{C}^\sharp$, then we denote by $\mathrm{Fun}^{\mathrm{mcc}}_f(K, \mathcal{E})$ the pullback

$$
\begin{array}{ccc}
\mathrm{Fun}^{\mathrm{mcc}}_f(K, \mathcal{E}) & \longrightarrow & \mathrm{Fun}^{\mathrm{mcc}}(K, \mathcal{E}) \\
\downarrow & & \downarrow{\scriptstyle p_*} \\
\Delta^0 & \xrightarrow{\ f\ } & \mathrm{Fun}(K, \mathcal{C})
\end{array}
$$

If K is an ordinary simplicial set, then we write $\mathrm{Fun}^{\mathrm{cc}}(K, \mathcal{E})$, respectively $\mathrm{Fun}^{\mathrm{cc}}_f(K, \mathcal{E})$ for $\mathrm{Fun}^{\mathrm{mcc}}(K^\sharp, \mathcal{E})$, respectively $\mathrm{Fun}^{\mathrm{mcc}}_f(K^\sharp, \mathcal{E})$.

Remark 3.2.15

In [Lur09], what we denote by $\mathrm{Fun}^{\mathrm{mcc}}(K, \mathcal{E})$ is written as $\mathrm{Map}^{\flat}(K, \mathcal{E}^{\natural})$. Likewise, what we denote by $\mathrm{Fun}^{\mathrm{mcc}}_{f}(K, \mathcal{E})$ is denoted by $\mathrm{Map}^{\flat}_{K}(K, \mathcal{E}^{\natural})$.

The reason for this notation is that the category of marked simplicial sets is *cartesian closed*, which means that for every marked simplicial set K, the functor $K \times -$ admits a right adjoint, denoted by $X \mapsto X^{K}$. We define a simplicial set $\mathrm{Map}^{\flat}(K, X)$ whose n-simplices are given by $\mathrm{Hom}_{\mathrm{sSet}^{+}}((\Delta^{n})^{\flat} \times K, X)$. Likewise, we define a simplicial set $\mathrm{Map}^{\sharp}(K, X)$ whose n-simplices are given by $\mathrm{Hom}_{\mathrm{sSet}_{+}}((\Delta^{n})^{\sharp} \times K, X)$. With our previous notation, we have that $u(X^{K}) = \mathrm{Map}^{\flat}(K, X)$ and $m(X^{K}) = \mathrm{Map}^{\sharp}(K, X)$.

Proposition 3.2.16

Let $p \colon \mathcal{E} \to \mathcal{C}$ be a cocartesian fibration, let $i \colon K \to L$ be a marked anodyne map, and let $f \colon L \to \mathcal{C}^{\sharp}$ be a morphism. Then the induced map

$$\mathrm{Fun}^{\mathrm{mcc}}_{f}(L, \mathcal{E}) \longrightarrow \mathrm{Fun}^{\mathrm{mcc}}_{fi}(K, \mathcal{E})$$

is a trivial fibration.

Proof We need to show that any lifting problem

$$
\begin{array}{ccc}
S & \longrightarrow & \mathrm{Fun}^{\mathrm{mcc}}_{f}(L, \mathcal{E}) \\
\downarrow & \nearrow & \downarrow \\
T & \longrightarrow & \mathrm{Fun}^{\mathrm{mcc}}_{fi}(K, \mathcal{E})
\end{array}
$$

can be solved if $S \to T$ is a monomorphism of simplicial sets. Based on the definitions, this is the case if and only if the lifting problem

$$
\begin{array}{ccc}
S^{\flat} \times L \amalg_{S^{\flat} \times K} T^{\flat} \times K & \longrightarrow & \mathcal{E}^{\natural} \\
\downarrow & \nearrow & \downarrow \\
T^{\flat} \times L & \longrightarrow & \mathcal{C}^{\sharp}
\end{array}
$$

can be solved. By Lemma 3.2.12, the left vertical map is marked anodyne, so the claim follows. $\qquad\square$

We learned the following lemma, which will be important for us in the sequel, from Hoang Kim Nguyen's thesis, [Ngu18, Lemma 3.2.3].

Lemma 3.2.17

Let $K \to L$ be a left-anodyne map. Then the map $K^\sharp \to L^\sharp$ of marked simplicial sets is marked anodyne.

Proof First we claim that the set of monomorphisms $K \to L$ of simplicial sets such that $K^\sharp \to L^\sharp$ is marked anodyne is saturated. It hence suffices to show that for $0 \leq i < n$, the map $(\Lambda_0^n)^\sharp \to (\Delta^n)^\sharp$ is marked anodyne. We observe that $\mathrm{sk}_1(\Lambda_i^n) = \mathrm{sk}_1(\Delta^n)$ once n is at least 3. Thus for $n \geq 3$ and $0 < i < n$, we have a pushout

$$
\begin{array}{ccc}
(\Lambda_i^n)^\flat & \longrightarrow & (\Lambda_i^n)^\sharp \\
\downarrow & & \downarrow \\
(\Delta^n)^\flat & \longrightarrow & (\Delta^n)^\sharp
\end{array}
$$

which shows that the right vertical map is marked anodyne. Likewise, there is a pushout

$$
\begin{array}{ccc}
(\Lambda_0^n)^{s\flat} & \longrightarrow & (\Lambda_0^n)^\sharp \\
\downarrow & & \downarrow \\
(\Delta^n)^{s\flat} & \longrightarrow & (\Delta^n)^\sharp
\end{array}
$$

so that the right vertical map is again marked anodyne. It remains to treat the cases $n < 3$. The case $n = 1$ is clear, so it remains to treat the case $n = 2$, for which we need to discuss the cases $i = 0$ and $i = 1$. There are pushouts

$$
\begin{array}{ccccc}
(\Lambda_0^2)^\flat & \longrightarrow & (\Lambda_0^2)^{s\flat} & \longrightarrow & (\Lambda_0^2)^\sharp \\
\downarrow & & \downarrow & & \downarrow \\
(\Delta^2)^\flat & \longrightarrow & (\Delta^2)^{s\flat} & \longrightarrow & (\Delta^2)^\flat \amalg_{(\Lambda_0^2)^\flat} (\Lambda_0^2)^\sharp
\end{array}
$$

so that the very right vertical map is marked anodyne. By Corollary 3.2.11 the further map

$$
(\Delta^2)^\flat \amalg_{(\Lambda_0^2)^\flat} (\Lambda_0^2)^\sharp \to (\Delta^2)^\sharp
$$

is also marked anodyne, so that the map $(\Lambda_0^2)^\sharp \to (\Delta^2)^\sharp$ is marked anodyne as well. For the remaining case, we have the pushout

$$
\begin{array}{ccc}
(\Lambda_1^2)^\flat & \longrightarrow & (\Lambda_1^2)^\sharp \\
\downarrow & & \downarrow \\
(\Delta^2)^\flat & \longrightarrow & (\Lambda_1^2)^\sharp \amalg_{(\Lambda_1^2)^\flat} (\Delta^2)^\flat
\end{array}
$$

so that the right vertical map is marked anodyne. By definition, the map

$$
(\Lambda_1^2)^\sharp \amalg_{(\Lambda_1^2)^\flat} (\Delta^2)^\flat \to (\Delta^2)^\sharp
$$

is marked anodyne as well, so the lemma is proven.   □

With this lemma at hand, we have the following immediate consequence which will be very important for us later.

Corollary 3.2.18

Let $p\colon \mathcal{E} \to \mathcal{C}$ be a cocartesian fibration, $i\colon K \to L$ a left-anodyne map of simplicial sets and $f\colon L \to \mathcal{C}$ a morphism. Then the induced map

$$
\operatorname{Fun}_f^{cc}(L, \mathcal{E}) \longrightarrow \operatorname{Fun}_{fi}^{cc}(K, \mathcal{E})
$$

is a trivial fibration.

Proof This is the special case of Proposition 3.2.16 where the marked left-anodyne map is $i^\sharp \colon K^\sharp \to L^\sharp$, using Lemma 3.2.17.   □

If $f\colon \Delta^1 \to \mathcal{C}$ is a morphism, then the ∞-category $\operatorname{Fun}_f^{cc}(\Delta^1, \mathcal{E})$ parametrizes all p-cocartesian lifts of a given morphism in $\mathcal{C}$. We thus find the following result.

Corollary 3.2.19

Let $p\colon \mathcal{E} \to \mathcal{C}$ be a cocartesian fibration and $f\colon \Delta^1 \to \mathcal{C}$ a morphism from x to y in $\mathcal{C}$. Then the map $\operatorname{Fun}_f^{cc}(\Delta^1, \mathcal{E}) \to \mathcal{E}_x$ given by evaluating at $\{0\}$ is a trivial fibration. In particular, the simplicial set

$$
\operatorname{Fun}_f^{cc}(\Delta^1, \mathcal{E}) \times_{\mathcal{E}_x} \{z\}
$$

is a contractible Kan complex for every object z of $\mathcal{E}_x$.

Proof We consider the pullback diagram

$$\operatorname{Fun}_f^{\mathrm{cc}}(\Delta^1, \mathcal{E}) \times_{\mathcal{E}_x} \{z\} \longrightarrow \operatorname{Fun}_f^{\mathrm{cc}}(\Delta^1, \mathcal{E})$$

$$\Delta^0 \xrightarrow{\quad z \quad} \operatorname{Fun}_x^{\mathrm{cc}}(\Delta^0, \mathcal{E}) \simeq \mathcal{E}_x$$

where the right vertical map is a trivial fibration because it is obtained by restriction along $\{0\} \to \Delta^1$ (which is left-anodyne so that we may apply Corollary 3.2.18). Therefore, the left vertical map is also a trivial fibration. □

In the remaining part of this section, we want to indicate how one can make use of marked simplicial sets to study Dwyer–Kan localizations.

In what follows, we will always view ∞-categories as a cocartesian fibration over Δ^0, so that $\mathcal{C}^\natural$ denotes the marked simplicial set $\mathcal{C}$ whose marked edges are all equivalences of $\mathcal{C}$.

Definition 3.2.20

Let X, Y be marked simplicial sets. A morphism $f \colon X \to Y$ is called a *marked equivalence* if for any ∞-category $\mathcal{C}$, the induced map.

$$\operatorname{Fun}^{\mathrm{mcc}}(Y, \mathcal{C}) \to \operatorname{Fun}^{\mathrm{mcc}}(X, \mathcal{C})$$

is an equivalence of ∞-categories.

Example Let $\mathcal{C}$ be an ∞-category, S a set of morphisms containing all equivalences, and $\mathcal{C}[S^{-1}]$ a localization. Then the map $(\mathcal{C}, S) \to \mathcal{C}[S^{-1}]^\natural$ is a marked equivalence.

Example Let $A \to B$ be a marked left-anodyne map. Then $A \to B$ is a marked equivalence. This is a special case of Proposition 3.2.16: We need to show that for every ∞-category $\mathcal{E}$, the restriction functor

$$\operatorname{Fun}^{\mathrm{mcc}}(B, \mathcal{E}) \to \operatorname{Fun}^{\mathrm{mcc}}(A, \mathcal{E})$$

is a Joyal equivalence. In fact, it is a trivial fibration, because $\operatorname{Fun}^{\mathrm{mcc}}(B, \mathcal{E}) = \operatorname{Fun}_*^{\mathrm{mcc}}(B, \mathcal{E})$ for the cocartesian fibration $\mathcal{E} \to \Delta^0$ and the canonical map $* \colon B \to \Delta^0$ (likewise for A instead of B).

As in Sect. 1.3, we expect to be able to factor any map into a marked anodyne map followed by a map which satisfies the RLP with respect to marked anodyne maps. For maps of the form $X \to \Delta^0$, we find that the resulting map $\mathcal{C} \to \Delta^0$ is an inner fibration, and the marked edges of $\mathcal{C}$ are precisely the equivalences. We will not prove the following theorem in this book.

Theorem 3.2.21
There exists a simplicial model structure on marked simplicial sets whose cofibrations are monomorphisms, whose equivalences are marked equivalences, and where fibrant objects are precisely ∞-categories whose marked edges are all equivalences.

Corollary 3.2.22
A Dwyer–Kan localization of $\mathcal{C}$ along S may thus be thought of as a fibrant replacement of $(\mathcal{C}, S)$ in this model structure on marked simplicial sets.

Lemma 3.2.23
Let $(\mathcal{C}, S)$ be a marked ∞-category and let $\mathcal{E}$ be an ∞-category. Then there is an equivalence $\mathrm{Fun}^{\mathrm{mcc}}((\mathcal{C}, S), \mathcal{E}) \simeq \mathrm{Fun}(\mathcal{C}[S^{-1}], \mathcal{E})$ of ∞-categories.

Proof This follows immediately from the definition of localizations.   □

More generally, we have the following result.

Lemma 3.2.24
Let $(\mathcal{C}, S)$ and $(\mathcal{D}, T)$ be ∞-categories equipped with sets of maps, viewed as marked simplicial sets. A map $f \colon (\mathcal{C}, S) \to (\mathcal{D}, T)$ of marked simplicial sets is a marked equivalence if and only if the induced map

$$\bar{f} \colon \mathcal{C}[S^{-1}] \xrightarrow{\simeq} \mathcal{D}[T^{-1}]$$

on Dwyer–Kan localizations is an equivalence.

Proof Let $\mathcal{E}$ be an auxiliary ∞-category. Consider the commutative diagram

$$
\begin{array}{ccc}
\mathrm{Fun}^{\mathrm{mcc}}((\mathcal{D}, T), \mathcal{E}) & \longrightarrow & \mathrm{Fun}^{\mathrm{mcc}}((\mathcal{C}, S), \mathcal{E}) \\
\downarrow & & \downarrow \\
\mathrm{Fun}(\mathcal{D}[T^{-1}], \mathcal{E}) & \longrightarrow & \mathrm{Fun}(\mathcal{C}[S^{-1}], \mathcal{E})
\end{array}
$$

where the vertical maps are equivalences by Lemma 3.2.23. Hence the upper horizontal map is an equivalence if and only if the lower one is. Since $\mathcal{E}$ is an arbitrary ∞-category, the lemma follows.   □

We conclude this section with the following results describing certain Dwyer–Kan localizations of 1-categories to the coherent nerve of related Kan-enriched categories. In more detail, for a simplicial model category M, we let M^c denote the category of cofibrant objects (viewed as an ordinary category), and we let M° denote the category of cofibrant-fibrant objects (viewed as a Kan-enriched category induced from the simplicial model structure on M). In [Lur17], Lurie constructs a functor $N(M^c) \to N(M^\circ)$ which induces a functor $N(M^c)[W^{-1}] \to N(M^\circ)$, where W denotes the collection of weak equivalences of the model structure.

Theorem 3.2.25

Let M be a simplicial model category. Then the canonical functor $N(M^c)[W^{-1}] \to N(M^\circ)$ is an equivalence of ∞-categories.

Corollary 3.2.26

There are equivalences of ∞-categories $\mathrm{sSet}[\mathrm{we}^{-1}] \simeq \mathrm{Spc}$ and $\mathrm{sSet}[\mathrm{Joy}^{-1}] \simeq \mathrm{Cat}_\infty$.

Proof The Kan–Quillen model structure on simplicial sets is simplicial and every object is cofibrant. The cofibrant-fibrant objects are the Kan complexes, and the induced Kan-enriched simplicial category is the one defining Spc, so the first claim follows from Theorem 3.2.25. In order to see the second claim, we first claim that there is an equivalence

$$\mathrm{sSet}[\mathrm{Joy}^{-1}] \simeq \mathrm{sSet}_+[\mathrm{me}^{-1}].$$

Indeed, the fibrant replacement functor determines an equivalence $\mathrm{Cat}^1_\infty[\mathrm{Joy}^{-1}] \simeq \mathrm{sSet}[\mathrm{Joy}^{-1}]$, see Lemma 2.4.9. Likewise, $\mathrm{sSet}_+[\mathrm{me}^{-1}]$ is equivalent to the localization of the fibrant objects at the marked equivalences. Now, the subcategory of fibrant objects of sSet_+ is isomorphic to Cat^1_∞, the inverse functors being "forgetting the marking" on the one hand, and marking all equivalences on the other hand. Moreover, a map between ∞-categories, viewed as a map of marked simplicial sets, is a marked equivalence if and only if it is a Joyal equivalence by Lemma 3.2.24. Therefore, we also obtain an equivalence $\mathrm{Cat}^1_\infty[\mathrm{Joy}^{-1}] \simeq \mathrm{sSet}_+[\mathrm{me}^{-1}]$. Now we apply Theorem 3.2.25 to the model structure on marked simplicial sets as in Theorem 3.2.21. In a last step, one needs to observe that the simplicial enrichment coming from the simplicial model structure on marked simplicial sets (the simplicial mapping objects are given by $\mathrm{Map}^\sharp(X, Y)$, see Remark 3.2.15) is the same enrichment as used to define Cat_∞ in Definition 2.1.13.   □

Remark 3.2.27

Instead of appealing to the general result on simplicial model categories, one can also directly appeal to a dual version of [Lur17, Proposition 1.3.4.7] (making use of path objects as opposed to cylinder objects) and deduce Corollary 3.2.26 in this manner. We point out that [Lur17, Proposition 1.3.4.7] enters the proof of Theorem 3.2.25.

3.3   Straightening-Unstraightening

In this section, we want to formulate and discuss an important correspondence: the Grothendieck construction. The main theorem which we will formulate and explain to some extend is Lurie's straightening-unstraightening equivalence, which compares the ∞-category $\mathrm{Fun}(\mathcal{C}, \mathrm{Cat}_\infty)$ of functors from a fixed ∞-category $\mathcal{C}$ to the ∞-category of ∞-categories with an ∞-category of cocartesian fibrations over $\mathcal{C}$. We will explain in detail how to associate a functor $\mathcal{C} \rightarrow \mathrm{Cat}_\infty$ to a cocartesian fibration over $\mathcal{C}$, and we will briefly state how this construction is part of an equivalence of ∞-categories as described above. Also, we will give some conceptual picture of how this equivalence is meant to arise, and give some references to where this process is analyzed. We note that the strategy for constructing a functor $\mathcal{C} \rightarrow \mathrm{Cat}_\infty$ from a cocartesian fibration over $\mathcal{C}$ is borrowed from [Hau17], where the variant for right fibrations is carried out.

We begin with an informal construction. Consider a cocartesian fibration $p \colon \mathcal{E} \rightarrow \mathcal{C}$. From this, we can extract the following data:

(1) for each object x of $\mathcal{C}$, we have the ∞-category $\mathcal{E}_x$;
(2) for each morphism $f \colon x \rightarrow y$ in $\mathcal{C}$ and an object z in $\mathcal{E}_x$, we can choose a p-cocartesian lift $z \rightarrow w$ of f, which we will denote by $w = f_!(z)$;
(3) given a further object z' in $\mathcal{E}_x$ and a morphism $\alpha \colon z \rightarrow z'$, we can choose another p-cocartesian lift $z' \rightarrow w' = f_!(z')$ and obtain a diagram

$$
\begin{array}{ccc}
z & \longrightarrow & z' \\
\downarrow & & \downarrow \\
f_!(z) & \dashrightarrow & f_!(z')
\end{array}
$$

Since $z \rightarrow f_!(z)$ is p-cocartesian, the space of dashed arrows making the diagram commutative is contractible. We will denote any such dashed arrow by $f_!(\alpha)$.

Summarizing, associated to a cocartesian fibration $p \colon \mathcal{E} \to \mathcal{C}$, we wish to find a functor $\mathcal{C} \to \mathrm{Cat}_\infty$ which sends an object x to the fibre $\mathcal{E}_x = p^{-1}(X)$ and a morphism $f \colon$ from x to y to a functor $f_! \colon \mathcal{E}_x \to \mathcal{E}_y$ as indicated above.

> **Proposition 3.3.1**
>
> *Given a cocartesian fibration $p \colon \mathcal{E} \to \mathcal{C}$ and a morphism $f \colon \Delta^1 \to \mathcal{C}$ from x to y, there exists a functor $\mathcal{E}_x \times \Delta^1 \to \mathcal{E}$ whose restriction to every object z of $\mathcal{E}_x$ provides a p-cocartesian morphism $\alpha \colon z \to z'$ over f. Restricting this functor to $\mathcal{E}_x \times \{1\}$ yields a functor $f_! \colon \mathcal{E}_x \to \mathcal{E}_y$.*

Proof We begin by constructing, for each cocartesian fibration $p \colon \mathcal{E} \to \mathcal{C}$ and each morphism $f \colon \Delta^1 \to \mathcal{C}$ from x to y, a functor $f_! \colon \mathcal{E}_x \to \mathcal{E}_y$. For this, recall that in this situation, the canonical map $\mathrm{Fun}_f^{cc}(\Delta^1, \mathcal{E}) \to \mathcal{E}_x$ given by taking the source of a morphism is a trivial fibration. Choosing a section of this trivial fibration produces the composite

$$f_! \colon \mathcal{E}_x \to \mathrm{Fun}_f^{cc}(\Delta^1, \mathcal{E}) \to \mathcal{E}_y$$

where the latter map is given by taking the target of a morphism. Furthermore, we find that the first map is adjoint to a map

$$\mathcal{E}_x \times \Delta^1 \to \mathcal{E}$$

with the following properties: Its restriction to $\mathcal{E}_x \times \{1\}$ is given by $f_!$, it makes the diagram

$$
\begin{array}{ccc}
\mathcal{E}_x \times \Delta^1 & \longrightarrow & \mathcal{E} \\
\downarrow & & \downarrow {\scriptstyle p} \\
\Delta^1 & \xrightarrow{\ f\ } & \mathcal{C}
\end{array}
$$

commute, and furthermore, for each object z in $\mathcal{E}_x$, the resulting morphism $\Delta^1 \to \mathcal{E}$ is a p-cocartesian morphism with source equal to z and target equal to $f_!(z)$. $\qquad\square$

Next, we wish to show that the association $f \mapsto f_!$ is functorial in f. For this, we consider a 2-simplex $\sigma \colon \Delta^2 \to \mathcal{C}$ inside $\mathcal{C}$, which exhibits h as a composition of f and g. Then in the diagram

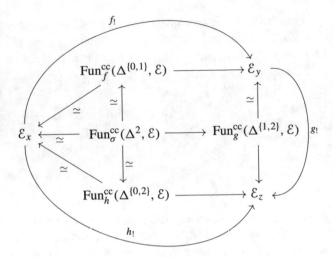

the maps labelled with $\simeq$ are trivial fibrations, because they arise as restrictions along left-anodyne maps. This shows that there is a natural isomorphism between $g_! f_!$ and $h_!$.

In the sequel, we shall explain the general construction. For this, we fix a cocartesian fibration $p\colon \mathcal{E} \to \mathcal{C}$.

Lemma 3.3.2

Associating to $\sigma\colon \Delta^n \to \mathcal{C}$ the ∞-category $\mathrm{Fun}^{cc}_\sigma(\Delta^n, \mathcal{E})$ extends to a functor $\Theta(p)\colon \Delta^{op}_{/\mathcal{C}} \to \mathrm{sSet}$.

Proof We need to show that a commutative diagram

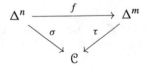

induces a well-defined functorial map $\mathrm{Fun}^{cc}_\tau(\Delta^m, \mathcal{E}) \to \mathrm{Fun}^{cc}_\sigma(\Delta^n, \mathcal{E})$. But this is clear from the definition. $\qquad\square$

Definition 3.3.3

Let X be a simplicial set. We denote by W_X the set of all morphisms $f\colon [n] \to [m]$ in $\Delta_{/X}$ such that $f(0) = 0$.

Lemma 3.3.4
For a cocartesian fibration $p\colon \mathcal{E} \to \mathcal{C}$, the functor $\Theta(p)$ sends any morphism in $W_\mathcal{E}$ to a Joyal equivalence.

Proof Consider a morphism f in $\Delta_{/\mathcal{C}}$, represented by the composite $\Delta^n \xrightarrow{f} \Delta^m \xrightarrow{\sigma} \mathcal{C}$. We will write $\tau = \sigma f$. By assumption, the composite

$$\Delta^{\{0\}} \to \Delta^n \xrightarrow{f} \Delta^m$$

picks out the object 0 in Δ^m. Therefore, we have a commutative diagram

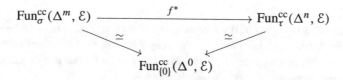

where both diagonal maps are trivial fibrations by Corollary 3.2.18, because the inclusion $\Delta^{\{0\}} \to \Delta^k$ is left-anodyne for any $k \geq 0$. Hence, the map f^* is a Joyal equivalence as claimed. $\square$

Corollary 3.3.5
For every cocartesian fibration $p\colon \mathcal{E} \to \mathcal{C}$, we obtain a functor

$$\Theta(p)\colon \mathrm{N}(\Delta_{/\mathcal{C}}^{\mathrm{op}})[W_\mathcal{E}^{-1}] \longrightarrow \mathrm{Cat}_\infty.$$

Proof By the previous lemma, $\Theta(p)$ induces a functor between the localizations

$$\mathrm{N}(\Delta_{/\mathcal{C}}^{\mathrm{op}})[W_\mathcal{E}^{-1}] \longrightarrow \mathrm{sSet}[\mathrm{Joy}^{-1}],$$

and the latter admits a further functor to Cat_∞ (which is in fact an equivalence by Corollary 3.2.26). $\square$

Lemma 3.3.6
Let X be a simplicial set. Then there is a canonical map of simplicial sets $\mathrm{N}(\Delta_{/X}^{\mathrm{op}}) \to X$ called the initial vertex map.

Proof Recall that a k-simplex of the nerve is given by a sequence

$$[n_0] \xrightarrow{\alpha_1} [n_1] \xrightarrow{\alpha_2} [n_2] \xrightarrow{\alpha_0} \cdots \xrightarrow{\alpha_k} [n_k]$$

together with a map $\sigma \colon \Delta^{n_k} \to X$. We observe that the association $\alpha \colon [k] \to [n_k]$ given by sending 0 to 0 and i to $\alpha_k \circ \cdots \circ \alpha_{k-i+1}(0)$ is a map of linearly ordered sets. Hence, we obtain a k-simplex of X by the composite $\Delta^k \xrightarrow{\alpha} \Delta^{n_k} \xrightarrow{\sigma} X$. It is straightforward to check that this association defines a map of simplicial sets as needed. □

Lemma 3.3.7

The initial vertex map sends all morphisms in W_X to degenerate edges of X.

Proof Recall that a 1-simplex in W_X is represented by the composite $[n] \xrightarrow{f} [m] \xrightarrow{\sigma} X$, where the map f sends 0 to 0. The resulting 1-simplex of X is given by restricting the map σ along the map $[1] \to [m]$ given by sending 0 to 0 and 1 to $f(0) = 0$. This is a degenerate edge in $[m]$, and it thus remains degenerate after applying the map σ. □

The following result is due to Joyal and Dwyer–Kan, and has also been proved by Stevenson, see [Ste17, Theorem 1.3].

Theorem 3.3.8

For every ∞-category $\mathcal{C}$, the initial vertex map induces a Joyal equivalence

$$N(\Delta^{op}_{/\mathcal{C}})[W_{\mathcal{C}}^{-1}] \xrightarrow{\simeq} \mathcal{C}.$$

Proof The proof will consist of two steps: First, one shows that the claim can be reduced to the case $\mathcal{C} = \Delta^n$, and then one has to show the claim in this case.

As a very first step, we need a slightly more general version of the above-said, namely we want to show that these maps make sense for an arbitrary simplicial set X instead of $\mathcal{C}$. Clearly, the initial vertex map defines a map $\Delta^{op}_{/X} \to X$. Now, for the moment, let us define for a marked simplicial set (X, S) a simplicial set $L(X, S)$ by the pushout

$$
\begin{array}{ccc}
\coprod_{f \in S} \Delta^1 & \longrightarrow & X \\
\downarrow & & \downarrow \\
\coprod_{f \in S} J & \longrightarrow & L(X, S)
\end{array}
$$

Observe that if $X = \mathcal{C}$ is an ∞-category, then $L(X, S)$ is Joyal-equivalent to $\mathcal{C}[S^{-1}]$, which was defined by choosing an inner-anodyne map $L(\mathcal{C}, S) \to \mathcal{C}[S^{-1}]$ to obtain an ∞-category. Since the initial vertex map takes a morphism in W_X to a degenerate edge in X, one can clearly extend the corresponding map $\Delta^1 \to X$ over the inclusion $\Delta^1 \to J$. In particular, we obtain an induced map $L(\mathrm{N}(\Delta^{\mathrm{op}}_{/X}), W_X) \to X$, and we claim that this map is a Joyal equivalence for every simplicial set X. Once this is shown, the theorem is proved by the above observation.

We denote the functor from simplicial sets to marked simplicial sets, sending X to $(\mathrm{N}(\Delta^{\mathrm{op}}_{/X}), W_X)$, by F. We will use the following properties, whose verification we leave as an exercise:

(1) The functor LF preserves colimits.
(2) The functor LF preserves monomorphisms.
(3) The initial vertex maps assemble into a natural transformation $LF \Rightarrow \mathrm{id}$.

Let us now suppose that the theorem is shown for $\mathcal{C} = \Delta^n$ and let X be an arbitrary simplicial set. We write X as the colimit over its skeleta $\mathrm{sk}_n(X)$ and obtain the map

$$L(\mathrm{N}(\Delta^{\mathrm{op}}_{/X}), W_X) \cong \operatorname*{colim}_n L(\mathrm{N}(\Delta^{\mathrm{op}}_{/\mathrm{sk}_n(X)}), W_{\mathrm{sk}_n(X)}) \to \operatorname*{colim}_n \mathrm{sk}_n(X) \cong X,$$

where we use the initial vertex map in each step. We leave it as an exercise to show that the required diagrams commute. If we can show that each initial vertex map

$$L(\mathrm{N}(\Delta^{\mathrm{op}}_{/\mathrm{sk}_n(X)}), W_{\mathrm{sk}_n(X)}) \to \mathrm{sk}_n(X)$$

is a Joyal equivalence, then the same applies to the above map by yet another exercise.

Next, we perform an induction over the dimension n. The induction start forces X to be a disjoint union of Δ^0's, and since the initial vertex map commutes with disjoint unions, this map is a Joyal equivalence by assumption.

For the induction step, we consider the pushout

$$
\begin{array}{ccc}
\coprod\limits_{J_n} \partial\Delta^n & \longrightarrow & \mathrm{sk}_{n-1}(X) \\
\downarrow & & \downarrow \\
\coprod\limits_{J_n} \Delta^n & \longrightarrow & \mathrm{sk}_n(X)
\end{array}
$$

and recall that the initial vertex map commutes with colimits. By induction and assumption, the initial vertex map is a Joyal equivalence on the corners except the lower right corner. However, since the functor $L(\mathrm{N}(\Delta^{\mathrm{op}}_{/(-)}), W_{(-)})$ preserves colimits and monomorphisms, it follows from Exercise 100 that this map is also a Joyal equivalence.

Now we have reached the second step of the proof, where we need to show the statement of the theorem for Δ^n. Observe that in this case, the initial vertex map

$$N(\Delta^{op}_{/\Delta^n}) \to \Delta^n$$

is the map induced on nerves of the functor $\Delta^{op}/\Delta^n \to [n]$ sending $f \colon [m] \to [n]$ to $f(0)$. Its effect on morphisms is given as follows: Given a composite $[k] \to [m] \to [n]$, where the composite is g and the latter map is f, we need to find a morphism in $[n]$ from $f(0)$ to $g(0)$. In other words, we need to show that $f(0) \leq g(0)$. But we have $g(0) = f(h(0)) \geq f(0)$, because both f and h are monotonically increasing.

Next, we construct a functor in the other direction $[n] \to \Delta^{op}_{/\Delta^n}$ as follows: The object i of $[n]$ is sent to the map $\Delta^{\{i,...,n\}} \to \Delta^n$. Clearly, if $i \leq j$, then there is a commutative diagram

$$
\begin{array}{ccc}
\Delta^{\{i,...,n\}} & \longrightarrow & \Delta^n \\
\uparrow & \nearrow & \\
\Delta^{\{j,...,n\}} & &
\end{array}
$$

so we get a functor as desired. One can immediately check that its composition with the initial vertex map is given by the identity. Let us now consider the composite

$$\Delta^{op}_{/\Delta^n} \xrightarrow{IV} [n] \xrightarrow{i} \Delta^{op}_{/\Delta^n}.$$

We claim that there is a canonical natural transformation from this composite $i \circ IV$ to the identity functor. Indeed, the composite is given by sending $f \colon [m] \to [n]$ to the canonical inclusion $\{f(0), \ldots, n\} \to [n]$, where the canonical commutative triangle

$$
\begin{array}{ccc}
\{0, \ldots, m\} & \xrightarrow{\ f\ } & \{0, \ldots, n\} \\
\ \downarrow{f} & \nearrow & \\
\{f(0), \ldots, n\} & &
\end{array}
$$

provides the components of this natural transformation (left vertical map). We observe that these components are all contained in the set W_{Δ^n}.

This construction provides a map of simplicial sets

$$\Delta^1 \to \mathrm{Fun}(N(\Delta^{op}_{/\Delta^n}), N(\Delta^{op}_{/\Delta^n}))$$

restricting to $i \circ IV$ on 0 and to the identity on 1. Post-composing with the localization map $N(\Delta^{op}_{/\Delta^n}) \to N(\Delta^{op}_{/\Delta^n})[W_{\Delta^n}^{-1}]$ and recalling that $i \circ IV$ sends W_{Δ^n} to equivalences, we obtain a map

$$\Delta^1 \to \mathrm{Fun}(N(\Delta^{op}_{/\Delta^n})[W_{\Delta^n}^{-1}], N(\Delta^{op}_{/\Delta^n})[W_{\Delta^n}^{-1}]).$$

This map is a natural equivalence due to the fact that the components of the above transformation are contained in the set W_{Δ^n}. We have thus constructed functors

$$N([n]) \to N(\Delta^{\mathrm{op}}_{/\Delta^n})[W^{-1}_{\Delta^n}] \to N([n]) \to N(\Delta^{\mathrm{op}}_{/\Delta^n})[W^{-1}_{\Delta^n}]$$

such that both composites are naturally equivalent to the identity functor. Hence, the initial vertex map is a Joyal equivalence as desired. □

> **Corollary 3.3.9**
>
> *Let* $p \colon \mathcal{E} \to \mathcal{C}$ *be a cocartesian fibration. By inverting the above equivalence, we obtain a functor*
>
> $$\mathcal{C} \xleftarrow{\simeq} N(\Delta^{\mathrm{op}}_{/\mathcal{C}})[W^{-1}_{\mathcal{C}}] \longrightarrow \mathrm{Cat}_\infty,$$
>
> *called the* straightening *of the cocartesian fibration* p.

We conclude this section with the straightening-unstraightening equivalence of Lurie (without giving a proof of it). Informally, it says that the straightening construction of Corollary 3.3.9 induces an equivalence of suitable ∞-categories. In order to make a precise statement, let us denote by $\mathrm{CoCart}(\mathcal{C})$ the subcategory of the slice $(\mathrm{Cat}_\infty)_{/\mathcal{C}}$ on objects which are cocartesian fibrations $\mathcal{E} \to \mathcal{C}$ and whose morphisms are the morphisms of cocartesian fibrations according to Definition 3.1.26.

> **Theorem 3.3.10**
>
> *For every* ∞-*category* $\mathcal{C}$, *there is an equivalence of* ∞-*categories*
>
> $$\mathrm{CoCart}(\mathcal{C}) \simeq \mathrm{Fun}(\mathcal{C}, \mathrm{Cat}_\infty).$$
>
> *On objects, this equivalence implements our previous construction.*

Definition 3.3.11

Consider the ∞-category $\mathcal{C} = \mathrm{Cat}_\infty$ and the identity functor. By Theorem 3.3.10, this corresponds to a cocartesian fibration over Cat_∞, called *the universal cocartesian fibration*. It is a functor $(\mathrm{Cat}_\infty)_{*/\!\!/} \to \mathrm{Cat}_\infty$, where $(\mathrm{Cat}_\infty)_{*/\!\!/}$ is an ∞-category whose objects are pairs $(\mathcal{C}, x)$, with x being an object of $\mathcal{C}$, and morphisms from $(\mathcal{C}, x)$ to $(\mathcal{D}, y)$ consist of pairs (F, α), with $F \colon \mathcal{C} \to \mathcal{D}$ and $\alpha \colon y \to Fx$ being a morphism in $\mathcal{D}$.

Remark 3.3.12

Constructing the ∞-category $(\mathrm{Cat}_\infty)_{*/\!/}$ is not easy, since it involves the composition in an ∞-category which is not strict. There are ways to work around this issue, but we will not get into the details here, see [RV18, Remark 6.1.19]. The idea is to consider the coherent nerve of the simplicial category of ∞-categories $\mathrm{N}(\mathrm{Cat}^1_\infty)$ without passing to the groupoids of the functor categories. This is a simplicial set, and one can form its slice under the point. The result is a map of simplicial sets $\mathrm{N}(\mathrm{Cat}^1_\infty)_{\Delta^0/\!/} \to \mathrm{N}(\mathrm{Cat}^1_\infty)$. By construction, there is also a functor $\mathrm{Cat}_\infty \to \mathrm{N}(\mathrm{Cat}^1_\infty)$, and the pullback of the slice projection turns out to be a cocartesian fibration.

Remark 3.3.13

In general, one would like to have, for each cocartesian fibration $\mathcal{E} \to \mathcal{D}$ and each ∞-category $\mathcal{C}$, a functor $\mathrm{Fun}(\mathcal{C}, \mathcal{D}) \to \mathrm{CoCart}(\mathcal{C})$ given on objects by pulling back the given cocartesian fibration. Then the statement that $\mathcal{E} \to \mathcal{D}$ is universal translates into the property that this functor is an equivalence of ∞-categories. Such a construction is done in [RV18, Theorem 6.1.13].

Remark 3.3.14

By means of the universal cocartesian fibration, we can also say what the equivalence

$$\mathrm{CoCart}(\mathcal{C}) \simeq \mathrm{Fun}(\mathcal{C}, \mathrm{Cat}_\infty)$$

does to a functor $F \colon \mathcal{C} \to \mathrm{Cat}_\infty$: It sends it to the pulled-back cocartesian fibration $F^*p \colon F^*(\mathrm{Cat}_\infty)_{*/\!/} \to \mathcal{C}$.

The following lemma will be useful to us soon.

Lemma 3.3.15

Let $f \colon \mathcal{C} \to \mathcal{D}$ be a fully faithful functor between ∞-categories. Then for any simplicial set K, the induced functor $\mathrm{Fun}(K, \mathcal{C}) \to \mathrm{Fun}(K, \mathcal{D})$ is again fully faithful.

Proof We first observe that a functor f is fully faithful if and only if the diagram

$$
\begin{array}{ccc}
\mathrm{Fun}(\Delta^1, \mathcal{C}) & \longrightarrow & \mathrm{Fun}(\Delta^1, \mathcal{D}) \\
\downarrow & & \downarrow \\
\mathcal{C} \times \mathcal{C} & \longrightarrow & \mathcal{D} \times \mathcal{D}
\end{array}
$$

is homotopy-cartesian. When applying the functor $\mathrm{Fun}(K, -)$ to this diagram, note that it preserves homotopy-cartesian diagrams. By using the equivalence $\mathrm{Fun}(K, \mathrm{Fun}(\Delta^1, \mathcal{C})) \cong \mathrm{Fun}(\Delta^1, \mathrm{Fun}(K, \mathcal{C}))$, the lemma follows. $\qquad\square$

Recall from Corollary 2.5.38 that the canonical functor $\mathrm{Spc} \to \mathrm{Cat}_\infty$ is fully faithful. From Lemma 3.3.15, it follows that for any ∞-category $\mathcal{C}$, the functor

$$\mathrm{Fun}(\mathcal{C}, \mathrm{Spc}) \to \mathrm{Fun}(\mathcal{C}, \mathrm{Cat}_\infty)$$

is also fully faithful. In particular, under the above equivalence, the ∞-category $\mathrm{Fun}(\mathcal{C}, \mathrm{Spc})$ must correspond to some full subcategory of $\mathrm{CoCart}(\mathcal{C})$. This subcategory is given by the following theorem.

Theorem 3.3.16

For every ∞-category $\mathcal{C}$, the straightening–unstraightening equivalence restricts to an equivalence

$$\mathrm{LFib}(\mathcal{C}) \simeq \mathrm{Fun}(\mathcal{C}, \mathrm{Spc}),$$

where $\mathrm{LFib}(\mathcal{C})$ denotes the full subcategory of the slice $(\mathrm{Cat}_\infty)_{/\mathcal{C}}$ on left fibrations.

Proof Under the straightening-unstraightening equivalence, the functor $\mathcal{C} \to \mathrm{Spc} \to \mathrm{Cat}_\infty$ corresponds to a cocartesian fibration $\mathcal{E} \to \mathcal{C}$ whose fibres are ∞-groupoids. Hence $\mathcal{E} \to \mathcal{C}$ is a left fibration by the dual version of Proposition 3.1.23. Since any morphism $\mathcal{E} \to \mathcal{E}'$ over $\mathcal{C}$ preserves cocartesian edges (since every edge is cocartesian by Lemma 3.1.22), this is in fact the *full* subcategory of the slice category as claimed. $\qquad\square$

The following result is almost a corollary of the above theorem.

Theorem 3.3.17

Let X be an ∞-groupoid. Then the straightening–unstraightening equivalence restricts to an equivalence

$$\mathrm{Spc}_{/X} \simeq \mathrm{Fun}(X, \mathrm{Spc}).$$

Proof By Theorem 3.3.16, we need to show that there is a canonical equivalence $\mathrm{LFib}(X) \simeq \mathrm{Spc}_{/X}$. Since X is an ∞-groupoid, any left fibration $\mathcal{E} \to X$ is in fact a Kan fibration. In particular, $\mathcal{E}$ is itself a Kan complex. We thus find that the category $\mathrm{LFib}(X)$ is the full subcategory of the slice $(\mathrm{Cat}_\infty)/X$ whose objects consist of Kan fibrations. We obtain the following diagram:

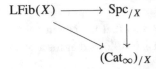

The functor $\mathrm{Spc}_{/X} \to (\mathrm{Cat}_\infty)/X$ is fully faithful, because the functor $\mathrm{Spc} \to \mathrm{Cat}_\infty$ is fully faithful by Corollary 2.5.38, see Corollary 3.3.19. It follows that the functor $\mathrm{LFib}(X) \to \mathrm{Spc}_{/X}$ is also fully faithful. Furthermore, any map $Y \to X$ between ∞-groupoids is equivalent to a Kan fibration. This implies that the inclusion $\mathrm{LFib}(X) \to \mathrm{Spc}_{/X}$ is essentially surjective and fully faithful, and thus an equivalence as needed.   □

Before we can establish Corollary 3.3.19 which is required to complete the above proof, we will need the following analysis of the mapping spaces in slice ∞-categories. This statement will also be used again later. The dual version is [Lur09, Lemma 5.5.5.12].

Proposition 3.3.18
Let $\mathcal{C}$ be an ∞-category, and let $f\colon x \to z$ and $g\colon y \to z$ be morphisms in $\mathcal{C}$, viewed as objects of $\mathcal{C}_{/z}$. Then the diagram

$$
\begin{array}{ccc}
\mathrm{map}_{\mathcal{C}_{/z}}(f, g) & \longrightarrow & \mathrm{map}_{\mathcal{C}}(x, y) \\
\downarrow & & \downarrow{\scriptstyle g_*} \\
\Delta^0 & \xrightarrow{\ f\ } & \mathrm{map}_{\mathcal{C}}(x, z)
\end{array}
$$

is homotopy-cartesian.

Proof Recall that the map $g_*\colon \mathrm{map}_{\mathcal{C}}(x, y) \to \mathrm{map}_{\mathcal{C}}(y, z)$ is constructed as follows. We have the two canonical restriction functors

$$\mathcal{C}_{/y} \xleftarrow{\ \simeq\ } \mathcal{C}_{/g} \to \mathcal{C}_{/z},$$

the first of which is an equivalence. Inverting this equivalence and taking fibres over x in $\mathcal{C}$, we obtain

$$\mathcal{C}_{/y} \times_{\mathcal{C}} \{x\} \simeq \mathcal{C}_{/g} \times_{\mathcal{C}} \{x\} \to \mathcal{C}_{/z} \times_{\mathcal{C}} \{x\},$$

where the first and the last term are given by $\mathrm{map}_{\mathcal{C}}^R(x, y)$ and $\mathrm{map}_{\mathcal{C}}^R(x, z)$, respectively. We then consider the diagram

$$
\begin{array}{ccccc}
\mathcal{C}_{/g} \times_{\mathcal{C}_{/z}} \{f\} & \longrightarrow & \mathcal{C}_{/g} \times_{\mathcal{C}} \{x\} & \longrightarrow & \mathcal{C}_{/g} \\
\downarrow & & \downarrow & & \downarrow \\
\Delta^0 & \xrightarrow{\ f\ } & \mathcal{C}_{/z} \times_{\mathcal{C}} \{x\} & \longrightarrow & \mathcal{C}_{/z}
\end{array}
$$

where both the right square and the big square are pullbacks. Hence all squares are pullbacks. Furthermore, since the very right vertical map is a right fibration, the middle vertical map is a right fibration whose target is an ∞-groupoid. Hence, the middle vertical map is a Kan fibration and models the map given by post-composition with g. It remains to show that there is a canonical equivalence

$$
\mathrm{map}_{\mathcal{C}_{/z}}(f, g) \simeq \mathcal{C}_{/g} \times_{\mathcal{C}_{/z}} \{f\}.
$$

For this, we observe that $(\mathcal{C}_{/z})_{/g} \cong \mathcal{C}_{/g}$, so the claim follows. $\qquad\square$

Corollary 3.3.19

Let $\mathcal{C} \subseteq \mathcal{D}$ be a full subcategory and let z be an object of $\mathcal{C}$. Then the canonical functor $\mathcal{C}_{/z} \to \mathcal{D}_{/z}$ is again fully faithful.

Proof Let $f \colon x \to z$ and $g \colon y \to z$ be objects of $\mathcal{C}_{/z}$. We wish to show that the map

$$
\mathrm{map}_{\mathcal{C}_{/z}}(f, g) \to \mathrm{map}_{\mathcal{D}_{/z}}(f, g)
$$

is a homotopy equivalence. By Proposition 3.3.18, it suffices to prove that in the diagram

$$
\begin{array}{ccc}
\mathrm{map}_{\mathcal{C}}(x, y) & \longrightarrow & \mathrm{map}_{\mathcal{D}}(x, y) \\
\downarrow & & \downarrow \\
\mathrm{map}_{\mathcal{C}}(x, z) & \longrightarrow & \mathrm{map}_{\mathcal{D}}(x, z)
\end{array}
$$

both horizontal maps are equivalences. But this follows from the assumption that f is fully faithful. $\qquad\square$

Limits, Colimits, and Quillen's Theorem A

4

4.1 Terminal and Initial Objects

In this section, we will discuss terminal and initial objects, as a warm-up for the later notion of limits and colimits in ∞-categories. We will give several standard equivalent characterizations for an object being terminal or initial, respectively.

Definition 4.1.1

Let $\mathcal{C}$ be an ∞-category. An object x is said to be *initial* if for all objects y of $\mathcal{C}$, the mapping space $\mathrm{map}_{\mathcal{C}}(x, y)$ is contractible. Likewise, x is said to be *terminal* if it is initial in $\mathcal{C}^{\mathrm{op}}$, i.e., if for all other objects y, the mapping space $\mathrm{map}_{\mathcal{C}}(y, x)$ is contractible.

It will be useful to have the following characterizations.

Lemma 4.1.2

Let $\mathcal{C}$ be an ∞-category and x in $\mathcal{C}$ an object. Then the following conditions are equivalent:

(1) x is terminal.
(2) The functor $\mathcal{C}_{/x} \to \mathcal{C}$ is a trivial fibration.
(3) For every $n \geq 1$, every lifting problem

(continued)

© The Author(s), under exclusive license to Springer Nature Switzerland AG 2021
M. Land, *Introduction to Infinity-Categories*, Compact Textbooks in Mathematics,
https://doi.org/10.1007/978-3-030-61524-6_4

Lemma 4.1.2 (continued)

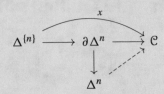

admits a solution.

Proof In order to show that (1) and (2) are equivalent, we consider the diagram

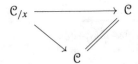

and we wish to show that the horizontal map is a trivial fibration. We already know that it is a right fibration, so it suffices to show that it is a Joyal equivalence if and only if x is terminal. By Theorem 3.1.27, this map is a Joyal equivalence if and only if it is a Joyal equivalence fibrewise, which amounts to saying that, for all objects y of $\mathcal{C}$, the canonical map $\mathrm{map}_{\mathcal{C}}^{R}(y, x) \to \Delta^0$ is a Joyal equivalence.

In order to see that (2) and (3) are equivalent, we observe that the map

$$\partial\Delta^{n-1} \star \Delta^0 \amalg_{\partial\Delta^{n-1}\star\emptyset} \Delta^{n-1} \star \emptyset \longrightarrow \Delta^{n-1} \star \Delta^0$$

is isomorphic to the map

$$\partial\Delta^n \to \Delta^n.$$

Hence the lifting problem

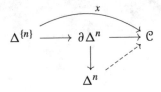

is equivalent to the lifting problem

$$
\begin{array}{ccc}
\partial\Delta^{n-1} & \longrightarrow & \mathcal{C}_{/x} \\
\downarrow & \nearrow & \downarrow \\
\Delta^{n-1} & \longrightarrow & \mathcal{C}
\end{array}
$$

Therefore, the lemma follows. □

The following result tells us that initial and terminal objects, if they exist, are unique up to contractible choices.

Proposition 4.1.3
Let $\mathcal{C}$ be an ∞-category and let $\mathcal{C}_{\text{term}}$ be the full subcategory spanned by all terminal objects. Then $\mathcal{C}_{\text{term}}$ is either empty or a contractible Kan complex.

Proof Suppose that $\mathcal{C}_{\text{term}}$ is not empty. We need to show that any lifting problem

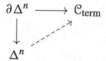

has a solution. If $n = 0$, a solution exists by the assumption that $\mathcal{C}_{\text{term}}$ is not empty. If $n \geq 1$, we use (3) of Lemma 4.1.2, which is possible since, in particular, the object $\Delta^{\{n\}}$ of $\partial \Delta^n$ is mapped to a terminal object. □

Lemma 4.1.4
Let $\mathcal{C}$ be an ∞-category. Then an object x of $\mathcal{C}$ is initial if and only if the map $x \colon \Delta^0 \to \mathcal{C}$ is left-anodyne. Dually, x is terminal if the map $\Delta^0 \to \mathcal{C}$ is right-anodyne.

Proof We prove the statement for initial objects; the case for terminal objects is obtained by passing to opposite categories. We first observe that for any monomorphism $S \to T$, the map $S \star \Delta^0 \to T \star \Delta^0$ is right-anodyne, and the map $\Delta^0 \star S \to \Delta^0 \star T$ is left-anodyne: To see this, it suffices to treat the case where $S \to T$ is a boundary inclusion $\partial \Delta^n \to \Delta^n$, in which case the maps in question become $\Lambda_{n+1}^{n+1} \to \Delta^{n+1}$ and $\Lambda_0^{n+1} \to \Delta^{n+1}$. Now let us assume that x is initial. By the version of Lemma 4.1.2 for initial objects, we find that $\mathcal{C}_{x/} \to \mathcal{C}$ is a trivial fibration. We choose a section $s \colon \mathcal{C} \to \mathcal{C}_{x/}$ and consider the diagram

$$
\begin{array}{ccccc}
\Delta^0 & \longrightarrow & \Delta^0 \star \Delta^0 & \longrightarrow & \Delta^0 \\
\downarrow{\scriptstyle x} & & \downarrow{\scriptstyle \Delta^0 \star x} & & \downarrow{\scriptstyle x} \\
\mathcal{C} & \longrightarrow & \Delta^0 \star \mathcal{C} & \xrightarrow{\ \hat{s}\ } & \mathcal{C}
\end{array}
$$

where $\hat{s}$ is the adjoint map of s. It follows that both horizontal composites are the identity. Therefore, the map $\Delta^0 \xrightarrow{x} \mathcal{C}$ is a retract of the map $\Delta^0 \star x$, which is left-anodyne by our first observation.

Conversely, assume that $\Delta^0 \xrightarrow{x} \mathcal{C}$ is left-anodyne. We can consider the diagram

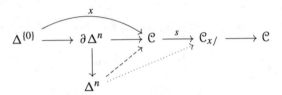

and find a dashed arrow s making the diagram commute. We will show that x is initial by establishing (3) of Lemma 4.1.2. We thus consider a diagram

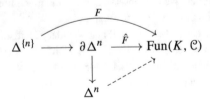

By construction, we have $s(x) = \mathrm{id}_x$, so it is an initial object of $\mathcal{C}_{x/}$ by Exercise 135. Hence the dotted arrow exists, and thus also a dashed arrow.   □

Proposition 4.1.5

Let $\mathcal{C}$ be an ∞-category and let K be a simplicial set. Suppose to be given a functor $F: K \to \mathcal{C}$ such that for all objects x of K, the object $F(x)$ is initial, respectively terminal, in $\mathcal{C}$. Then F is initial, respectively terminal, in $\mathrm{Fun}(K, \mathcal{C})$.

Proof We show the case of terminal objects. We need to prove that for all $n \geq 1$, any lifting problem

$$
\begin{array}{ccc}
\Delta^{\{n\}} \longrightarrow \partial\Delta^n & \xrightarrow{\hat{F}} & \mathrm{Fun}(K, \mathcal{C}) \\
\downarrow & \nearrow & \\
\Delta^n & &
\end{array}
$$

with the arc labeled F from $\Delta^{\{n\}}$ to $\mathrm{Fun}(K, \mathcal{C})$

admits a solution. By adjunction, this corresponds to the lifting problem

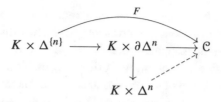

where now by assumption the further restriction of the top horizontal composite along any object $x \colon \Delta^0 \to K$ is a terminal object of $\mathcal{C}$. We consider the filtration $F_k(K) = \mathrm{sk}_k(K \times \Delta^n) \cup K \times \partial\Delta^n$, and we wish to solve the following extension problem inductively:

$$
\begin{array}{ccc}
\coprod_{i\in I(k)} \partial\Delta^k & \xrightarrow{\ a_i\ } F_{k-1}(K) \xrightarrow{\ f_{k-1}\ } \mathcal{C} \\
\downarrow & \qquad\downarrow \qquad \nearrow \\
\coprod_{i\in I(k)} \Delta^k & \longrightarrow F_k(K)
\end{array}
$$

For this, it suffices to observe that for all $i \in I(k)$, the composite $f_{k-1} \circ a_i$ sends the vertex $\{k\}$ to a terminal object in $\mathcal{C}$. By definition, $I(k)$ consists of those non-degenerate k-simplices of $K \times \Delta^n$ which are not contained in $K \times \partial\Delta^n$, in other words, it consists in particular of pairs $(\alpha_i, \beta_i) \in K_k \times \Delta^n_k$ such that $b \colon [k] \to [n]$ is surjective. In particular, β_i sends the object $\{k\}$ to $\{n\}$. Hence $f_{k-1}(a_i(\{k\})) = \hat{F}(\alpha_i(\{k\}, \{n\}))$, which is a terminal object by assumption. Hence the dashed arrow exists. By passing to the colimit over k, the proposition follows. □

The converse of Proposition 4.1.5 is almost true, as the following lemma reveals.

Lemma 4.1.6
Suppose that $\mathcal{C}_{\mathrm{term}}$ is not empty. Then any terminal object of $\mathrm{Fun}(K, \mathcal{C})$ *takes values in* $\mathcal{C}_{\mathrm{term}}$.

Proof Let x be a terminal object and consider the constant functor c_x with value x. By Proposition 4.1.5, c_x is a terminal object of $\mathrm{Fun}(K, \mathcal{C})$. By Proposition 4.1.3, any other terminal object T of $\mathrm{Fun}(K, \mathcal{C})$ is equivalent to c_x. In particular, for any object k of K, we have that $T(k)$ is equivalent to $c_x(k)$, and is therefore terminal. □

4.2   The Yoneda Lemma for ∞-Categories

In this section, we will prove the ∞-categorical version of the Yoneda lemma. Using the straightening-unstraightening equivalence which we discussed in Sect. 3.3, we construct a bivariant mapping-space functor $\mathcal{C}^{\mathrm{op}} \times \mathcal{C} \to \mathrm{Spc}$ by means of the twisted arrow category, following Lurie's construction in [Lur17]. We then formulate and sketch a proof of the Yoneda lemma in the language of left fibrations. Additionally, we indicate how one can deduce from this lemma that the functor $\mathcal{C} \to \mathrm{Fun}(\mathcal{C}^{\mathrm{op}}, \mathrm{Spc})$ associated with the bivariant mapping-space functor is fully faithful, which is why this functor is called the *Yoneda embedding*.

To begin, we wish to show that for every ∞-category $\mathcal{C}$, there is a Yoneda functor $\mathcal{C} \to \mathrm{Fun}(\mathcal{C}^{\mathrm{op}}, \mathrm{Spc})$ which should send an object x of $\mathcal{C}$ to the functor $y \mapsto \mathrm{map}_{\mathcal{C}}(y, x)$. Afterwards, we will indicate that this functor is fully faithful, which is the ∞-categorical version of the Yoneda lemma. The fact that $\mathrm{map}_{\mathcal{C}}(-, x)$ is a functor for every single x is something we already know.

Definition 4.2.1

Let x be an object of an ∞-category $\mathcal{C}$. Then the functor $\mathcal{C}^{/x} \to \mathcal{C}$ is a right fibration and hence, by Theorem 3.3.16 (applied to $\mathcal{C}^{\mathrm{op}}$), it is equivalently given by a functor $\mathcal{C}^{\mathrm{op}} \to \mathrm{Spc}$ sending y to $\mathcal{C}^{/x} \times_{\mathcal{C}} \{y\} \simeq \mathrm{map}_{\mathcal{C}}(y, x)$. We shall denote this functor by $\mathrm{map}_{\mathcal{C}}(-, x)$.

Our task now is to specify that the functors $\mathrm{map}_{\mathcal{C}}(-, x)$ are functorial in x. We begin with a construction.

Lemma 4.2.2

The association $[n] \mapsto [n] \star [n]^{\mathrm{op}} \cong [2n + 1]$ extends to a functor $\Delta \to \Delta$. In particular, sending $[n]$ to $\Delta^n \star (\Delta^n)^{\mathrm{op}}$ is a cosimplicial object in simplicial sets, i.e., a functor $\Delta \to \mathrm{sSet}$.

Definition 4.2.3

Let $\mathcal{C}$ be an ∞-category. We define its *twisted arrow category* $\mathrm{Tw}(\mathcal{C})$ as the simplicial set

$$\mathrm{Tw}(\mathcal{C})_n = \mathrm{Hom}_{\mathrm{sSet}}(\Delta^n \star (\Delta^n)^{\mathrm{op}}, \mathcal{C}),$$

where the simplicial structure comes from the cosimplicial object $[n] \mapsto \Delta^n \star (\Delta^n)^{\mathrm{op}}$. The inclusions $\Delta^n \to \Delta^n \star (\Delta^n)^{\mathrm{op}} \leftarrow (\Delta^n)^{\mathrm{op}}$ determine a functor

$$\mathrm{Tw}(\mathcal{C}) \longrightarrow \mathcal{C} \times \mathcal{C}^{\mathrm{op}}.$$

The proof for the following proposition is taken from [Lur17, Proposition 5.2.1.3].

Proposition 4.2.4

For an ∞-category $\mathcal{C}$, *the functor*

$$\mathrm{Tw}(\mathcal{C}) \longrightarrow \mathcal{C} \times \mathcal{C}^{\mathrm{op}}$$

is a right fibration. In particular, $\mathrm{Tw}(\mathcal{C})$ *is again an* ∞-*category.*

Proof Let $0 < k \leq n$, and consider a lifting problem

$$
\begin{array}{ccc}
\Lambda_n^k & \longrightarrow & \mathrm{Tw}(\mathcal{C}) \\
\downarrow & \nearrow & \downarrow \\
\Delta^n & \longrightarrow & \mathcal{C} \times \mathcal{C}^{\mathrm{op}}
\end{array}
$$

Unravelling the definition of the twisted arrow category, this corresponds to the lifting problem

$$
\begin{array}{ccc}
K & \longrightarrow & \mathcal{C} \\
\downarrow & \nearrow & \downarrow \\
\Delta^{2n+1} & \longrightarrow & \Delta^0
\end{array}
$$

where K is the sub-simplicial set of Δ^{2n+1} consisting of those faces σ which satisfy any of the following three properties:

(1) σ is contained in $\Delta^{\{0,\ldots,n\}} \subseteq \Delta^{2n+1}$.
(2) σ is contained in $\Delta^{\{n+1,\ldots,2n+1\}} \subseteq \Delta^{2n+1}$.
(3) There exists $j \neq k$, with $0 \leq j \leq n$, such that neither j nor $2n+1-j$ is a vertex of σ.

Since $\mathcal{C}$ is an ∞-category, it suffices to show that the inclusion $K \to \Delta^{2n+1}$ is inner-anodyne. We call a simplex σ primary, if it is not contained in K and its vertices are contained in the set $\{k, \ldots, 2n+1\}$; correspondingly, we call σ secondary if it is not contained in K and not primary. We let S be the set containing the following simplices τ of Δ^{2n+1}:

(1) τ is primary and k is not a vertex of τ;
(2) τ is secondary and $2n+1-k$ is not a vertex of τ.

Given a simplex τ in S, we let τ' be the simplex obtained by adding the vertex k if τ is primary, and by adding $2n+1-k$ if τ is secondary. We observe that each simplex of Δ^{2n+1} is either contained in K, in S, or is of the form τ' for a unique τ in S: If it is neither contained in K or S, then it must be either primary and contain k as vertex, or be secondary and contain

$2n + 1 - k$ as vertex, and in either case, one can remove the vertex k or $2n + 1 - k$ and obtain a primary or secondary simplex as needed.

We now choose an ordering $\{\sigma_1, \sigma_2, \ldots, \sigma_m\}$ of S with the following two properties:

(1) If $p \leq q$, then $\dim(\sigma_p) \leq \dim(\sigma_q)$.
(2) If $p \leq q$ and $\dim(\sigma_p) = \dim(\sigma_q)$ and σ_q is primary, then σ_p is primary as well.

For $0 \leq q \leq m$, we let K_q be the sub-simplicial set of Δ^{2n+1} obtained from K by adding the simplices σ_p and σ_p' for $1 \leq p \leq q$. Clearly, we have $K_m = \Delta^{2n+1}$, so we obtain a filtration

$$K \to K_1 \to K_2 \to \cdots \to K_m = \Delta^{2n+1},$$

and it will suffice to show that for each q, the map $K_{q-1} \to K_q$ is inner-anodyne. Since K_q is obtained from K_{q-1} by adding the simplices σ_q and σ_q', and since σ_q' contains σ_q, it suffices to show that there is a pushout

$$
\begin{array}{ccc}
\Lambda_j^d & \longrightarrow & K_{q-1} \\
\downarrow & & \downarrow \\
\Delta^d & \xrightarrow{\;\sigma_q'\;} & K_q
\end{array}
$$

where d is the dimension of σ_q' and $0 < j < d$. For this, we need to check which of the faces of σ_q' are already contained in K_{q-1}, and argue that precisely one inner face is not contained in K_{q-1}. $\qquad\square$

Definition 4.2.5

For every ∞-category, we denote by $\mathrm{map}_{\mathcal{C}}(-, -)\colon \mathcal{C}^{\mathrm{op}} \times \mathcal{C} \to \mathrm{Spc}$ the functor associated with the right fibration $\mathrm{Tw}(\mathcal{C}) \to \mathcal{C} \times \mathcal{C}^{\mathrm{op}}$.

Remark 4.2.6

We could also go a different route: If we only wanted to construct a mapping functor $\mathrm{map}_{\mathcal{C}}(-, -)\colon \mathcal{C}^{\mathrm{op}} \times \mathcal{C} \to \mathrm{Spc}$, we could observe that by definition $\mathrm{Spc} = \mathrm{N}(\mathrm{Kan})$. Therefore, such a functor is equivalently given by a simplicial functor $\mathfrak{C}[\mathcal{C}^{\mathrm{op}} \times \mathcal{C}] \to \mathrm{Kan}$. We could then consider the composite

$$\mathfrak{C}[\mathcal{C}^{\mathrm{op}} \times \mathcal{C}] \to \mathfrak{C}[\mathcal{C}]^{\mathrm{op}} \times \mathfrak{C}[\mathcal{C}] \to \mathrm{Kan},$$

where the first part is given by the canonical map and the second part by the simplicial mapping space followed by a functorial Kan-replacement. The confirmation that this version of a bivariant mapping-space functor is (at least pointwise) equivalent to our approach is given by [Lur09, Theorem 1.1.5.13].

> **Lemma 4.2.7**
>
> *For an ∞-category $\mathcal{C}$ and every object x of $\mathcal{C}$, there is a canonical commutative diagram*
>
> $$\begin{array}{ccc} \mathcal{C}_{/x} & \longrightarrow & \mathrm{Tw}(\mathcal{C}) \\ \downarrow & & \downarrow \\ \mathcal{C} \times \{x\} & \longrightarrow & \mathcal{C} \times \mathcal{C}^{\mathrm{op}} \end{array}$$
>
> *This diagram is homotopy-cartesian. In other words, the induced map $\mathcal{C}_{/x} \to \mathrm{Tw}(\mathcal{C})_x$ is a Joyal equivalence between right fibrations over $\mathcal{C}$.*

Proof We recall that the n-simplices of $\mathrm{Tw}(\mathcal{C})_x$ are given by those maps $\Delta^n \star (\Delta^n)^{\mathrm{op}} \to \mathcal{C}$ whose restriction along the inclusion $(\Delta^n)^{\mathrm{op}} \to \Delta^n \star (\Delta^n)^{\mathrm{op}}$ are the map which is constant at the object x of $\mathcal{C}$. Then we define an auxiliary simplicial set $\mathcal{M}$, whose n-simplices are given by maps $\Delta^n \star \Delta^0 \star (\Delta^n)^{\mathrm{op}} \to \mathcal{C}$, whose restriction along $\emptyset \star \Delta^0 \star (\Delta^n)^{\mathrm{op}} \to \Delta^n \star \Delta^0 \star (\Delta^n)^{\mathrm{op}}$ is constant at x. The obvious inclusions define maps

$$\mathcal{C}_{/x} \leftarrow \mathcal{M} \to \mathrm{Tw}(\mathcal{C})_x,$$

and the left map admits a section induced from the map $\Delta^0 \star (\Delta^n)^{\mathrm{op}} \to \Delta^0$. It hence suffices to show that both of the above maps are Joyal equivalences. This is done by showing that both maps are trivial fibrations, i.e., that they satisfy the lifting property with respect to boundary inclusions. Let us start with the case $\mathcal{M} \to \mathcal{C}_{/x}$, by considering a lifting problem

$$\begin{array}{ccc} \partial \Delta^n & \longrightarrow & \mathcal{M} \\ \downarrow & \nearrow & \downarrow \\ \Delta^n & \longrightarrow & \mathcal{C}_{/x} \end{array}$$

Unravelling the definitions shows that this is equivalently given by a lifting problem

$$\begin{array}{ccc} K & \longrightarrow & \mathcal{C} \\ \downarrow & \nearrow & \downarrow \\ \Delta^n \star \Delta^0 \star \Delta^n & \longrightarrow & \Delta^0 \end{array}$$

where K denotes the smallest sub-simplicial set containing $\Delta^n \star \Delta^0 \star \emptyset$, $\emptyset \star \Delta^0 \star \Delta^n$, and $\Delta^I \star \Delta^0 \star (\Delta^I)^{\mathrm{op}}$, for every proper subset $I \subseteq [n]$. By Theorem 2.5.14, it suffices to show that the inclusion $K \to \Delta^n \star \Delta^0 \star \Delta^n$ is a Joyal equivalence. First, we claim that the composite

$$\Delta^n \star \Delta^0 \amalg_{\Delta^0} \Delta^0 \star \Delta^n \to K \to \Delta^n \star \Delta^0 \star \Delta^n$$

is a Joyal equivalence. In order to see this, we consider the diagram

$$
\begin{array}{ccc}
I^{n+1} \amalg_{\Delta^0} I^{n+1} & \longrightarrow & I^{2n+2} \\
\downarrow & & \downarrow \\
\Delta^{n+1} \amalg_{\Delta^0} \Delta^{n+1} & \longrightarrow & \Delta^{2n+2}
\end{array}
$$

where the upper horizontal map is an isomorphism and the lower horizontal map is the map under investigation. It hence suffices to show that the vertical maps are Joyal equivalences. For the right-hand side, this follows from the fact that the spine inclusion is inner-anodyne by Proposition 1.3.22 and thus a Joyal equivalence by Corollary 2.2.13. For the left vertical map, we argue likewise for the maps $I^{n+1} \to \Delta^{n+1}$ and then use Corollary 2.5.8 to conclude that the map on pushouts is also a Joyal equivalence. It now suffices to show that the map

$$
\Delta^n \star \Delta^0 \amalg_{\Delta^0} \Delta^0 \star \Delta^n \to K
$$

is a Joyal equivalence. We denote by K_0 the sub-simplicial set of K spanned by the simplices of the form $\Delta^I \star \Delta^0 \star (\Delta^I)^{\mathrm{op}}$, with $I \subseteq [n]$ a proper subset. Since the diagram

$$
\begin{array}{ccc}
\partial \Delta^n \star \Delta^0 \amalg_{\Delta^0} \Delta^0 \star \partial \Delta^n & \xrightarrow{\ j\ } & K_0 \\
\downarrow & & \downarrow \\
\Delta^n \star \Delta^0 \amalg_{\Delta^0} \Delta^0 \star \Delta^n & \longrightarrow & K
\end{array}
$$

is a pushout, it suffices to show that the map j is a Joyal equivalence. But this map is a colimit of maps of the form

$$
\Delta^I \star \Delta^0 \amalg_{\Delta^0} \Delta^0 \star (\Delta^I)^{\mathrm{op}} \to \Delta^I \star \Delta^0 \star (\Delta^I)^{\mathrm{op}}
$$

which we have just argued to be Joyal equivalences. We thus conclude that the map $\mathcal{M} \to \mathcal{C}_{/x}$ is a Joyal equivalence.

It remains to prove that the map $\mathcal{M} \to \mathrm{Tw}(\mathcal{C})_x$ is also a Joyal equivalence. Since the proof is very similar in spirit to the one of Proposition 4.2.4, we will refrain from spelling out the details and refer to [Lur17, Proposition 5.2.1.10] instead.   □

Corollary 4.2.8
For every object x, the composite $\mathcal{C} \times \{x\} \to \mathcal{C} \times \mathcal{C}^{\mathrm{op}} \to \mathrm{Spc}$ is equivalent to the functor $\mathrm{map}_{\mathcal{C}}(-, x)$.

Proof The first functor is the one associated to the right fibration $\mathrm{Tw}(\mathcal{C})_x \to \mathcal{C}$, whereas the other one is associated to the right fibration $\mathcal{C}^{/x} \to \mathcal{C}$. By Lemma 4.2.7 and

Proposition 2.5.27, these right fibrations are equivalent. Hence the claim follows from the straightening-unstraightening equivalence (Theorem 3.3.16). $\qquad\square$

Definition 4.2.9

Let $\mathcal{C}$ be an ∞-category. The functor $\mathrm{map}_{\mathcal{C}}(-,-)\colon \mathcal{C}^{\mathrm{op}} \times \mathcal{C} \to \mathrm{Spc}$ is adjoint to a functor $Y\colon \mathcal{C} \to \mathrm{Fun}(\mathcal{C}^{\mathrm{op}}, \mathrm{Spc})$ which we call the *Yoneda functor*.

The following result is the ∞-categorical version of the Yoneda lemma.

Proposition 4.2.10

Let $F\colon \mathcal{C} \to \mathrm{Spc}$ be a functor and let x be an object of $\mathcal{C}$. Then the canonical map

$$\mathrm{map}_{\mathrm{Fun}(\mathcal{C}, \mathrm{Spc})}(\mathrm{map}_{\mathcal{C}}(x,-), F) \to F(x)$$

given by evaluation at the identity is an equivalence.

Proof We let $p\colon \mathcal{E} \to \mathcal{C}$ be the left fibration corresponding to the functor F. Under the straightening-unstraightening equivalence from Theorem 3.3.16,

$$\mathrm{Fun}(\mathcal{C}, \mathrm{Spc}) \simeq \mathrm{LFib}(\mathcal{C}) \overset{\mathrm{full}}{\subseteq} \mathrm{Cat}_{\infty/\mathcal{C}},$$

the left-hand mapping spaces correspond to $\mathrm{map}_{\mathrm{Cat}_{\infty/\mathcal{C}}}(\mathcal{C}_{x/}, \mathcal{E})$. We claim that there is a canonical equivalence

$$\mathrm{map}_{\mathrm{Cat}_{\infty/\mathcal{C}}}(\mathcal{C}_{x/}, \mathcal{E}) \simeq \mathrm{Fun}_q(\mathcal{C}_{x/}, \mathcal{E}),$$

where $q\colon \mathcal{C}_{x/} \to \mathcal{C}$ is the canonical forgetful functor and $\mathrm{Fun}_q(\mathcal{C}_{x/}, \mathcal{E})$ is as in Definition 3.2.14 without the superscripts.

Taking this for granted for the moment, we need to show that the map

$$\mathrm{Fun}_q(\mathcal{C}_{x/}, \mathcal{E}) \to \mathrm{Fun}_{\mathrm{id}_x}(\Delta^0, \mathcal{E}) \cong \mathcal{E}_x$$

is an equivalence. For this purpose, recall that by construction of the straightening-unstraightening equivalence, there is a canonical equivalence $\mathcal{E}_x \simeq F(x)$. Here, the map is induced by the canonical map $\Delta^0 \to \mathcal{C}_{x/}$ specifying the identity of x. By Exercise 135, the identity of x is an initial object of $\mathcal{C}_{x/}$, so that the map $\Delta^0 \to \mathcal{C}_{x/}$ is left-anodyne by Lemma 4.1.4. It then follows from Corollary 3.2.18 that the map in question is a trivial fibration and thus a Joyal equivalence: Since $p\colon \mathcal{E} \to \mathcal{C}$ is a left fibration, we find that $\mathrm{Fun}_q(\mathcal{C}_{x/}, \mathcal{E}) = \mathrm{Fun}_q^{\mathrm{cc}}(\mathcal{C}_{x/}, \mathcal{E})$, and likewise for Δ^0 instead of $\mathcal{C}_{x/}$.

It remains to prove the claim about the mapping spaces in $\mathrm{Cat}_\infty/\mathcal{C}$. For this, we invoke Proposition 3.3.18, which states that the following diagram is homotopy-cartesian:

$$
\begin{array}{ccc}
\mathrm{map}_{\mathrm{Cat}_\infty/\mathcal{C}}(q, p) & \longrightarrow & \mathrm{map}_{\mathrm{Cat}_\infty}(\mathcal{C}_{x/}, \mathcal{E}) \\
\downarrow & & \downarrow{\scriptstyle p_*} \\
\Delta^0 & \xrightarrow{\ q\ } & \mathrm{map}_{\mathrm{Cat}_\infty}(\mathcal{C}_{x/}, \mathcal{C})
\end{array}
$$

We now recall from Theorem 2.5.35 that the mapping spaces in Cat_∞ are canonically equivalent to the groupoid cores of the functor categories. Hence, the right vertical map in the above diagram identifies with the left vertical map in the diagram

$$
\begin{array}{ccc}
\mathrm{Fun}(\mathcal{C}_{x/}, \mathcal{E})^\simeq & \longrightarrow & \mathrm{Fun}(\mathcal{C}_{x/}, \mathcal{E}) \\
\downarrow{\scriptstyle p_*} & & \downarrow{\scriptstyle p_*} \\
\mathrm{Fun}(\mathcal{C}_{x/}, \mathcal{C})^\simeq & \longrightarrow & \mathrm{Fun}(\mathcal{C}_{x/}, \mathcal{C})
\end{array}
$$

Since p is a right fibration, p_* is also a right fibration by Theorem 1.3.37. Thus, by Proposition 2.1.3 the map p_* is conservative, so that the diagram is cartesian by Exercise 78. Pasting together the two diagrams, we find that the square

$$
\begin{array}{ccc}
\mathrm{map}_{\mathrm{Cat}_\infty/\mathcal{C}}(q, p) & \longrightarrow & \mathrm{Fun}(\mathcal{C}_{x/}, \mathcal{E}) \\
\downarrow & & \downarrow{\scriptstyle p_*} \\
\Delta^0 & \xrightarrow{\ q\ } & \mathrm{Fun}(\mathcal{C}_{x/}, \mathcal{C})
\end{array}
$$

is homotopy-cartesian. Since $\mathrm{Fun}_q(\mathcal{C}_{x/}, \mathcal{E})$ is the pullback of this diagram, we find that the canonical map

$$
\mathrm{map}_{\mathrm{Cat}_\infty/\mathcal{C}}(q, p) \to \mathrm{Fun}_q(\mathcal{C}_{x/}, \mathcal{E})
$$

is a Joyal equivalence as claimed. $\qquad\qquad\qquad\qquad\qquad\qquad\qquad\qquad\qquad\qquad\square$

With this result, we almost find the usual consequence of the Yoneda lemma, namely that the Yoneda functor is fully faithful. Henceforth, it will be called the *Yoneda embedding*.

Proposition 4.2.11
Let $\mathcal{C}$ be an ∞-category. Then the Yoneda functor $Y\colon \mathcal{C} \to \mathrm{Fun}(\mathcal{C}^{\mathrm{op}}, \mathrm{Spc})$ is fully faithful.

Proof We simply calculate that the evaluation map

$$\mathrm{map}_{\mathrm{Fun}(\mathcal{C}^{\mathrm{op}},\mathrm{Spc})}(\mathrm{map}_{\mathcal{C}}(-,x),\mathrm{map}_{\mathcal{C}}(-,y)) \longrightarrow \mathrm{map}_{\mathcal{C}}(x,y)$$

is an equivalence by Proposition 4.2.10 and claim that the composite

$$\mathrm{map}_{\mathcal{C}}(x,y) \xrightarrow{Y} \mathrm{map}_{\mathrm{Fun}(\mathcal{C}^{\mathrm{op}},\mathrm{Spc})}(\mathrm{map}_{\mathcal{C}}(-,x),\mathrm{map}_{\mathcal{C}}(-,y)) \longrightarrow \mathrm{map}_{\mathcal{C}}(x,y)$$

is also an equivalence. Informally, it is clearly the identity, but a formal proof requires more effort, which is why we will not work it out here. (See the remark below for more explanations.) In sum, we obtain that the Yoneda functor induces an equivalence on mapping spaces, which proves the proposition. □

Remark 4.2.12

A different way of defining the bivariant mapping-space functor was suggested to us by Fabian Hebestreit: One can consider the source-target functor $\mathrm{Fun}(\Delta^1, \mathcal{C}) \to \mathcal{C} \times \mathcal{C}$ as a functor over $\mathcal{C}$ via the source map and the projection onto the first factor, both of which are cartesian fibrations. In addition, the source-target map preserves cartesian morphisms, so the diagram provides a map

$$\Delta^1 \longrightarrow \mathrm{Cart}(\mathcal{C}) \simeq \mathrm{Fun}(\mathcal{C}^{\mathrm{op}}, \mathrm{Cat}_\infty).$$

Adjoining, we obtain a functor

$$\mathcal{C}^{\mathrm{op}} \longrightarrow \mathrm{Fun}(\Delta^1, \mathrm{Cat}_\infty)$$

which, unravelling the definitions, sends x to the projection functor $\mathcal{C}^{x/} \to \mathcal{C}$. Since this is a left fibration, and since left fibrations are a full subcategory of $\mathrm{Fun}(\Delta^1, \mathrm{Cat}_\infty)$, we obtain a functor

$$\mathcal{C}^{\mathrm{op}} \longrightarrow \mathrm{LFib}(\mathcal{C}) \simeq \mathrm{Fun}(\mathcal{C}, \mathrm{Spc}),$$

which is another candidate for the Yoneda functor, as it sends x to the left fibration $\mathcal{C}^{x/} \to \mathcal{C}$ which represents the functor $\mathrm{map}_{\mathcal{C}}(x, -)$. The advantage of this approach is that it can be shown from the definitions that this Yoneda functor $\mathcal{C}^{\mathrm{op}} \to \mathrm{Fun}(\mathcal{C}, \mathrm{Spc})$ is fully faithful, see Hebestreit's lecture notes [Heb20, Yoneda's lemma, adjunctions, and (co)limits]. Unfortunately, neither of the following two statements are obvious:

(1) The functor $\mathcal{C}^{\mathrm{op}} \times \mathcal{C} \to \mathrm{Spc}$ as constructed above is equivalent to the one of Definition 4.2.9 constructed via the twisted arrow category.
(2) The functor $\mathcal{C}^{\mathrm{op}} \times \mathcal{C} \to \mathrm{Spc}$ as constructed above is equivalent to its dual version, where we view the source-target map as a map of cocartesian fibrations, using the target map and the second projection.

It turns out that both statements are true, but their proofs require more tools.

4.3   Limits and Colimits

In this section, we will give a definition of (co)limits of diagrams in ∞-categories which was suggested to us by A. Krause and T. Nikolaus and relate it to the usual notion of (co)limits, i.e., phrased in terms of initial and terminal objects of certain slice categories. We will prove that (co)limits, if they exist, are unique up to a contractible space of choices, and show that an ∞-category admits small (co)limits if and only if it admits small (co)products and pullbacks, respectively pushouts. Afterwards, we will use these findings to indicate how one can show that the ∞-categories Spc of spaces and Cat_∞ of ∞-categories have all small limits and colimits. We will conclude the section with a selection of properties that (co)limits have.

The following definition of colimits is taken from Krause–Nikolaus.

Definition 4.3.1

Let $F\colon K \to \mathcal{C}$ be a functor and let x be an object of $\mathcal{C}$. We define a simplicial set $\mathrm{Map}_\mathcal{C}(F, x)$ by the pullback

$$\begin{array}{ccc}
\mathrm{Map}_\mathcal{C}(F, x) & \longrightarrow & \mathrm{Fun}(K \star \Delta^0, \mathcal{C}) \\
\downarrow & & \downarrow \\
\Delta^0 & \xrightarrow{\ (F,x)\ } & \mathrm{Fun}(K, \mathcal{C}) \times \mathcal{C}
\end{array}$$

where the right vertical map is given by restriction along the canonical inclusion $K \cup \{\infty\} \subseteq K \star \Delta^0$.

Lemma 4.3.2

In the above situation, $\mathrm{Map}_\mathcal{C}(F, x)$ is an ∞-groupoid. If $F = y\colon \Delta^0 \to \mathcal{C}$, then $\mathrm{Map}_\mathcal{C}(y, x) = \mathrm{map}_\mathcal{C}(y, x)$.

Proof Restriction along a monomorphism is a conservative inner fibration, which is stable under pullbacks. The second part is clear from the definition, since $\Delta^0 \star \Delta^0 = \Delta^1$.   □

Proposition 4.3.3

Let $F\colon K \to \mathcal{C}$ be a functor and $i\colon L \to K$ be a map of simplicial sets. Then for all objects x of $\mathcal{C}$, i induces a map $\mathrm{Map}_\mathcal{C}(F, x) \to \mathrm{Map}_\mathcal{C}(Fi, x)$. If i is right-anodyne, then this map is a homotopy equivalence.

Proof For the proof of the first statement, we observe that there is a commutative diagram

$$
\begin{array}{ccccc}
\Delta^0 & \xrightarrow{(F,x)} & \mathrm{Fun}(K, \mathcal{C}) \times \mathcal{C} & \longleftarrow & \mathrm{Fun}(K \star \Delta^0, \mathcal{C}) \\
\Big\| & & \Big\downarrow & & \Big\downarrow \\
\Delta^0 & \xrightarrow{(Fi,x)} & \mathrm{Fun}(L, \mathcal{C}) \times \mathcal{C} & \longleftarrow & \mathrm{Fun}(L \star \Delta^0, \mathcal{C})
\end{array}
$$

which induces the map of interest on pullbacks. In order to prove that this map is a homotopy equivalence if i is right-anodyne, we claim that the diagram

$$
\begin{array}{ccc}
\mathrm{Fun}(K \star \Delta^0, \mathcal{C}) & \longrightarrow & \mathrm{Fun}(L \star \Delta^0, \mathcal{C}) \\
\Big\downarrow & & \Big\downarrow \\
\mathrm{Fun}(K \cup \{\infty\}, \mathcal{C}) & \longrightarrow & \mathrm{Fun}(L \cup \{\infty\}, \mathcal{C})
\end{array}
$$

is homotopy-cartesian. This is seen by calculating the pullback given by

$$
\mathrm{Fun}(L \star \Delta^0 \amalg_{L \cup \{\infty\}} K \cup \{\infty\}, \mathcal{C}).
$$

The comparison map is then induced by the canonical inclusion

$$
L \star \Delta^0 \amalg_{L \cup \{\infty\}} K \cup \{\infty\} \to K \star \Delta^0,
$$

which is inner-anodyne since $L \to K$ is right-anodyne, see Lemma 1.4.22. Therefore, the comparison map is a Joyal equivalence as needed. It follows that the diagram

$$
\begin{array}{ccc}
\mathrm{Map}_{\mathcal{C}}(F, x) & \longrightarrow & \mathrm{Map}_{\mathcal{C}}(Fi, x) \\
\Big\downarrow & & \Big\downarrow \\
\Delta^0 & =\!\!=\!\!=\!\!= & \Delta^0
\end{array}
$$

is also homotopy-cartesian, so that the upper map is a homotopy equivalence as desired. $\quad\square$

Definition 4.3.4

Let $F\colon K \to \mathcal{C}$ be a functor and $\bar{F}\colon K \star \Delta^0 \to \mathcal{C}$ a cone over F. We say that $\bar{F}$ is a *colimit cone* if for all objects x of $\mathcal{C}$, the canonical map

$$
\mathrm{Map}_{\mathcal{C}}(\bar{F}, x) \to \mathrm{Map}_{\mathcal{C}}(F, x)
$$

is a homotopy equivalence.

Remark 4.3.5

Since the map $\{\infty\} \to K \star \Delta^0$ is right-anodyne, we find that for any cone $\bar{F}: K \star \Delta^0 \to \mathcal{C}$, the canonical map

$$\mathrm{Map}_{\mathcal{C}}(\bar{F}, x) \to \mathrm{Map}_{\mathcal{C}}(\bar{F}(\infty), x) = \mathrm{map}_{\mathcal{C}}(\bar{F}(\infty), x)$$

is a homotopy equivalence. Hence, $\bar{F}$ is a colimit cone if and only if, for all objects x of $\mathcal{C}$, the above maps assemble to a homotopy equivalence

$$\mathrm{Map}_{\mathcal{C}}(F, x) \simeq \mathrm{map}_{\mathcal{C}}(\bar{F}(\infty), x).$$

Definition 4.3.6

Dually, for a functor $F: K \to \mathcal{C}$, one defines a simplicial set $\mathrm{Map}_{\mathcal{C}}(x, F)$ as the pullback

$$\begin{array}{ccc} \mathrm{Map}_{\mathcal{C}}(x, F) & \longrightarrow & \mathrm{Fun}(\Delta^0 \star K, \mathcal{C}) \\ \downarrow & & \downarrow \\ \Delta^0 & \xrightarrow{(x,F)} & \mathcal{C} \times \mathrm{Fun}(K, \mathcal{C}) \end{array}$$

As before, a map $i: L \to K$ induces a map $\mathrm{Map}_{\mathcal{C}}(x, F) \to \mathrm{Map}_{\mathcal{C}}(x, Fi)$ which is a homotopy equivalence if i is left-anodyne. A cocone $\bar{F}: \Delta^0 \star K \to \mathcal{C}$ of F is then called a *limit cocone* if the canonical map

$$\mathrm{Map}_{\mathcal{C}}(x, \bar{F}) \to \mathrm{Map}_{\mathcal{C}}(x, F)$$

is a homotopy equivalence for all x in $\mathcal{C}$.

Remark 4.3.7

Let us compare this definition with the definition of initial and terminal objects. The goal is to see that an initial object is a colimit of the empty functor $\emptyset \to \mathcal{C}$: A cone over the empty functor is simply a functor $y: \Delta^0 \to \mathcal{C}$. Furthermore, $\mathrm{Map}_{\mathcal{C}}(\emptyset, x) \cong \Delta^0$, and $\mathrm{Map}_{\mathcal{C}}(y, x) = \mathrm{map}_{\mathcal{C}}(y, x)$ by Lemma 4.3.2. Thus we find that an object y, viewed as a cone over the empty functor, is a colimit cone if and only if, for all objects x, the mapping space $\mathrm{map}_{\mathcal{C}}(y, x)$ is Joyal-equivalent to Δ^0, i.e. is contractible. Thus, a colimit cone over the empty functor is precisely an initial object. Likewise, a limit cocone over the empty functor is precisely a terminal object.

Example A colimit over a set (viewed as a discrete category) is called a coproduct. A limit over a set is called a product.

Example A colimit of $\Lambda_0^2 \to \mathcal{C}$ is called a pushout. A limit of $\Lambda_2^2 \to \mathcal{C}$ is called a pullback. Notice that $\Lambda_0^2 \star \Delta^0 \cong \Delta^1 \times \Delta^1$ and likewise that $\Delta^0 \star \Lambda_2^2 \cong \Delta^1 \times \Delta^1$.

In Lemma 4.1.2, we characterized initial and terminal objects in terms of certain maps between slices to be trivial fibrations. We will now head towards a similar description for general colimits and limits. To get started, we have the following lemma.

Lemma 4.3.8

For any two simplicial sets K and S, there is a canonical isomorphism

$$K \diamond S \cong [(K \diamond \Delta^0) \times S] \amalg_{K \times S} K$$

compatible with the maps from K and S.

Proof Consider the diagram

$$K \times \Delta^1 \times S \longleftarrow K \times \partial\Delta^1 \times S \longrightarrow K \times S \amalg S$$
$$\| \qquad\qquad \| \qquad\qquad \downarrow$$
$$K \times \Delta^1 \times S \longleftarrow K \times \partial\Delta^1 \times S \longrightarrow K \amalg S$$

whose pushouts are given by $(K \diamond \Delta^0) \times S$ and $K \diamond S$, respectively. We thus find that the right of the small squares in the diagram

$$K \times S \longrightarrow K \times S \amalg S \longrightarrow (K \diamond \Delta^0) \times S$$
$$\downarrow \qquad\qquad \downarrow \qquad\qquad \downarrow$$
$$K \longrightarrow K \amalg S \longrightarrow K \diamond S$$

is a pushout. The left square is a pushout by inspection, hence the combined square is a pushout as well. $\qquad\square$

Lemma 4.3.9
Let $F \colon K \to \mathcal{C}$ *be a functor and let* x *be an object of* $\mathcal{C}$. *Then the diagrams*

$$
\begin{array}{ccc}
\mathrm{Map}_{\mathcal{C}}(F, x) & \longrightarrow & \mathcal{C}_{F/} \\
\downarrow & & \downarrow \\
\Delta^0 & \xrightarrow{\ x\ } & \mathcal{C}
\end{array}
\qquad
\begin{array}{ccc}
\mathrm{Map}_{\mathcal{C}}(x, F) & \longrightarrow & \mathcal{C}_{/F} \\
\downarrow & & \downarrow \\
\Delta^0 & \xrightarrow{\ x\ } & \mathcal{C}
\end{array}
$$

are homotopy-cartesian.

Proof We argue for the left-hand square, the other case is analogous. We first show that there is a pullback diagram

$$
\begin{array}{ccc}
\mathcal{C}^{F/} & \longrightarrow & \mathrm{Fun}(K \diamond \Delta^0, \mathcal{C}) \\
\downarrow & & \downarrow \\
\Delta^0 & \xrightarrow{\ F\ } & \mathrm{Fun}(K, \mathcal{C})
\end{array}
$$

For the time being, let us call the pullback $\Phi(F)$. For a simplicial set S, a map to $\mathcal{C}^{F/}$ corresponds to a map $K \diamond S \to \mathcal{C}$, whereas a map to $\Phi(F)$ corresponds to a map

$$
(K \diamond \Delta^0) \times S \amalg_{K \times S} K \to \mathcal{C}.
$$

Thus we conclude this first part of the proof by Lemma 4.3.8.

Next, we define a simplicial set $\mathcal{C}_{F/\!/}$ by the pullback

$$
\begin{array}{ccc}
\mathcal{C}_{F/\!/} & \longrightarrow & \mathrm{Fun}(K \star \Delta^0, \mathcal{C}) \\
\downarrow & & \downarrow \\
\Delta^0 & \xrightarrow{\ F\ } & \mathrm{Fun}(K, \mathcal{C})
\end{array}
$$

From the canonical map $K \diamond \Delta^0 \to K \star \Delta^0$, we obtain maps

$$
\mathcal{C}_{F/\!/} \to \mathcal{C}^{F/} \leftarrow \mathcal{C}_{F/}
$$

and we claim that both are Joyal equivalences: The right map was dealt with in Proposition 2.5.27, and for the left map, this follows again from the fact that the map $K \diamond \Delta^0 \to K \star \Delta^0$ is a Joyal equivalence, and that the pullbacks involving $\mathcal{C}^{F/}$ and $C_{F/\!/}$ are invariant under Joyal equivalences by Lemma 2.5.7. We then consider the diagram

$$\begin{array}{ccccc}
\mathrm{Map}_{\mathcal{C}}(F, x) & \longrightarrow & \mathcal{C}_{F/\!/} & \longrightarrow & \mathrm{Fun}(K \star \Delta^0, \mathcal{C}) \\
\downarrow & & \downarrow & & \downarrow \\
\Delta^0 & \longrightarrow & \mathcal{C} & \longrightarrow & \mathrm{Fun}(K, \mathcal{C}) \times \mathcal{C} \\
\downarrow & & & & \downarrow \\
\Delta^0 & & \longrightarrow & & \mathrm{Fun}(K, \mathcal{C})
\end{array}$$

consisting of pullback diagrams. The claim then follows from the fact that there is a Joyal equivalence $\mathcal{C}_{F/\!/} \simeq \mathcal{C}^{F/}$. $\qquad\qquad\square$

> **Remark 4.3.10**
>
> One could define a variant of $\mathrm{Map}_{\mathcal{C}}(F, x)$ by using the fat join instead of the ordinary join. The resulting ∞-groupoid $\widetilde{\mathrm{Map}}_{\mathcal{C}}(F, x)$ is canonically equivalent to $\mathrm{Map}_{\mathcal{C}}(F, x)$, and we will freely exchange the two whenever it is useful. The proof of Lemma 4.3.9 then shows that for this variant, the following diagram is a pullback:
>
> $$\begin{array}{ccc}
> \widetilde{\mathrm{Map}}_{\mathcal{C}}(F, x) & \longrightarrow & \mathcal{C}^{F/} \\
> \downarrow & & \downarrow \\
> \Delta^0 & \longrightarrow & \mathcal{C}
> \end{array}$$

The following theorem states that our definition of limits and colimits coincides with the definition that is usually given, e.g., in [Lur09].

> **Theorem 4.3.11**
>
> *Let $F\colon K \to \mathcal{C}$ be a diagram in an ∞-category $\mathcal{C}$. A cone $\bar{F}\colon K \star \Delta^0 \to \mathcal{C}$ of F is a colimit cone if and only if it is an initial object of $\mathcal{C}_{F/}$. Dually, a cocone $\bar{F}\colon \Delta^0 \star K \to \mathcal{C}$ of F is a limit cocone if and only if it is a terminal object of $\mathcal{C}_{/F}$.*

Proof A cocone $\bar{F}$ gives rise to a commutative diagram

of left fibrations. By Exercise 142, $\bar{F}$ is an initial object if and only if the horizontal map is a trivial fibration, which is the case if and only if it is a Joyal equivalence. By Theorem 3.1.27, this is the case if and only if the induced map on fibres over objects x of $\mathcal{C}$ is a homotopy equivalence. By Lemma 4.3.9, the induced map on fibres is equivalent to the map

$$\mathrm{Map}_{\mathcal{C}}(\bar{F}, x) \to \mathrm{Map}_{\mathcal{C}}(F, x).$$

We thus find that the map $\mathcal{C}_{\bar{F}/} \to \mathcal{C}_{F/}$ is a Joyal equivalence if and only if $\bar{F}$ is a colimit cone of F. The argument for limits is the same, using the dual version of Lemma 4.3.9.   □

The following result will be very useful for our later purposes.

Proposition 4.3.12
Let $\mathcal{C}$ be an ∞-category and consider a pushout of simplicial sets

$$
\begin{array}{ccc}
A & \xrightarrow{\ i\ } & B \\
\downarrow & & \downarrow \\
C & \longrightarrow & K
\end{array}
$$

where the map i is a monomorphism. Let $F \colon K \to \mathcal{C}$ be a functor, and denote by F_A, F_B, and F_C its restriction to A, B and C, respectively. Then, for each object x of $\mathcal{C}$, the diagram

$$
\begin{array}{ccc}
\mathrm{Map}_{\mathcal{C}}(F, x) & \longrightarrow & \mathrm{Map}_{\mathcal{C}}(F_B, x) \\
\downarrow & & \downarrow \\
\mathrm{Map}_{\mathcal{C}}(F_C, x) & \longrightarrow & \mathrm{Map}_{\mathcal{C}}(F_A, x)
\end{array}
$$

is homotopy-cartesian.

Proof We may replace $\mathrm{Map}_{\mathcal{C}}(F, x)$ by $\widetilde{\mathrm{Map}}_{\mathcal{C}}(F, x)$, i.e. the version using the fat slice. First, we observe that the diagram is a pullback, because $\widetilde{\mathrm{Map}}_{\mathcal{C}}(F, x)$ is defined as a fibre which commutes with pullbacks, and both functors $\mathrm{Fun}(-, \mathcal{C})$ and $\mathrm{Fun}(-, \mathcal{C}) \times \mathcal{C}$ send pushouts to pullbacks. Then we can use the fact that $- \diamond \Delta^0$ also preserves pushouts. It hence suffices to show that the map $\widetilde{\mathrm{Map}}_{\mathcal{C}}(F_B, x) \to \widetilde{\mathrm{Map}}_{\mathcal{C}}(F_A, x)$ is a Kan fibration. By a previous remark, there is a pullback diagram

$$
\begin{array}{ccc}
\widetilde{\mathrm{Map}}_{\mathcal{C}}(F_B, x) & \longrightarrow & \mathcal{C}^{F_B/} \\
\downarrow & & \downarrow \\
\widetilde{\mathrm{Map}}_{\mathcal{C}}(F_A, x) & \longrightarrow & \mathcal{C}^{F_A/}
\end{array}
$$

so it suffices to recall that the map $\mathcal{C}^{F_B/} \to \mathcal{C}^{F_A/}$ is a left fibration and that its pullback is therefore a left fibration between Kan complexes, and thus a Kan fibration. $\qquad\square$

Again, we find that (co)limits, if they exist, are unique up to a contractible space of choices.

Lemma 4.3.13
Let $p\colon K \to \mathcal{C}$ be a diagram. Let $(\mathcal{C}_{p/})^{\mathrm{colim}} \subseteq \mathcal{C}_{p/}$ and $(\mathcal{C}_{/p})^{\mathrm{lim}} \subseteq \mathcal{C}_{/p}$ be the full subcategories spanned by colimit cones and limit cocones, respectively. Then $(\mathcal{C}_{p/})^{\mathrm{colim}}$, respectively $(\mathcal{C}_{/p})^{\mathrm{lim}}$, are either empty or contractible Kan complexes.

Proof This is merely a reformulation of the case for initial and terminal objects, see Proposition 4.1.3. $\qquad\square$

The following result will be required later on when we show that the formation of colimits (if possible) is a functor.

Proposition 4.3.14
Let K be a simplicial set, $F\colon K \to \mathcal{C}$ a functor, and x an object of an ∞-category $\mathcal{C}$. Then there is a canonical isomorphism of simplicial sets

$$\widetilde{\mathrm{Map}}_{\mathcal{C}}(F, x) \simeq \mathrm{map}_{\mathrm{Fun}(K,\mathcal{C})}(F, \mathrm{const}_x)$$

and therefore a canonical homotopy equivalence

$$\mathrm{Map}_{\mathcal{C}}(F, x) \simeq \mathrm{map}_{\mathrm{Fun}(K,\mathcal{C})}(F, \mathrm{const}_x)$$

Proof The second assertion follows from the canonical homotopy equivalence $\mathrm{Map}_{\mathcal{C}}(F, x) \simeq \widetilde{\mathrm{Map}}_{\mathcal{C}}(F, x)$ of Remark 4.3.10. To see the first assertion, we claim that the diagram

$$
\begin{array}{ccc}
\mathcal{C}^{F/} & \longrightarrow & \mathrm{Fun}(K, \mathcal{C})^{F/} \\
\downarrow & & \downarrow \\
\mathcal{C} & \xrightarrow{\mathrm{const}} & \mathrm{Fun}(K, \mathcal{C})
\end{array}
$$

is cartesian, where we view F as a functor $\Delta^0 \to \mathrm{Fun}(K, \mathcal{C})$ as well. Then, under the assumption that our claim is true, pulling back along a map $x\colon \Delta^0 \to \mathcal{C}$, we obtain an isomorphism

$$\widetilde{\mathrm{Map}}_{\mathcal{C}}(F, x) \xrightarrow{\;\cong\;} \mathrm{map}_{\mathrm{Fun}(K,\mathcal{C})}(F, \mathrm{const}_x)$$

as desired. In order to see the claim, note that a map $X \to \mathcal{C}^{F/}$ corresponds to a map $X \diamond K \to \mathcal{C}$ whose restriction to K is F. Likewise, a map to the pullback of the above diagram corresponds to a map

$$(X \diamond \Delta^0) \times K \amalg_{X \times K} X \to \mathcal{C},$$

restricting also appropriately. Then the claim follows again from Lemma 4.3.8.   □

Lemma 4.3.15
Let $\mathcal{C}$ be an ∞-category and $\bar{\mathrm{id}} \colon \mathcal{C} \star \Delta^0 \to \mathcal{C}$ a cone over the identity of $\mathcal{C}$. Then $\bar{\mathrm{id}}$ is a colimit cone if and only if $\bar{\mathrm{id}}(\infty)$ is a terminal object. In particular, $\mathcal{C}$ has a terminal object if and only if the identity functor has a colimit.

Proof Suppose that t is a terminal object of $\mathcal{C}$. Consider the composite $\{t\} \to \mathcal{C} \to \mathcal{C}$, where the latter functor is the identity. We obtain an induced functor on slices $\mathcal{C}_{\mathrm{id}/} \to \mathcal{C}_{t/}$ which is a trivial fibration, since the inclusion $\{t\} \to \mathcal{C}$ is right-anodyne by Lemma 4.1.4. Since $\mathcal{C}_{t/}$ has an initial object, the same applies to $\mathcal{C}_{\mathrm{id}/}$. Hence by Theorem 4.3.11, t is a colimit of the identity functor. Conversely, let us assume that the identity has a colimit cone $\bar{\mathrm{id}}$. We need to show that $x = \bar{\mathrm{id}}(\infty)$ is a terminal object. For this, we will allude to Lemma 4.1.2 and consider a lifting problem

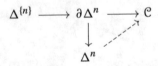

where the upper composite is given by the object x. By applying the functor $- \star \Delta^0$ to this diagram, we obtain

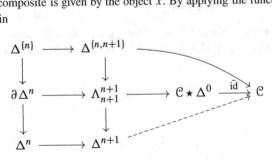

The composite $\Delta^{\{n,n+1\}} \to \mathcal{C}$ is a morphism in $\mathcal{C}$ from x to x, and we claim that this morphism is an equivalence (the justification of this claim will follow momentarily). Hence a dashed arrow exists by Joyal's lifting theorem Theorem 2.1.8.

As for the proof of the claim, we will show that the above morphism extends to a morphism in $\mathcal{C}_{\mathrm{id}/}$ from $\bar{\mathrm{id}}$ to itself. Since $\bar{\mathrm{id}}$ is a colimit cone, it is an initial object of $\mathcal{C}_{\mathrm{id}/}$, so that all of its endomorphisms are invertible by Proposition 4.1.3. Now, we consider the composite

$$\mathcal{C} \star \Delta^1 \cong \mathcal{C} \star \Delta^0 \star \Delta^0 \xrightarrow{\overline{\mathrm{id}} \star \Delta^0} \mathcal{C} \star \Delta^0 \xrightarrow{\overline{\mathrm{id}}} \mathcal{C}$$

which determines a 1-simplex in $\mathcal{C}_{\mathrm{id}}$, whose source and target are $\overline{\mathrm{id}}$. It hence suffices to show that the induced map on cone points is the one considered above. By construction, we thus have to analyze the map

$$\Delta^1 \to \mathcal{C} \star \Delta^1 \to \mathcal{C} \star \Delta^0 \to \mathcal{C},$$

and we have to observe that the map $\Delta^1 \to \mathcal{C} \star \Delta^0$ sends 0 to $\overline{\mathrm{id}}(\infty) = x$ and 1 to the cone point ∞. Therefore, it is given by the map

$$\Delta^1 \cong \Delta^0 \star \Delta^0 \xrightarrow{x \star \mathrm{id}} \mathcal{C} \star \Delta^0.$$

Composing this map with $\overline{\mathrm{id}}$ is precisely the bended map in the above diagram. □

The following proposition is a refined version of Proposition 4.3.12.

Proposition 4.3.16

Let $\mathcal{C}$ be an ∞-category. Consider a pushout of simplicial sets

$$
\begin{array}{ccc}
A & \xrightarrow{i} & B \\
\downarrow & & \downarrow \\
C & \longrightarrow & K
\end{array}
$$

where the map i is a monomorphism. Let $F\colon K \to \mathcal{C}$ be a functor, and denote by F_A, F_B, and F_C its restriction to A, B and C, respectively. Then the square

$$
\begin{array}{ccc}
\mathcal{C}^{F/} & \longrightarrow & \mathcal{C}^{F_B/} \\
\downarrow & & \downarrow \\
\mathcal{C}^{F_C/} & \longrightarrow & \mathcal{C}^{F_A/}
\end{array}
$$

is a cartesian and homotopy-cartesian square of left fibrations over $\mathcal{C}$.

Remark 4.3.17

Upon passing to fibres over objects x of $\mathcal{C}$, we obtain the cartesian and homotopy-cartesian square of Proposition 4.3.12.

We now interpret the definitions and properties of colimits as follows. First, we observe that the left fibration $\mathcal{C}_{F/} \to \mathcal{C}$ corresponds to a functor $\mathcal{C} \to \mathrm{Spc}$ by the straightening-unstraightening equivalence. Since it takes the value $\mathrm{Map}_{\mathcal{C}}(F, x)$ on an object x in $\mathcal{C}$, we simply denote this functor by $\mathrm{Map}_{\mathcal{C}}(F, -)$.

Definition 4.3.18

A functor $\mathcal{C} \to \mathrm{Spc}$ is called *representable* if it is equivalent to the functor $\mathrm{map}_{\mathcal{C}}(x, -)$ for some x in $\mathcal{C}$. Any such x will be called a *representing object*. Equivalently, a functor is representable by x, if the associated left fibration $\mathcal{E} \to \mathcal{C}$ is equivalent to the left fibration $\mathcal{C}_{x/} \to \mathcal{C}$.

Proposition 4.3.19

Let $F: K \to \mathcal{C}$ be a functor. Then F admits a colimit if and only if the functor $\mathrm{Map}_{\mathcal{C}}(F, -)$ is representable, and any representing object is a colimit of F.

Proof Suppose that F admits a colimit cone $\bar{F}: K \star \Delta^0 \to \mathcal{C}$ and let $x = \bar{F}(\infty)$. As we have previously seen, this implies that both functors

$$\mathcal{C}_{F/} \leftarrow \mathcal{C}_{\bar{F}/} \to \mathcal{C}_{x/}$$

are equivalences of left fibrations over $\mathcal{C}$: For the right map, this is always the case, and the left one is an equivalence if and only if $\bar{F}$ is a colimit cone.

It hence remains to show that if $\mathrm{Map}_{\mathcal{C}}(F, -)$ is representable by an object x, then F admits a colimit. By the assumption, we find an equivalence $\mathcal{C}_{x/} \simeq \mathcal{C}_{F/}$ of left fibrations over $\mathcal{C}$. Pick an initial object of $\mathcal{C}_{x/}$, e.g., the identity of x. Under the above equivalence, this object is transported to an initial object of $\mathcal{C}_{F/}$. Any such object is a colimit cone, whose colimit point is x by construction.   □

Corollary 4.3.20

Every diagram $F: \Delta^n \to \mathcal{C}$ admits both a limit and a colimit, where a limit is given by evaluation on 0, and a colimit is given by evaluation on n.

Proof It suffices to recall that these inclusions induce equivalences of left/right fibrations $\mathcal{C}^{F/} \simeq \mathcal{C}^{F(n)/}$ and $\mathcal{C}^{/F} \simeq \mathcal{C}^{/F(0)}$, respectively, so that both $\mathcal{C}^{F/}$ and $\mathcal{C}^{/F}$ are representable left fibrations or right fibrations, respectively.   □

Proposition 4.3.21

Let $F \colon K \to \mathcal{C}$ be a functor, let $i \colon L \to K$ be a map of simplicial sets, let $\bar{F}$ be a colimit cone of F, and let $\overline{Fi}$ be a colimit cone of Fi. Furthermore, let $G \colon \Delta^1 \to \mathcal{C}^{Fi/}$ be a map with $G(0) = \overline{Fi}$ and $G(1) = \bar{F}i$, let g be the composite $\Delta^1 \to \mathcal{C}^{Fi/} \to \mathcal{C}$, and let $g(0) = x$ and $g(1) = y$. Then there is a commutative diagram

$$
\begin{array}{ccccccc}
\mathcal{C}^{F/} & \longrightarrow & \mathcal{C}^{Fi/} & \xleftarrow{\;\simeq\;} & \mathcal{C}^{\overline{Fi}/} & \xrightarrow{\;\simeq\;} & \mathcal{C}^{x/} \\
{\scriptstyle\simeq}\big\uparrow & & \big\uparrow & & \big\uparrow & & \nearrow \\
\mathcal{C}^{\bar{F}/} & \longrightarrow & \mathcal{C}^{\bar{F}i/} & \xleftarrow{\;\simeq\;} & (\mathcal{C}^{Fi/})^{G/} & & \\
{\scriptstyle\simeq}\big\downarrow & & \big\downarrow & & \big\downarrow & \nearrow & \\
\mathcal{C}^{y/} & =\!=\!= & \mathcal{C}^{y/} & \xleftarrow{\;\simeq\;} & \mathcal{C}^{g/} & &
\end{array}
$$

and y and x are a colimit of F and Fi, respectively. The resulting map

$$
\mathcal{C}^{y/} \to \mathcal{C}^{x/}
$$

is given by precomposition with the map g, which we call the induced map $x = \mathrm{colim}_L\, Fi \to \mathrm{colim}_K\, F = y$.

Remark 4.3.22

The obvious dual situation for limiting cocones holds as well.

Observation 4.3.23

Informally, we summarize the above situation by saying that there is a commutative diagram

$$
\begin{array}{ccc}
\mathcal{C}^{F/} & \longrightarrow & \mathcal{C}^{Fi/} \\
{\scriptstyle\simeq}\big\downarrow & & \big\downarrow{\scriptstyle\simeq} \\
\mathcal{C}^{y/} & \xrightarrow{\;g\;} & \mathcal{C}^{x/}
\end{array}
$$

where the vertical maps are the equivalences coming from the fact that y is a colimit of F and x is a colimit of Fi. Of course, this diagram does not actually commute, but it commutes up to an invertible natural transformation. On the other hand, we could also take this diagram as a definition for the morphism $g \colon x \to y$. Namely, by inverting the map $\mathcal{C}^{F/} \to \mathcal{C}^{y/}$, we obtain a functor $\mathcal{C}^{y/} \to \mathcal{C}^{x/}$ which makes the diagram commute. By the Yoneda lemma, such a functor is equivalently given by an object of $\mathcal{C}^{x/} \times_{\mathcal{C}} \{y\} \simeq$

(continued)

4.3.23 (continued)

$\mathrm{map}_{\mathcal{C}}(x, y)$, which is going to be a morphism equivalent to the morphism g of the previous proposition. The advantage of this approach is that it shows how the morphism g canonically lifts to a morphism between cones over Fi.

Lemma 4.3.24

Let I be a discrete category and let $F \colon I \to \mathcal{C}$ be a diagram, i.e., a collection x_i of objects of $\mathcal{C}$. Then there is a canonical equivalence

$$\mathrm{Map}_{\mathcal{C}}(F, y) \simeq \prod_{i \in I} \mathrm{map}_{\mathcal{C}}(x_i, y)$$

between functors $\mathcal{C} \to \mathrm{Spc}$ in y.

Proof The following diagram is a pushout:

$$
\begin{array}{ccc}
I \times \{0\} \sqcup I \times \{1\} & \longrightarrow & \coprod_I \Delta^1 \\
\downarrow & & \downarrow \\
I \sqcup \{\infty\} & \longrightarrow & I \star \Delta^0
\end{array}
$$

Thus in the following diagram, both small squares are pullbacks:

$$
\begin{array}{ccccc}
\mathcal{C}_{F/} & \longrightarrow & \mathrm{Fun}(I \star \Delta^0, \mathcal{C}) & \longrightarrow & \prod_I \mathrm{Fun}(\Delta^1, \mathcal{C}) \\
\downarrow & & \downarrow & & \downarrow \\
\mathcal{C} & \xrightarrow{(F,\mathrm{id})} & \mathrm{Fun}(I, \mathcal{C}) \times \mathcal{C} & \xrightarrow{(s,\Delta)} & \prod_I \mathcal{C} \times \mathcal{C}
\end{array}
$$

Therefore, the big square is also a pullback.

Since pullbacks commute with arbitrary products, we find that there is an isomorphism

$$\mathcal{C}_{F/} \cong \prod_I \mathcal{C}^{x_i/}$$

of left fibrations. This implies the lemma. $\square$

Corollary 4.3.25

Let $F: \Lambda_0^2 \to \mathcal{C}$ *be a diagram, depicted as*

$$x \xrightarrow{\ f\ } y$$
$$\downarrow g$$
$$z$$

and let t *be another object. Then there is a homotopy-cartesian square of* ∞-*groupoids*

$$
\begin{array}{ccc}
\mathrm{Map}_{\mathcal{C}}(F, t) & \longrightarrow & \mathrm{map}_{\mathcal{C}}(y, t) \\
\downarrow & & \downarrow \\
\mathrm{map}_{\mathcal{C}}(z, t) & \longrightarrow & \mathrm{map}_{\mathcal{C}}(x, t)
\end{array}
$$

which is natural in t.

Proof By Proposition 4.3.16 and the fact that Λ_0^2 is the pushout $\Delta^1 \amalg_{\Delta^0} \Delta^1$, we find that the diagram

$$
\begin{array}{ccc}
\mathcal{C}^{F/} & \longrightarrow & \mathcal{C}^{f/} \\
\downarrow & & \downarrow \\
\mathcal{C}^{g/} & \longrightarrow & \mathcal{C}^{x/}
\end{array}
$$

is cartesian and homotopy-cartesian and consists of left fibrations over $\mathcal{C}$. Furthermore, we have $\mathcal{C}^{f/} \simeq \mathcal{C}^{y/}$ and $\mathcal{C}^{g/} \simeq \mathcal{C}^{z/}$. Passing to fibres over a point t, we obtain the statement. $\square$

Proposition 4.3.26

Let $\mathcal{C}$ *be an* ∞-*category and* K *a simplicial set, written as a pushout* $B \amalg_A C$, *where the map* $A \to B$ *is a monomorphism. Let* $F: K \to \mathcal{C}$ *be a diagram. Suppose that* $F_{|B}$ *has a colimit* y, $F_{|A}$ *has a colimit* x *and* $F_{|C}$ *has a colimit* y. *If* $\mathcal{C}$ *has pushouts, then a pushout* $y \amalg_x z$ *is a colimit of* F.

Proof In the sequel, we shall follow the reasoning of [Lur09, 4.4.2.2]. Let $G: \Lambda_0^2 \to \mathcal{C}$ be the diagram given by $y \leftarrow x \to z$, where the maps originate from Proposition 4.3.21, and let $\bar{G}: \Delta^1 \times \Delta^1 \to \mathcal{C}$ be a colimit cone of G. We wish to show that $w = \bar{G}(1, 1)$ is a colimit of $F: K \to \mathcal{C}$. By Proposition 4.3.19, we have to show that there is an equivalence of functors between $\mathrm{map}_{\mathcal{C}}(w, -)$ and $\mathrm{Map}_{\mathcal{C}}(F, -)$. In terms of left fibrations, we need to show that there is an equivalence of left fibrations $\mathcal{C}^{w/} \simeq \mathcal{C}^{F/}$. We have $\mathcal{C}^{w/} \simeq \mathcal{C}^{G/}$, since w is a colimit of G. Furthermore, there is a homotopy-cartesian square

$$
\begin{array}{ccc}
\mathcal{C}^{G/} & \longrightarrow & \mathcal{C}^{y/} \\
\downarrow & & \downarrow \\
\mathcal{C}^{z/} & \longrightarrow & \mathcal{C}^{x/}
\end{array}
$$

Since x, y, and z are colimits themselves, we find that there is commutative diagram

$$
\begin{array}{ccccc}
\mathcal{C}^{F_B/} & \longrightarrow & \mathcal{C}^{F_A/} & \longleftarrow & \mathcal{C}^{F_C/} \\
\downarrow{\scriptstyle \simeq} & & \downarrow{\scriptstyle \simeq} & & \downarrow{\scriptstyle \simeq} \\
\mathcal{C}^{y/} & \longrightarrow & \mathcal{C}^{x/} & \longleftarrow & \mathcal{C}^{z/}
\end{array}
$$

where the vertical comparison maps are Joyal equivalences (see Observation 4.3.23). Hence, there is an induced equivalence on homotopy pullbacks

$$
\mathcal{C}^{G/} \simeq \mathcal{C}^{F/},
$$

so that $\mathcal{C}^{F/}$ is indeed representable, with w a representing object. $\square$

Proposition 4.3.27

Let $F \colon K \to \mathcal{C}$ be a functor, let $K = \operatorname*{colim}_{i \geq 0} K_i$ be an $\mathbb{N}$-indexed decomposition with each map $K_i \to K_{i+1}$ a monomorphism, and let F_i be the restriction of F to K_i. Suppose that, for all i, the functor F_i admits a colimit and that $\mathcal{C}$ admits colimits over 1-dimensional simplicial sets. Then F admits a colimit.

Proof First, we find that there is an isomorphism of left fibrations $\mathcal{C}^{F/} \to \lim_{i \geq 0} \mathcal{C}^{F_i/}$, because for every simplicial set X, the functor $X \diamond -$ preserves connected colimits. By assumption, all $\mathcal{C}^{F_i}$ are corepresentable left fibrations, say $\mathcal{C}^{F_i/} \simeq \mathcal{C}^{x_i/}$. We thus obtain canonical maps $\mathcal{C}^{x_{i-1}/} \to \mathcal{C}^{x_i/}$ which make the comparison diagrams commute. By the Yoneda lemma, all of these maps corresponds to morphisms $\alpha_i \colon x_{i-1} \to x_i$, and they assemble into a functor $G \colon I^\infty \to \mathcal{C}$. Since $I^\infty = \operatorname{colim}_n I^n$, we find that $\mathcal{C}^{G/} \to \lim_{i \geq 0} \mathcal{C}^{G_i}$. Since I^n has a terminal object, we find that $\mathcal{C}^{G_i} \simeq \mathcal{C}^{G(i)/} = \mathcal{C}^{x_i/}$. We obtain commutative diagrams

$$
\begin{array}{ccccc}
\mathcal{C}^{F_i/} & \longrightarrow & \mathcal{C}^{x_i/} & \longleftarrow & \mathcal{C}^{G_i/} \\
\downarrow & & \downarrow & & \downarrow \\
\mathcal{C}^{F_{i-1}/} & \longrightarrow & \mathcal{C}^{x_{i-1}/} & \longleftarrow & \mathcal{C}^{G_{i-1}/}
\end{array}
$$

where all horizontal maps are Joyal equivalences, and the outer vertical maps are isofibrations. We thus find that the induced map on vertical limits is a Joyal equivalence $\mathcal{C}^{G/} \simeq \mathcal{C}^{F/}$. Since I^∞ is 1-dimensional, we find that $\mathcal{C}$ admits I^∞-indexed colimits. Hence $\mathcal{C}^{G/}$ is corepresentable, and thus the same applies to $\mathcal{C}^{F/}$. Therefore, F admits a colimit.   □

Proposition 4.3.28

If an ∞-category $\mathcal{C}$ admits small coproducts and pushouts, then it admits all small colimits.

Proof We follow the reasoning of [Lur09, 4.4.2.6]. Firs, we show that $\mathcal{C}$ admits colimits indexed over finite-dimensional simplicial sets K by induction over the dimension of K. If K is zero-dimensional, it is simply a discrete set, so that colimits over it are coproducts and hence exist by assumption. Now suppose that K is n-dimensional, and consider its skeletal pushout

$$
\begin{array}{ccc}
\coprod_{i \in I} \partial \Delta^n & \longrightarrow & \mathrm{sk}_{n-1}(K) \\
\downarrow & & \downarrow \\
\coprod_{i \in I} \Delta^n & \longrightarrow & K
\end{array}
$$

For a functor $F \colon K \to \mathcal{C}$, it suffices by Proposition 4.3.26 to argue that each restriction of F to any of the other three corners admits a colimit. For $\mathrm{sk}_{n-1}(K)$ and $\coprod_{i \in I} \partial \Delta^n$, this follows by the induction hypothesis. It remains to show that any functor $\coprod_{i \in I} \Delta^n \to \mathcal{C}$ admits a colimit. By Corollary 4.3.20, we know that every single functor $\Delta^n \to \mathcal{C}$ admits a colimit. By the same argument, we find that the restriction along all terminal objects provides a Joyal equivalence

$$
\mathcal{C}^{\coprod_{i \in I} \Delta^n /} \overset{\sim}{\to} \mathcal{C}^{I/},
$$

because the coproduct of right-anodyne maps is again right-anodyne. It hence suffices to argue that $\mathcal{C}^{I/}$ is equivalent to a representable left fibration, which again follows from the assumption that $\mathcal{C}$ admits coproducts.

Based on this first step, we can then use Proposition 4.3.27 to conclude that $\mathcal{C}$ admits K-shaped colimits for all small simplicial sets K by writing K as the colimit over its skeleta. □

In the presence of finite coproducts, having pushouts is in fact equivalent to having coequalizers, as the following lemma reveals.

Lemma 4.3.29

If an ∞-category $\mathcal{C}$ admits finite coproducts, then it admits pushouts if and only if it admits coequalizers. In particular, if $\mathcal{C}$ admits small coproducts and coequalizers, then it admits all small colimits.

Proof The coequalizer category is the pushout of the diagram $\Delta^1 \leftarrow \partial\Delta^1 \to \Delta^1$. Hence by Proposition 4.3.26, $\mathcal{C}$ admits coequalizers if it admits pushouts and finite coproducts. In order to see the converse, one argues similarly as in Proposition 4.3.26. Namely, suppose that K' is the coequalizer of two morphisms $L \to K$ of simplicial sets, and suppose that $K' \to \mathcal{C}$ is a functor such that the restrictions to K and to L admit a colimit. Then the coequalizer of the colimits is a colimit of the functor $K' \to \mathcal{C}$, see [Lur09, 4.4.3.1] or Exercise 143. Based on this finding, we observe that there is a coequalizer diagram

$$\Delta^0 \rightrightarrows \Delta^1 \amalg \Delta^1 \longrightarrow \Lambda_0^2$$

where the two maps are the two inclusions as vertex 0. From this, we can deduce that colimits over Λ_0^2 are given by a coequalizer of two maps between colimits indexed over Δ^0 and $\Delta^1 \amalg \Delta^1$, which exist if $\mathcal{C}$ admits finite coproducts. $\qquad\square$

Definition 4.3.30

Let $f\colon \mathcal{C} \to \mathcal{D}$ be a functor between ∞-categories and let $F\colon K \to C$ be a diagram. Suppose that F admits a colimit in $\mathcal{C}$. We say that f preserves this colimit, if for some (and hence any) colimit cone $\bar{F}\colon K \star \Delta^0 \to \mathcal{C}$, the resulting diagram $K \star \Delta^0 \to \mathcal{C} \to \mathcal{D}$ is a colimit cone over fF. Furthermore,

we say that F preserves K-shaped colimits, if for every functor $F\colon K \to \mathcal{C}$ which admits a colimit, F preserves this colimit.

Remark 4.3.31

A functor $f\colon \mathcal{C} \to \mathcal{D}$ thus preserves K-shaped colimits, if for every functor $F\colon K \to \mathcal{C}$, the induced functor $\mathcal{C}^{F/} \to \mathcal{D}^{fF/}$ preserves initial objects.

Proposition 4.3.32

Let $F\colon \mathcal{C} \to \mathcal{D}$ be a functor between ∞-categories. Then F preserves small colimits if and only if it preserves small coproducts and pushouts. The same holds true if one replaces pushouts by coequalizers.

Proof Exercise 144. $\qquad\square$

We will collect the following properties of colimits without proof (but we recommend that the reader tries to prove some of them as an exercise).

Proposition 4.3.33

If $\mathcal{D}$ is (co)complete and K is a small simplicial set, then $\mathrm{Fun}(K, \mathcal{D})$ is again (co)complete and colimits are calculated pointwise. In other words, for every object x of K, the evaluation functor $\mathrm{Fun}(K, \mathcal{D}) \to \mathcal{D}$ preserves (co)limits.

Proposition 4.3.34

Let $\mathcal{C}$ be a (co)complete ∞-category and $p \colon K \to \mathcal{C}$ a diagram. Then $\mathcal{C}^{/F}$ admits colimits and the functor $\mathcal{C}^{/F} \to \mathcal{C}$ preserves colimits. Dually, $\mathcal{C}^{F/}$ admits limits and $\mathcal{C}^{F/} \to \mathcal{C}$ preserves limits.

Proof [Lur09, 1.2.13.8]. □

Proposition 4.3.35

Let $\mathcal{C}$ be a cocomplete ∞-category and let $F \colon K \to \mathcal{C}$ be a diagram. Then $\mathcal{C}^{F/}$ is again cocomplete. Dually, if $\mathcal{C}$ is complete, then $C^{/F}$ is complete. The forgetful functors, however, do not preserve these (co)limits in general.

Proposition 4.3.36

Let $f \colon \mathcal{E} \to \mathcal{C}$ be a left fibration and let K be a weakly contractible simplicial set. Then f preserves K-shaped colimits. Likewise, right fibrations preserve contractible limits.

Proof [Lur09, 4.4.2.8 & 4.4.2.9]. □

For the following theorem, we will later give an independent proof which makes use of the straightening-unstraightening equivalence. The following proof has the advantage, however, that we can see that limits and colimits are given as expected.

Theorem 4.3.37

The ∞-categories Cat_∞ and Spc admit all small limits and colimits.

Proof By Proposition 4.3.28, it suffices to show that these ∞-categories admit small products, coproducts, pullbacks and pushouts. Coproducts and products are quite easy, as we have seen earlier. We show that Cat_∞ admits pullbacks, all other cases are similar in flavour. For this purpose, consider a diagram $F \colon \Lambda_2^2 \to \mathrm{Cat}_\infty$ given by two functors

$$\mathcal{D} \xrightarrow{p} \mathcal{C} \xleftarrow{f} \mathcal{C}'.$$

Without loss of generality, we may assume that $\mathcal{D} \to \mathcal{C}$ is an isofibration. We let $\mathcal{D}'$ be the pullback of the above diagram of simplicial sets and let $\bar{F} \colon \Delta^1 \times \Delta^1 \to \mathrm{Cat}_\infty$ be the whole pullback diagram. We wish to show that $\mathrm{Cat}_\infty^{/F}$ is a representable right fibration. By the dual argument of Corollary 4.3.25, we know that there is a cartesian and homotopy-cartesian square

$$
\begin{array}{ccc}
\mathrm{Cat}_\infty^{/F} & \longrightarrow & \mathrm{Cat}_\infty^{/f} \\
\downarrow & & \downarrow \\
\mathrm{Cat}_\infty^{/p} & \longrightarrow & \mathrm{Cat}_\infty^{/\mathcal{C}}
\end{array}
$$

Furthermore, there is a canonical map $\mathrm{Cat}_\infty^{/\mathcal{D}'} \simeq \mathrm{Cat}_\infty^{/\bar{F}} \to \mathrm{Cat}_\infty^{/F}$, and we wish to show that this functor is an equivalence. We will show that it is essentially surjective and fully faithful. As for the essential surjectivity, an object of $\mathcal{C}/F$ can be represented (up to equivalence) by a commutative diagram of ∞-categories

$$
\begin{array}{ccc}
\mathcal{E} & \longrightarrow & \mathcal{D} \\
\downarrow & & \downarrow{\scriptstyle p} \\
\mathcal{C}' & \xrightarrow{f} & \mathcal{C}
\end{array}
$$

By the universal property of the pullback $\mathcal{D}'$, $\mathcal{E}$ comes with a unique map to $\mathcal{D}'$, which provides the resulting object of $\mathrm{Cat}_\infty^{/\mathcal{D}'}$. In order to see the full faithfulness, it suffices to show that for any two objects $\mathcal{E}$, $\mathcal{E}'$ of $\mathrm{Cat}_\infty^{/\mathcal{D}'}$, the following diagram of mapping spaces is homotopy-cartesian:

$$
\begin{array}{ccc}
\mathrm{map}_{\mathrm{Cat}_\infty^{/\mathcal{D}'}}(\mathcal{E}, \mathcal{E}') & \longrightarrow & \mathrm{map}_{\mathrm{Cat}_\infty^{/\mathcal{C}'}}(\mathcal{E}, \mathcal{E}') \\
\downarrow & & \downarrow \\
\mathrm{map}_{\mathrm{Cat}_\infty^{/\mathcal{D}}}(\mathcal{E}, \mathcal{E}') & \longrightarrow & \mathrm{map}_{\mathrm{Cat}_\infty^{/\mathcal{C}}}(\mathcal{E}, \mathcal{E}')
\end{array}
$$

But this follows from the description of mapping spaces in slice-categories, from Proposition 3.3.18, from the fact that $\mathcal{D}' \cong \mathcal{D} \times_{\mathcal{C}} \mathcal{C}'$, and from the fact that $\mathcal{D} \to \mathcal{C}$ is an isofibration. $\qquad\square$

4.4 Cofinal and Coinitial Functors

In this section, we discuss the notion of cofinal and coinitial functors. Informally, they are functors $f : I \to J$ which induce equivalences on colimits and limits, respectively, for any choice of diagrams $J \to \mathcal{C}$. Using the notion of cofinal and coinitial functors, we will discuss the notion of smooth and proper functors, and we will show that left fibrations are smooth and that right fibrations are proper. We will then use this result to give a simple proof of the ∞-categorical version of Quillen's Theorem A, which states that cofinal functors can be detected by their behaviour on slice categories (Theorem 4.4.20).

Definition 4.4.1

Let $f : K \to L$ be a map of simplicial sets and $p : X \to L$ an inner fibration. As before, we define an ∞-category by the pullback

$$
\begin{array}{ccc}
\mathrm{Fun}_f(K, X) & \longrightarrow & \mathrm{Fun}(K, X) \\
\downarrow & & \downarrow{\scriptstyle p_*} \\
\Delta^0 & \xrightarrow{\ f\ } & \mathrm{Fun}(K, L)
\end{array}
$$

If $f = \mathrm{id} : L \to L$, then we simply write $\mathrm{Fun}_L(L, X)$ instead of $\mathrm{Fun}_{\mathrm{id}}(K, X)$.

Definition 4.4.2

Let $f : K \to L$ be a map of simplicial sets. Then f is called *cofinal* if for all right fibrations $p : X \to Y$, the canonical map

$$
\mathrm{Fun}_L(L, X) \to \mathrm{Fun}_f(K, X)
$$

induced by f is a Joyal equivalence. Likewise, it is called *coinitial* if for all left fibrations $p : X \to Y$, the canonical map

$$
\mathrm{Fun}_L(L, X) \to \mathrm{Fun}_f(K, X)
$$

is a Joyal equivalence.

Remark 4.4.3

By construction, $\mathrm{Fun}_f(K, X)$ is an ∞-groupoid if $p : X \to L$ is a right fibration or a left fibration, since in this case its canonical map to Δ^0 is a right fibration or a left fibration and hence a Kan fibration. In our previous notation, the fact that $p : \mathcal{E} \to \mathcal{C}$

(continued)

4.4.3 (continued)

is a left fibration between ∞-categories was written as $\mathrm{Fun}_f^{cc}(\mathcal{C}, \mathcal{E}^\flat)$, because for left fibrations, any morphism in $\mathcal{E}$ is p-cocartesian.

Proposition 4.4.4

Let $f\colon K \to L$ and $g\colon L \to M$ be maps of simplicial sets.

(1) If f is cofinal, then gf is cofinal if and only if g is.

(2) If f is cofinal, then f is a weak equivalence.

(3) If f is a monomorphism, then f is cofinal if and only if it is right-anodyne.

Proof We prove (1) first. For this purpose, consider a right fibration $\mathcal{D} \to M$ and the diagram

$$
\begin{array}{ccccc}
\mathrm{Fun}_M(M, \mathcal{D}) & \longrightarrow & \mathrm{Fun}_g(L, \mathcal{D}) & \xleftarrow{\ \cong\ } & \mathrm{Fun}_L(L, g^*(\mathcal{D})) \\
& \searrow & \downarrow & & \downarrow \\
& & \mathrm{Fun}_{gf}(K, \mathcal{D}) & \xleftarrow{\ \cong\ } & \mathrm{Fun}_f(K, g^*(\mathcal{D}))
\end{array}
$$

where the right horizontal maps are isomorphisms, and the rightmost vertical map is a Joyal equivalence by the assumption that f is cofinal and the fact that pullbacks of right fibrations are right fibrations. We thus conclude the statement by the 3-for-2 property.

In order to show (2), it suffices to prove that for any Kan complex X, the canonical map $\mathrm{Fun}(L, X) \to \mathrm{Fun}(K, X)$ is a homotopy equivalence. Consider the map $L \times X \to L$ which is a Kan fibration, and in particular a right fibration. Then we find that $\mathrm{Fun}_L(L, L \times X) \cong \mathrm{Fun}(L, X)$ and that $\mathrm{Fun}_f(K, L \times X) \cong \mathrm{Fun}(K, X)$ as needed.

In order to see that a right-anodyne map is cofinal, we claim that the restriction map which we have to analyze is a trivial fibration. Namely, by adjunction, it suffices to prove that for any monomorphism $S \to T$, the induced map

$$
S \times L \cup T \times K \to T \times L
$$

is right-anodyne as well, which was established in Lemma 1.3.31. Conversely, suppose that f is a cofinal monomorphism and let $X \to Y$ be a right fibration, and consider a lifting problem to see whether f is right-anodyne. By pulling back, we may assume that $Y = L$ and get a diagram where we wish to show the existence of the dashed arrow. By assumption, we know that the morphism

$$\text{Fun}_L(L, X) \to \text{Fun}_f(K, X)$$

is a Joyal equivalence. If we can show that it is in addition an isofibration, then it is a trivial fibration by Lemma 2.2.17 and hence surjective on 0-simplices (which shows the existence of the dashed arrow above). In order to see that it is indeed an isofibration, we first note that it is an inner fibration, using the facts that the pushout product of an inner-anodyne map with a monomorphism is inner-anodyne, and that right fibrations are inner fibrations in particular. It remains to show that any diagram

$$
\begin{array}{ccc}
\{1\} & \longrightarrow & \text{Fun}_L(L, X) \\
\downarrow & \nearrow & \downarrow \\
\Delta^1 & \longrightarrow & \text{Fun}_f(K, X)
\end{array}
$$

admits a dashed arrow making the diagram commutative. But this is true, since the right vertical map is conservative (as a functor between ∞-groupoids). Now we use the fact that the pushout product of a right-anodyne map and a monomorphism is again right-anodyne, so that a lift exists due to $X \to L$ being a right fibration. $\qquad\square$

Corollary 4.4.5

Let $f \colon K \to L$ and $g \colon L \to M$ be maps of simplicial sets.

(1) If f is coinitial, then gf is coinitial if and only if g is coinitial.

(2) If f is a monomorphism, then f is coinitial if and only if f is left-anodyne.

Proof This follows immediately from Exercise 145. $\qquad\square$

Corollary 4.4.6

Among monomorphisms, the left-anodyne and right-anodyne maps satisfy the right-cancellation property: If f and g are composable morphisms, and both f and gf are left-anodyne or right-anodyne, respectively, then the same applies to g.

Next, we want to prove an important characterization of cofinal maps, which builds up on the following lemma.

Lemma 4.4.7
Let $F\colon L \to \mathcal{C}$ be a diagram and x an object of $\mathcal{C}$. Then there is a canonical cartesian (and homotopy-cartesian) diagram as follows:

$$
\begin{array}{ccc}
\widetilde{\mathrm{Map}}_{\mathcal{C}}(F, x) & \longrightarrow & \mathrm{Fun}(L, \mathcal{C}^{/x}) \\
\downarrow & & \downarrow \\
\Delta^0 & \xrightarrow{\ \ F\ \ } & \mathrm{Fun}(L, \mathcal{C})
\end{array}
$$

Proof Again, for the time of the proof, let us call the pullback $\Phi(F)$. Note that a map from a simplicial set X to $\Phi(F)$ corresponds to a map

$$[(X \times L) \diamond \Delta^0] \amalg_{X \times L} L \to \mathcal{C}$$

whose restriction to Δ^0 is given by x and whose restriction to L is F. On the other hand, a map from X to $\widetilde{\mathrm{Map}}_{\mathcal{C}}(F, x)$ corresponds to a map

$$[X \times (L \diamond \Delta^0)] \amalg_{X \times (L \cup \{\infty\})} L \cup \{\infty\} \to \mathcal{C}$$

whose restriction to $L \cup \{\infty\}$ is the pair (F, x). We claim that there is an isomorphism of simplicial sets

$$[(X \times L) \diamond \Delta^0] \amalg_{X \times L} L \cong [X \times (L \diamond \Delta^0)] \amalg_{X \times (L \cup \{\infty\})} L \cup \{\infty\}.$$

For this, we calculate as follows:

$$
\begin{aligned}
[(X \times L) \diamond \Delta^0] \amalg_{X \times L} L &\cong \left[(X \times L \times \Delta^1) \amalg_{X \times L \times \partial \Delta^1} (X \times L) \amalg \Delta^0 \right] \amalg_{X \times L} L \\
&\cong (X \times L \times \Delta^1) \amalg_{X \times L \times \partial \Delta^1} \left((X \times L) \amalg \Delta^0 \amalg_{X \times L} L \right) \\
&\cong (X \times L \times \Delta^1) \amalg_{X \times L \times \partial \Delta^1} L \amalg \Delta^0 \\
&\cong (X \times L \times \Delta^1) \amalg_{X \times L \times \partial \Delta^1} \left(X \times (L \amalg \Delta^0) \amalg_{X \times (L \amalg \Delta^0)} L \amalg \Delta^0 \right) \\
&\cong \left((X \times L \times \Delta^1) \amalg_{X \times L \times \partial \Delta^1} X \times (L \amalg \Delta^0) \right) \amalg_{X \times (L \amalg \Delta^0)} L \amalg \Delta^0 \\
&\cong X \times (L \diamond \Delta^0) \amalg_{X \times (L \amalg \Delta^0)} L \amalg \Delta^0
\end{aligned}
$$

This shows the lemma, once we convince ourselves that the inclusions of Δ^0 and L correspond to each other (which is a simple matter of checking the maps).   □

> **Theorem 4.4.8**
>
> Let $f: K \to L$ be a map of simplicial sets. Then f is cofinal if and only if for each ∞-category $\mathcal{C}$ and each diagram $p: L \to \mathcal{C}$, the induced map $\mathcal{C}^{p/} \to \mathcal{C}^{pf/}$ is a Joyal equivalence.

Proof First, we assume that f is cofinal and show that in this case the map $\mathcal{C}^{p/} \to \mathcal{C}^{pf/}$ is a Joyal equivalence. Since this map is a map of left fibrations over $\mathcal{C}$, it suffices to show that the induced map on fibres over objects x of $\mathcal{C}$ is an equivalence. By Lemma 4.4.7, this map can be identified (up to homotopy equivalence) with the map $\mathrm{Fun}_p(L, \mathcal{C}^{/x}) \to \mathrm{Fun}_{pf}(K, \mathcal{C}^{/x})$, which is an equivalence because $\mathcal{C}^{/x} \to \mathcal{C}$ is a right fibration and f is cofinal. Conversely, assume that $\mathcal{C}^{p/} \to \mathcal{C}^{pf/}$ is a Joyal equivalence for any diagram $p: L \to \mathcal{C}$, and let $X \to L$ be a right fibration. By the straightening-unstraightening equivalence, there is a functor $p: L \to \mathrm{Spc}^{\mathrm{op}}$ whose pullback of the universal right fibration is equivalent to $X \to L$. Now we use the fact that the universal right fibration is given by $(\mathrm{Spc}_{*/})^{\mathrm{op}} \to \mathrm{Spc}^{\mathrm{op}}$, which is a representable right fibration due to $(\mathrm{Spc}_{*/})^{\mathrm{op}} \simeq (\mathrm{Spc}^{\mathrm{op}})_{/*}$. Thus, using Lemma 4.4.7, we find that

$$\mathrm{Fun}_L(L, X) \simeq \mathrm{Fun}_p(L, (\mathrm{Spc}_{*/})^{\mathrm{op}}) \simeq \mathrm{Map}_{\mathrm{Spc}^{\mathrm{op}}}(p, *) \simeq (\mathrm{Spc}^{\mathrm{op}})^{p/} \times_{\mathrm{Spc}^{\mathrm{op}}} \{*\},$$

and likewise that

$$\mathrm{Fun}_f(K, X) \simeq (\mathrm{Spc}^{\mathrm{op}})^{pf/} \times_{\mathrm{Spc}^{\mathrm{op}}} \{*\}.$$

The map which we have to investigate is the map induced by the map $(\mathrm{Spc}^{\mathrm{op}})^{p/} \to (\mathrm{Spc}^{\mathrm{op}})^{pf/}$ of left fibrations over $\mathrm{Spc}^{\mathrm{op}}$ by taking the fibre over $* \in \mathrm{Spc}^{\mathrm{op}}$. By assumption, this map is a Joyal equivalence, and thus the induced map on fibres is also a Joyal equivalence. □

> **Corollary 4.4.9**
>
> Any Joyal equivalence $f: K \to L$ is cofinal.

Proof By Theorem 4.4.8, it suffices to show that for each ∞-category $\mathcal{C}$ and each diagram, the induced map $\mathcal{C}^{p/} \to \mathcal{C}^{pf/}$ is a Joyal equivalence. For this, it suffices to show that for each further ∞-category $\mathcal{D}$, the induced map

$$\mathrm{Fun}(\mathcal{D}, \mathcal{C}^{p/}) \to \mathrm{Fun}(\mathcal{D}, \mathcal{C}^{pf/})$$

is an equivalence. By adjunction, this map is isomorphic to

$$\mathrm{Fun}_p(\mathcal{D} \diamond L, \mathcal{C}) \to \mathrm{Fun}_{pf}(\mathcal{D} \diamond K, \mathcal{C}),$$

which is in turn given by the induced map on pullbacks in the diagram

$$
\begin{array}{ccccc}
\Delta^0 & \xrightarrow{p} & \mathrm{Fun}(L, \mathcal{C}) & \longleftarrow & \mathrm{Fun}(\mathcal{D} \diamond L, \mathcal{C}) \\
\| & & \downarrow & & \downarrow \\
\Delta^0 & \xrightarrow{pf} & \mathrm{Fun}(K, \mathcal{C}) & \longleftarrow & \mathrm{Fun}(\mathcal{D} \diamond K, \mathcal{C})
\end{array}
$$

where the right horizontal maps are isofibrations and all vertical maps are Joyal equivalences. (For the right-hand side, this follows from the fact that $\mathcal{D} \diamond K \to \mathcal{D} \diamond L$ is again a Joyal equivalence by Corollary 2.5.17.) Therefore, the induced map on pullbacks is also a Joyal equivalence by Lemma 2.5.7.   □

Corollary 4.4.10

Let $f \colon K \to L$ be a cofinal map and let $p \colon L \to \mathcal{C}$ be a diagram with $\mathcal{C}$ an ∞-category. Then f admits a colimit if and only if pf admits a colimit, and in either case, f preserves this colimit.

Lemma 4.4.11

Let $f \colon K \to L$ be a map of simplicial sets. Then f is a trivial fibration if and only if it is a cofinal right fibration.

Proof The "only if" part follows from the fact that trivial fibrations are right fibrations and Joyal equivalences. In order to see the converse direction, we will show that the fibres are contractible and then allude to the dual version of Exercise 111. So let $K \to L$ be a cofinal right fibration. Then the map

$$\mathrm{Fun}_L(L, K) \to \mathrm{Fun}_f(K, K)$$

is an equivalence. The right-hand side contains the functor $\mathrm{id}_K \colon K \to K$, so there exists an object $\varphi \colon \Delta^0 \to \mathrm{Fun}_L(L, X)$ whose image in $\mathrm{Fun}_f(K, K)$ is equivalent to id_K. Spelling this out, we obtain that $f\varphi = \mathrm{id}_L$ and that there exists a 1-simplex $\Delta^1 \to \mathrm{Fun}_f(K, K)$ connecting id_K to φf. This corresponds to a commutative diagram

$$
\begin{array}{ccc}
\Delta^1 \times K & \xrightarrow{h} & K \\
\downarrow{\scriptstyle \mathrm{pr}} & & \downarrow{\scriptstyle f} \\
K & \xrightarrow{f} & L
\end{array}
$$

One can then restrict the map h to $\Delta^1 \times K_x$ for any 0-simplex x of L. The resulting map is easily seen to provide a homotopy between id_{K_x} and φf restricted to K_x. The latter map is constant, since f is constant on the fibres. Thus each fibre K_x is a contractible Kan complex.

□

Proposition 4.4.12

A map is cofinal if and only if it is a composite of a right-anodyne map followed by a trivial fibration.

Proof The "if" case is clear: Both right-anodyne maps and trivial fibrations are cofinal, and compositions of cofinal maps are cofinal. Conversely, given a cofinal map $f\colon K \to L$, we may factor it as a right-anodyne map $K \to K'$ followed by a right fibration $K' \to L$. Since right-anodyne maps are cofinal, we find that the right fibration $K' \to L$ is cofinal by Proposition 4.4.4, part (1). Then the claim follows from Lemma 4.4.11. □

Definition 4.4.13

Let $p\colon Y \to X$ be a map of simplicial sets. We call p *smooth*, if for every pullback diagram

$$
\begin{array}{ccc}
B & \xrightarrow{\ j\ } & Y \\
\downarrow & & \downarrow{\scriptstyle p} \\
A & \xrightarrow{\ i\ } & X
\end{array}
$$

with i being cofinal, the map j is again cofinal. Dually, it is called *proper* if for every such pullback diagram with i being coinitial, the map j is again coinitial.

Definition 4.4.14

A map of simplicial sets $p\colon Y \to X$ is called *universally smooth* if the pullback along any map $X' \to X$ is smooth. Likewise, it is called *universally proper* if the pullback along any map is proper.

Remark 4.4.15

A few words of warning are necessary concerning different notations in the literature. In [Lur09], what we call universally smooth is simply called smooth, and likewise for the terms proper and universally proper. The reason for our distinction is that universally proper maps are closed under pullback, whereas proper (in our terminology) maps are not closed under pullback, see Exercise 150. In [Cis19, Ngu18], what we

(continued)

4.4.15 (continued)
call universally smooth is called proper, and what we call universally proper is called
smooth. In fact, the terminology breaks a little earlier: What we call cofinal is called
final, what we call coinitial is called cofinal, and what we call smooth or proper does
not have a separate name in [Cis19, Ngu18].

For the following proposition, we follow the proof given in [Ngu18, 2.3.23] and
[Cis19].

Proposition 4.4.16
Consider a pullback diagram

$$
\begin{array}{ccc}
B & \xrightarrow{\ j\ } & Y \\
\downarrow & & \downarrow{\scriptstyle p} \\
A & \xrightarrow{\ i\ } & X
\end{array}
$$

where p is a left fibration and i is right-anodyne. Then the map j is again right-anodyne.

Proof The first step is to see that it suffices to prove the claim for i contained in a generating
set of right-anodyne maps. For this purpose, we claim that the set S of right-anodyne maps
satisfying the conclusion of the lemma is saturated, see Exercise 148. While this is easy to
see for compositions and retracts, one needs to work a little harder to see that these maps
are closed under pushouts. By Corollary 1.3.36, it is then enough to show that the maps
$\{1\} \times \Delta^n \cup \Delta^1 \times \partial \Delta^n \to \Delta^1 \times \Delta^n$ are contained in this setright-anodyne. Also observe that the
set S has the right-cancellation property, since according to Exercise 148 all right-anodyne
maps have this propertyright-anodyne. Thus it suffices to show that for any simplicial set K,
the map

$$
\{1\} \times K \to \Delta^1 \times K
$$

is contained in S. For this, consider the diagram

$$
\begin{array}{ccc}
\{1\} \times \partial \Delta^n & \longrightarrow & \Delta^1 \times \partial \Delta^n \\
\downarrow & & \downarrow \\
\{1\} \times \Delta^n & \longrightarrow & \{1\} \times \Delta^n \cup \Delta^1 \times \partial \Delta^n \\
& & \qquad \searrow \\
& & \qquad\qquad \Delta^1 \times \Delta^n
\end{array}
$$

Using the claim, we find that the top horizontal map is contained in S. Therefore, since S is closed under pushouts, the same applies to the lower horizontal map. Also, the lower bended map is contained in S. Since S satisfies the right-cancellation property, the diagonal map is also contained in S.

Now we observe that the maps $\{1\} \times K \to \Delta^1 \times K$ are particular instances of right-anodyne extensions, namely they are right deformation retracts, which are shown to be closed under pullbacks along left fibrations in Exercise 149. □

Remark 4.4.17

By applying the opposite functor, we find that if p is a right fibration and i is left-anodyne, then j is again left-anodyne.

Remark 4.4.18

The conclusion of Proposition 4.4.16 holds more generally for cocartesian fibrations $p \colon Y \to X$, see [Lur09, Proposition 4.1.2.15].

Corollary 4.4.19

Left fibrations are universally smooth. In fact, cocartesian fibrations are universally smooth. Right fibrations, in fact cartesian fibrations, are universally proper.

Proof Since left and right fibrations are closed under pullbacks, it suffices to show that a left fibration is smooth. For this purpose, consider a diagram as in the definition of smooth maps, factor the map i as

$$A \xrightarrow{i'} A' \xrightarrow{p} X,$$

with i' right-anodyne and p a trivial fibration, and consider the enlarged diagram

$$
\begin{array}{ccccc}
B & \xrightarrow{j'} & B' & \xrightarrow{q} & Y \\
\downarrow & & \downarrow & & \downarrow \\
A & \longrightarrow & A' & \longrightarrow & X
\end{array}
$$

We find that q is a trivial fibration and that j' is right-anodyne by Proposition 4.4.16. Thus j is cofinal as a composition of cofinal maps. □

Using the fact that left fibrations are smooth and right fibrations are proper, we obtain a nice proof of an ∞-categorical version of Quillen's Theorem A. Note that this theorem provides yet another characterization of cofinality for maps whose target is an ∞-category.

Theorem 4.4.20
Let $f \colon \mathcal{C} \to \mathcal{D}$ be a map of simplicial sets with $\mathcal{D}$ an ∞-category. Then f is cofinal if and only if for all objects d of $\mathcal{D}$, the simplicial set $\mathcal{C}_{d/}$ is weakly contractible.

Proof First, assume that f is cofinal. In the pullback diagram

$$
\begin{array}{ccc}
\mathcal{C}_{d/} & \longrightarrow & \mathcal{D}_{d/} \\
\downarrow & & \downarrow \\
\mathcal{C} & \longrightarrow & \mathcal{D}
\end{array}
$$

the right vertical map is a left fibration, and thus smooth. It follows that $\mathcal{C}_{d/} \to \mathcal{D}_{d/}$ is cofinal and thus a weak equivalence. Since $\mathcal{D}_{d/}$ is weakly contractible (it has an initial object), the same applies to $\mathcal{C}_{d/}$.

For the converse direction, we consider a factorization of f as

$$
\mathcal{C} \xrightarrow{i} \mathcal{E} \xrightarrow{p} \mathcal{D}
$$

where i is right-anodyne and p is a right fibration. We want to show that p is a trivial fibration, so that f is cofinal by Proposition 4.4.12. For this purpose, consider the diagram

$$
\begin{array}{ccccc}
\mathcal{C}_{d/} & \longrightarrow & \mathcal{E}_{d/} & \longrightarrow & \mathcal{D}_{d/} \\
\downarrow & & \downarrow & & \downarrow \\
\mathcal{C} & \longrightarrow & \mathcal{E} & \longrightarrow & \mathcal{D}
\end{array}
$$

where all squares are pullbacks. Since the very right vertical map is a left fibration, the middle vertical map is also a left fibration. By Proposition 4.4.16, the map $\mathcal{C}_{d/} \to \mathcal{E}_{d/}$ is again right-anodyne and hence a weak equivalence, and hence $\mathcal{E}_{d/}$ is weakly contractible.

Now consider the diagram

$$
\begin{array}{ccccc}
\mathcal{E}_d & \longrightarrow & \mathcal{E}_{d/} & \longrightarrow & \mathcal{E} \\
\downarrow & & \downarrow & & \downarrow \\
\Delta^0 & \xrightarrow{\mathrm{id}_d} & \mathcal{D}_{d/} & \longrightarrow & \mathcal{D}
\end{array}
$$

Again, all squares are pullbacks. This time, the rightmost vertical map is a right fibration, hence the middle vertical map is also right fibration. Furthermore, the map $\Delta^0 \to \mathcal{D}_{d/}$ is left-anodyne by Lemma 4.1.4, since id_d is an initial object of $\mathcal{D}_{d/}$ (see Exercise 135). Hence, by the dual version of Proposition 4.4.16, the map $\mathcal{E}_d \to \mathcal{E}_{d/}$ is left-anodyne and thus a weak equivalence. Since $\mathcal{E}_{d/}$ is weakly contractible by the first step and $\mathcal{E}_d$ is an ∞-groupoid, this means that the fibres $\mathcal{E}_d$ of the right fibration $p \colon \mathcal{E} \to \mathcal{D}$ are contractible. Hence p is a trivial fibration, e.g., by Exercise 111. $\qquad\square$

Remark 4.4.21

Dually, a map of simplicial sets $F \colon \mathcal{C} \to \mathcal{D}$, with $\mathcal{D}$ an ∞-category, is coinitial if and only if for all objects d of $\mathcal{D}$, the simplicial set $\mathcal{C}_{/d}$ is weakly contractible.

To conclude this chapter, here is the actual statement which Quillen proved.

Corollary 4.4.22

Let $F \colon \mathcal{C} \to \mathcal{D}$ be a functor between ordinary categories. If all slices $\mathcal{C}_{d/}$ are weakly contractible, then the functor $\mathrm{N}(\mathcal{C}) \to \mathrm{N}(\mathcal{D})$ is a weak equivalence.

Proof By Theorem 4.4.20, the functor $\mathrm{N}(F) \colon \mathrm{N}(\mathcal{C}) \to \mathrm{N}(\mathcal{D})$ is cofinal, and hence a weak equivalence by Proposition 4.4.4. $\qquad\square$

Adjunctions and Adjoint Functor Theorems

5

5.1   Adjunctions

In this section, we will discuss adjunctions between ∞-categories. We will define them in the language of fibrations and show that they may equivalently be described by choosing a binatural transformation of bivariant mapping-space functors. We will give several sufficient criteria for a fixed functor $f\colon \mathcal{C} \to \mathcal{D}$ to admit an adjoint, similarly as in ordinary category theory, and discuss some examples. Furthermore, we will prove that (co)limits (if they exist) can be functorially formed by showing that their formation assembles into an adjoint of the diagonal or constant functor. Finally, we will discuss the notion of Bousfield localizations, which are those Dwyer–Kan localizations which admit a right adjoint, or equivalently those functors which admit a fully faithful right adjoint.

Definition 5.1.1

An *adjunction* is a bicartesian fibration $\mathcal{E} \to \Delta^1$, i.e., a functor which is both a cartesian and a cocartesian fibration.

Definition 5.1.2

Given an adjunction, we can use the straightening-unstraightening equivalence to obtain a functor $f\colon \mathcal{E}_0 \to \mathcal{E}_1$, classified by the cocartesian fibration, and also to obtain a functor $g\colon \mathcal{E}_1 \to \mathcal{E}_0$, classified by the underlying cartesian fibration of the adjunction. We refer to f as the left adjoint and to g as the right adjoint of the adjunction.

Remark 5.1.3

We say that a functor $f\colon \mathcal{C} \to \mathcal{D}$ admits a right adjoint, if there exists an adjunction $\mathcal{E} \to \Delta^1$ whose associated functor is equivalent to f, i.e., where one specifies equivalences $\mathcal{C} \simeq \mathcal{E}_0$ and $\mathcal{D} \simeq \mathcal{E}_1$, such that the composite $\mathcal{C} \simeq \mathcal{E}_0 \to \mathcal{E}_1 \simeq \mathcal{D}$ is equivalent to f. In general, an adjunction between two ∞-categories $\mathcal{C}$ and $\mathcal{D}$ hence consists of a bicartesian fibration $\mathcal{E} \to \Delta^1$ together with specified equivalences $\mathcal{C} \simeq \mathcal{E}_0$ and $\mathcal{D} \simeq \mathcal{E}_1$.

Remark 5.1.4

We can show directly that an adjunction $\mathcal{E} \to \Delta^1$ gives rise to an equivalence of spaces

$$\mathrm{map}_{\mathcal{E}_1}(f(x), z) \simeq \mathrm{map}_{\mathcal{E}_0}(x, g(z))$$

if x is an object of $\mathcal{E}_0$ and z is an object of $\mathcal{E}_1$. For this, we consider the spaces $\mathrm{map}_{\mathcal{E}}(x, z)$. Choosing a cartesian lift of the unique map $0 \to 1$ with target z, we find by Corollary 3.1.19 that there is a fibre sequence

$$\mathrm{map}_{\mathcal{E}_0}(x, g(z)) \to \mathrm{map}_{\mathcal{E}}(x, z) \to \mathrm{map}_{\Delta^1}(0, 1) \simeq *,$$

so that the first map is a homotopy equivalence. Likewise, choosing a cocartesian lift with domain x gives us a fibre sequence

$$\mathrm{map}_{\mathcal{E}_1}(f(x), z) \to \mathrm{map}_{\mathcal{E}}(x, z) \to \mathrm{map}_{\Delta^1}(0, 1) \simeq *.$$

Therefore, we find the desired equivalence as the zig-zag

$$\mathrm{map}_{\mathcal{E}_1}(f(x), z) \xrightarrow{\simeq} \mathrm{map}_{\mathcal{E}}(x, z) \xleftarrow{\simeq} \mathrm{map}_{\mathcal{E}_0}(x, g(z)).$$

Now we want to promote this result to a natural equivalence of functors $\mathcal{E}_0^{\mathrm{op}} \times \mathcal{E}_1 \to \mathrm{Spc}$.

Proposition 5.1.5

Let $\mathcal{E} \to \Delta^1$ be an adjunction, with $f\colon \mathcal{E}_0 \to \mathcal{E}_1$ and $g\colon \mathcal{E}_1 \to \mathcal{E}_0$ the associated functors. Then there is a natural equivalence of functors

$$\mathrm{map}_{\mathcal{E}_0}(-, g(-)) \simeq \mathrm{map}_{\mathcal{E}_1}(f(-), -).$$

Proof We claim that both functors are equivalent to the composite

$$\mathcal{E}_0^{\mathrm{op}} \times \mathcal{E}_1 \to \mathcal{E}^{\mathrm{op}} \times \mathcal{E} \to \mathrm{Spc},$$

where the latter is the bivariant mapping-space functor for the ∞-category $\mathcal{E}$. For this, we first notice that there is a natural transformation of functors $\tau_g : i_0 \circ g \to i_1$ and $\tau_f : i_0 \to i_1 \circ f$ which picks out the required (co)cartesian maps: One considers the diagrams

$$
\begin{array}{ccc}
\mathcal{E}_0 \times \{0\} & \longrightarrow & \mathcal{E} \\
\downarrow & {\scriptstyle \tau_f} \nearrow & \downarrow \\
\mathcal{E}_0 \times \Delta^1 & \longrightarrow & \Delta^1
\end{array}
\qquad
\begin{array}{ccc}
\mathcal{E}_1 \times \{1\} & \longrightarrow & \mathcal{E} \\
\downarrow & {\scriptstyle \tau_g} \nearrow & \downarrow \\
\mathcal{E}_1 \times \Delta^1 & \longrightarrow & \Delta^1
\end{array}
$$

where the dashed arrows exist because $\mathcal{E}^{\mathcal{E}_i} \to \Delta^{\mathcal{E}_i}$ is again bicartesian, so that we find such lifts as desired. We then consider the composite

$$\mathcal{E}_0^{\mathrm{op}} \times \mathcal{E}_1 \times \Delta^1 \to \mathcal{E}_0^{\mathrm{op}} \times \mathcal{E} \to \mathcal{E}^{\mathrm{op}} \times \mathcal{E} \to \mathrm{Spc},$$

which is a natural transformation from $\mathrm{map}_{\mathcal{E}}(-, g(-))$ to $\mathrm{map}_{\mathcal{E}}(-, -)$. It is a pointwise equivalence, since τ_g is a pointwise cartesian edge. Using that $\mathcal{E}_0 \to \mathcal{E}$ is fully faithful (Δ^1 has trivial spaces of self-maps) and that natural transformations which are pointwise equivalences are themselves equivalences (see Corollary 2.2.2), we conclude that τ_g induces a natural equivalence

$$\mathrm{map}_{\mathcal{E}_0}(-, g(-)) \simeq \mathrm{map}_{\mathcal{E}}(-, -) \colon \mathcal{E}_0^{\mathrm{op}} \times \mathcal{E}_1 \to \mathrm{Spc}.$$

Likewise, τ_f induces a natural equivalence

$$\mathrm{map}_{\mathcal{E}_1}(f(-), -) \simeq \mathrm{map}_{\mathcal{E}}(-, -) \colon \mathcal{E}_0^{\mathrm{op}} \times \mathcal{E}_1 \to \mathrm{Spc},$$

which shows the claim. □

Next, we wish to define a unit transformation and a counit transformation which are associated to an adjunction. As in the proof of Proposition 5.1.5, we consider the transformations τ_f and τ_g, and we observe that the two maps

$$
\begin{array}{ccc}
\mathcal{E}_1 \times \Delta^1 & \xrightarrow{\ \tau_g\ } & \mathcal{E} \\
{\scriptstyle g \times \mathrm{id}} \downarrow & \nearrow {\scriptstyle \tau_f} & \\
\mathcal{E}_0 \times \Delta^1 & &
\end{array}
$$

are identical when restricted to $\mathcal{E}_1 \times \{0\}$: By definition, τ_g restricts to the functor $i_0 \circ g$, whereas $\tau_f \circ (g \times \mathrm{id})$ restricts to the composite of g with the inclusion i_0.

Hence, these two maps combine to a map $\mathcal{E}_1 \times \Lambda_0^2 \to \mathcal{E}$ such that the restriction to $\mathcal{E}_1 \times \Delta^{\{0,1\}}$ is given by $\tau_f \circ (g \times \mathrm{id})$, and the restriction to $\mathcal{E}_1 \times \Delta^{\{0,2\}}$ is given by τ_g. This gives the top horizontal map in the diagram

$$
\begin{array}{ccc}
\mathcal{E}_1 \times \Lambda_0^2 & \longrightarrow & \mathcal{E} \\
\downarrow & \nearrow & \downarrow \\
\mathcal{E}_1 \times \Delta^2 & \longrightarrow & \Delta^1
\end{array}
$$

and the lower horizontal map is given by the composite $\mathcal{E}_1 \times \Delta^2 \to \Delta^2 \to \Delta^1$, where the latter map sends 0 to 0 and both 1 and 2 to 1. In order to see that the diagram commutes, it suffices to recall that both $p \circ \tau_f$ and $p \circ \tau_g$ are the projections. We now find that this lifting problem can be solved, since the composite

$$\mathcal{E}_1 \times \Delta^{\{0,1\}} \to \mathcal{E}_1 \times \Lambda_0^2 \to \mathcal{E}$$

is pointwise cocartesian. For an object z of $\mathcal{E}_1$, the resulting 2-simplex of $\mathcal{E}$ is given by

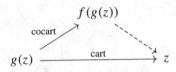

Finally, we see that the restriction $\mathcal{E}_1 \times \Delta^{\{1,2\}} \to \mathcal{E}_1 \times \Delta^2 \to \mathcal{E}$ factors through the inclusion $\mathcal{E}_1 \to \mathcal{E}$ by construction.

A similar constructions provides a functor $\mathcal{E}_0 \times \Delta^2 \to \mathcal{E}_0$ which can be depicted as follows:

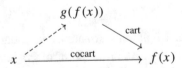

Definition 5.1.6

We refer to the resulting functor $\varepsilon \colon \mathcal{E}_1 \times \Delta^1 \to \mathcal{E}_1$ as the *counit of the adjunction*. Dually, we refer to the resulting functor $\eta \colon \mathcal{E}_0 \times \Delta^1 \to \mathcal{E}_0$ as the *unit of the adjunction*.

Remark 5.1.7

In an adjunction $\mathcal{E} \to \Delta^1$, the unique morphism $0 \to 1$ has both a cartesian and a cocartesian lift. In general, however, a cartesian lift does not need to be cocartesian and vice versa. In fact, this can be controlled very nicely.

Proposition 5.1.8

Let $p \colon \mathcal{E} \to \Delta^1$ be an adjunction, with a left adjoint $f \colon \mathcal{E}_0 \to \mathcal{E}_1$ and a right adjoint $g \colon \mathcal{E}_1 \to \mathcal{E}_0$. Then p-cartesian edges are p-cocartesian if and only if g is fully faithful. Conversely, p-cocartesian edges are p-cartesian if and only if f is fully faithful.
In particular, f and g are mutually inverse equivalences if and only if the set of p-cartesian edges equals the set of p-cocartesian edges.

Proof We first show that cartesian edges are cocartesian if and only if the counit is an equivalence. For this, consider the 2-simplex from above,

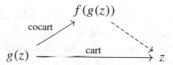

and assume that cartesian edges are cocartesian. By the dual version of Lemma 3.1.7, we find that the counit map $f(g(z)) \to z$ is also cocartesian and thus an equivalence for every z, because its image is invertible in Δ^1. Hence the counit is a natural equivalence. Conversely, if the counit is an equivalence, then it is also cocartesian, so that the cartesian edge $g(z) \to z$ is cocartesian as well (as a composition of cocartesian edges).

Next, we will show that g is fully faithful if and only if the counit is an equivalence. By construction, the diagram

$$
\begin{array}{ccc}
\mathrm{map}_{\mathcal{E}_1}(z, w) & \longrightarrow & \mathrm{map}_{\mathcal{E}_0}(g(z), g(w)) \\
\downarrow & & \downarrow \\
\mathrm{map}_{\mathcal{E}_1}(f(g(z)), w) & \longrightarrow & \mathrm{map}_{\mathcal{E}}(g(z), w)
\end{array}
$$

commutes. The lower horizontal map and the right vertical map are equivalences, because they are induced by post-composition with a cartesian edge, or by precomposition with a cocartesian edge, respectively. Hence, g is fully faithful if and only if the counit is an equivalence.

The argument for the unit is similar, and the "in particular" part follows from the fact that f and g are mutually inverse if and only if they are both fully faithful. $\qquad\square$

Definition 5.1.9

Let $f : \mathcal{C} \to \mathcal{D}$ be a functor between ∞-categories. Then f is said to admit a right adjoint if the cocartesian fibration $\mathcal{E} \to \Delta^1$ associated to f is cartesian. Conversely, $g : \mathcal{D} \to \mathcal{C}$ is said to admit a left adjoint if the cartesian fibration $\mathcal{E}' \to \Delta^1$ associated to g is cocartesian.

The following proposition provides a useful way of constructing functors.

Proposition 5.1.10

Let $f : \mathcal{C} \to \mathcal{D}$ be a functor between ∞-categories. Specify for each object x of $\mathcal{D}$ an object gx of $\mathcal{C}$ and maps $f(gx) \to x$ in $\mathcal{D}$. If the induced composite

$$\mathrm{map}_{\mathcal{C}}(z, gx) \xrightarrow{f} \mathrm{map}_{\mathcal{D}}(f(z), f(gx)) \xrightarrow{\varepsilon} \mathrm{map}_{\mathcal{D}}(f(z), x)$$

is an equivalence, then there exists a functor $g : \mathcal{D} \to \mathcal{C}$, sending x to gx, which is right-adjoint to f. Furthermore, the counit of the adjunction is then equivalent to the chosen map $f(gx) \to x$.

Proof Let $p : \mathcal{E} \to \Delta^1$ be the cocartesian fibration associated to the functor f. We want to show that p is cartesian. In other words, we must specify, for each object x of $\mathcal{E}_1 \simeq \mathcal{D}$, a p-cartesian morphism over the unique non-identity morphism of Δ^1. For this purpose, we consider the object gx of $\mathcal{E}_0 \simeq \mathcal{C}$ and choose a p-cocartesian morphism $gx \to f(gx)$. Composing this morphism with the specified morphism $f(gx) \to x$, we obtain a map $gx \to x$ over the unique non-identity morphism of Δ^1. If we can show that this map is p-cartesian, then the first part of the proposition follows. For this, note that a morphism $\alpha : u \to v$ in $\mathcal{E}$ over $0 \to 1$ is p-cartesian if and only if the map

$$\mathrm{map}_{\mathcal{E}}(w, u) \xrightarrow{\alpha_*} \mathrm{map}_{\mathcal{E}}(w, v)$$

is a homotopy equivalence for all $w \in \mathcal{E}_0$ (for instance using Remark 3.1.18). In other words, we must show that the composite

$$\mathrm{map}_{\mathcal{E}}(z, gx) \to \mathrm{map}_{\mathcal{E}}(z, f(gx)) \to \mathrm{map}_{\mathcal{E}}(z, x)$$

is an equivalence for all $z \in \mathcal{E}_0$. For this, we choose a p-cocartesian edge $z \to f(z)$ and consider the diagram

$$
\begin{array}{ccccc}
\mathrm{map}_{\mathcal{E}}(z, gx) & \longrightarrow & \mathrm{map}_{\mathcal{E}}(z, f(gx)) & \longrightarrow & \mathrm{map}_{\mathcal{E}}(z, x) \\
\| & & \simeq \uparrow & & \simeq \uparrow \\
\mathrm{map}_{\mathcal{E}}(z, gx) & \xrightarrow{f} & \mathrm{map}_{\mathcal{E}}(f(z), f(gx)) & \longrightarrow & \mathrm{map}_{\mathcal{E}}(f(z), x)
\end{array}
$$

where the middle vertical map and the right vertical map are equivalences, since $z \to f(z)$ is p-cocartesian and x and $f(gx)$ are objects of $\mathcal{E}_1$. The lower composite is an equivalence by assumption, hence the upper composite is also an equivalence. Therefore, the above constructed map $gx \to x$ is p-cartesian.

It remains to prove the claim about the counit of the adjunction. But this follows from the construction: Note that there is a 2-simplex in $\mathcal{E}$

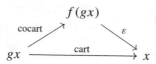

and we have just verified that any composite is a cartesian edge $gx \to x$. □

Remark 5.1.11

Likewise, specifying gx for each object x of $\mathcal{D}$ and maps $x \to f(gx)$ such that the composite

$$\mathrm{map}_{\mathcal{C}}(gx, z) \to \mathrm{map}_{\mathcal{D}}(f(gx), f(z)) \to \mathrm{map}_{\mathcal{D}}(x, f(z))$$

is an equivalence for all $z \in \mathcal{C}$ leads to a functor g which is left-adjoint to f, and the specified map is equivalent to the unit of the adjunction.

Proposition 5.1.12

The association $\mathcal{C} \mapsto \mathcal{C}[W^{-1}]$, where W consists of all morphisms of $\mathcal{C}$, extends to a left adjoint of the inclusion $\mathrm{Spc} \to \mathrm{Cat}_{\infty}$.

Proof By Proposition 5.1.10 and its variant for the existence of left adjoints, it suffices to specify the ∞-groupoid $\mathcal{C}[W^{-1}]$ for each ∞-category $\mathcal{C}$, together with the map $\mathcal{C} \to \mathcal{C}[W^{-1}]$, and to show that for each ∞-groupoid X, the composite

$$\mathrm{map}_{\mathrm{Spc}}(\mathcal{C}[W^{-1}], X) \to \mathrm{map}_{\mathrm{Cat}_{\infty}}(\mathcal{C}[W^{-1}], X) \to \mathrm{map}_{\mathrm{Cat}_{\infty}}(\mathcal{C}, X)$$

is an equivalence. As we have already seen, the functor $\mathrm{Spc} \to \mathrm{Cat}_{\infty}$ is fully faithful, so the first map is an equivalence. Next, we recall that for arbitrary ∞-categories $\mathcal{C}$ and $\mathcal{D}$, we have

$$\mathrm{map}_{\mathrm{Cat}_{\infty}}(\mathcal{C}, \mathcal{D}) \simeq \mathrm{Fun}(\mathcal{C}, \mathcal{D})^{\simeq}.$$

However, if $\mathcal{D}$ is an ∞-groupoid, then $\text{Fun}(\mathcal{C}, \mathcal{D})$ is also an ∞-groupoid, therefore it suffices to show that the map

$$\text{Fun}(\mathcal{C}[W^{-1}], X) \simeq \text{Fun}(\mathcal{C}, X)$$

is an equivalence for all ∞-groupoids. The universal property of the localization shows that the former is canonically equivalent to $\text{Fun}^{\simeq}(\mathcal{C}, X)$, the full subcategory of $\text{Fun}(\mathcal{C}, X)$ on functors sending all morphisms to equivalences. Since every morphism in X is an equivalence, the inclusion $\text{Fun}^{\simeq}(\mathcal{C}, X) \rightarrow \text{Fun}(\mathcal{C}, X)$ is an equivalence. With this, the proposition is shown.   □

The next proposition makes sure that if we already have a candidate for a right-adjoint functor, then it is really a right adjoint.

Proposition 5.1.13
Let $f \colon \mathcal{C} \rightarrow \mathcal{D}$ and $g \colon \mathcal{D} \rightarrow \mathcal{C}$ be functors, and let $\varepsilon \colon fg \rightarrow \text{id}$ be a natural transformation such that the induced map

$$\text{map}(x, g(y)) \rightarrow \text{map}(f(x), f(g(y))) \rightarrow \text{map}(f(x), y)$$

is an equivalence for all x and y. Then g is right-adjoint to f.

Proof By Proposition 5.1.10, there exists a functor g' which is right-adjoint to f, pointwise equivalent to g and such that the counit map is equivalent to the chosen map. We now need to show that g' is equivalent to g. For this, we first construct a natural transformation $g \rightarrow g'$ as follows: We recall that the functor $\text{Fun}(\mathcal{D}, \mathcal{C}) \rightarrow \text{Fun}(\mathcal{D}, \mathcal{P}(\mathcal{C}))$, given by post-composition with the Yoneda functor, is fully faithful; it hence suffices to construct an equivalence between the images of g and g', these images being given by the two functors

$$d \mapsto \begin{cases} \text{map}_{\mathcal{D}}(f(-), d) & \text{for } g', \\ \text{map}_{\mathcal{C}}(-, g(d)) & \text{for } g. \end{cases}$$

By Exercise 147, we find that these two functors are equivalent. Hence, we can deduce that g and g' are also equivalent.   □

Corollary 5.1.14
Let $f \colon \mathcal{C} \rightarrow \mathcal{D}$ and $g \colon \mathcal{D} \rightarrow \mathcal{C}$ be functors. Then g is right-adjoint to f if and only if there exist unit transformations and counit transformations that satisfy the triangle identities.

Proof The fact that the triangle identities are satisfied for an adjunction is left as an exercise to the reader. The converse follows from the previous proposition, since satisfying the triangle identities implies that the canonical map

$$\text{map}(x, g(y)) \to \text{map}(f(x), f(g(y))) \to \text{map}(f(x), y)$$

is an equivalence, and an inverse is given by the composite

$$\text{map}(fx, y) \to \text{map}(gfx, gy) \to \text{map}(x, gy).$$

□

Based on this result, we can prove the following result about the compatibility of adjunctions with Dwyer–Kan localizations and functor categories.

Proposition 5.1.15
Let $f: \mathcal{C} \to \mathcal{D}$ and $g: \mathcal{D} \to \mathcal{C}$. Suppose that $\mathcal{C}$ is equipped with a set S of morphisms and that $\mathcal{D}$ is equipped with a set T of morphisms. If $f(S) \subseteq T$ and $g(T) \subseteq S$, then there are induced functors $F: \mathcal{C}[S^{-1}] \to \mathcal{D}[T^{-1}]$ and $G: \mathcal{D}[T^{-1}] \to \mathcal{C}[S^{-1}]$. If furthermore f is left-adjoint to g, then F is left-adjoint to G.

Proof By Corollary 5.1.14, it suffices to construct transformations $FG \to \text{id}$ and $\text{id} \to GF$ which satisfy the triangle identities. We first construct the map $\text{id} \to GF$, which is a 1-simplex of $\text{Fun}(\mathcal{C}[S^{-1}], \mathcal{C}[S^{-1}])$ from id to the composition GF. We know that the restriction functor

$$\ell^*: \text{Fun}(\mathcal{C}[S^{-1}], \mathcal{C}[S^{-1}]) \to \text{Fun}(\mathcal{C}, \mathcal{C}[S^{-1}])$$

is fully faithful, so it suffices to construct the desired 1-simplex in the latter category, namely from ℓ to $GF \circ \ell$. There is also a functor

$$\text{Fun}(\mathcal{C}, \mathcal{C}) \to \text{Fun}(\mathcal{C}, \mathcal{C}[S^{-1}])$$

given by post-composition with the localization map $\ell: \mathcal{C} \to \mathcal{C}[S^{-1}]$. The unit of the adjunction η is a 1-simplex from id to gf in the former category, hence this functor gives rise to a 1-simplex in $\text{Fun}(\mathcal{C}, \mathcal{C}[S^{-1}])$ from ℓ to $\ell \circ gf$. By definition of G and F, there is an equivalence $\ell \circ gf \simeq GF \circ \ell$. Hence we find a transformation from ℓ to $GF \circ \ell$ as needed. Likewise, one obtains the counit transformation $FG \to \text{id}$. In order to see that the triangle identities are fulfilled, we consider the 2-simplex witnessing the triangle identity for f in the diagram

$$\Delta^2 \to \mathrm{map}_{\mathrm{Fun}(\mathcal{C},\mathcal{D})}(f, f)$$

$$\to \mathrm{map}_{\mathrm{Fun}(\mathcal{C},\mathcal{D}[T^{-1}])}(\ell_D \circ f, \ell_D \circ f)$$

$$\xleftarrow{\simeq} \mathrm{map}_{\mathrm{Fun}(\mathcal{C}[S^{-1}],\mathcal{D}[T^{-1}])}(F, F)$$

Note that the resulting 2-simplex of $\mathrm{map}(F, F)$ witnesses the triangle identity for F. The argument for G is similar. □

Proposition 5.1.16

Let $f\colon \mathcal{C} \to \mathcal{D}$ be a functor, K a simplicial set and $\mathcal{E}$ an auxiliary ∞-category. If f admits a right adjoint or a left adjoint, respectively, then so does $f_*\colon \mathrm{Fun}(K, \mathcal{C}) \to \mathrm{Fun}(K, \mathcal{D})$. If f admits a right adjoint or a left adjoint, respectively, then $f^*\colon \mathrm{Fun}(\mathcal{D}, \mathcal{E}) \to \mathrm{Fun}(\mathcal{C}, \mathcal{E})$ admits a left adjoint or a right adjoint, respectively.

Proof Let g be a right adjoint of f. We prove that g_* is right-adjoint to f_* and that g^* is left-adjoint to f^*. The other cases are similar. For this, let $\varepsilon\colon fg \to \mathrm{id}$ be the counit and $\eta\colon \mathrm{id} \to gf$ be the unit of the adjunction, viewed as morphisms

$$\Delta^1 \xrightarrow{\varepsilon} \mathrm{Fun}(\mathcal{D}, \mathcal{D}) \qquad \Delta^1 \xrightarrow{\eta} \mathrm{Fun}(\mathcal{C}, \mathcal{C}).$$

Then we can post-compose with the canonical functor $\mathrm{Fun}(\mathcal{C}, \mathcal{D}) \to \mathrm{Fun}(\mathcal{C}^K, \mathcal{D}^K)$ and obtain new transformations $\varepsilon_*\colon f_*g_* \to \mathrm{id}_*$ and $\eta_*\colon \mathrm{id}_* \to g_*f_*$. It is easy to check that these transformations satisfy the triangle identities, and therefore form an adjunction. Likewise, one can compose with the functor $\mathrm{Fun}(\mathcal{C}, \mathcal{D}) \to \mathrm{Fun}(\mathcal{E}^{\mathcal{D}}, \mathcal{E}^{\mathcal{C}})$ and obtain transformations $\varepsilon^*\colon (fg)^* = g^*f^* \to \mathrm{id}^*$ and $\eta^*\colon \mathrm{id}^* \to (gf)^* = f^*g^*$. Again, these transformations satisfy the triangle identities, thus g^* is left-adjoint to f^*. □

Corollary 5.1.17

The functor $\mathrm{Cat}_\infty \to \mathrm{Spc}$ given by taking the maximal sub-∞-groupoid is a right adjoint of the inclusion $\mathrm{Spc} \to \mathrm{Cat}_\infty$.

Proof The inclusion functor $\mathrm{Kan} \to \mathrm{Cat}_\infty^1$ has a right adjoint given by $\mathrm{Cat}_\infty^1 \to \mathrm{Kan}$, sending $\mathcal{C}$ to $\mathcal{C}^{\simeq}$. The inclusion sends homotopy equivalences to Joyal equivalences, and the maximal sub-groupoid functor sends Joyal equivalences to homotopy equivalences. Hence we may apply Proposition 5.1.15. □

Remark 5.1.18

Of course, one can also prove Corollary 5.1.17 by using Proposition 5.1.13. In this case, one has to show that for an ∞-groupoid X and an ∞-category $\mathcal{C}$, the canonical map

$$\mathrm{Fun}(X, \mathcal{C}^{\simeq}) \to \mathrm{Fun}(X, \mathcal{C})^{\simeq}$$

is a homotopy equivalence.

Proposition 5.1.19

Let $\ell\colon \mathcal{C} \to \mathcal{D}$ be a Dwyer–Kan localization. Suppose that ℓ admits a right adjoint r. Then r is fully faithful.

Proof We claim that the there is a commutative diagram in the ∞-category Cat_∞

$$
\begin{array}{ccc}
\mathcal{D} & \xrightarrow{\;\;r\;\;} & \mathcal{C} \\
\downarrow{\scriptstyle y_{\mathcal{D}}} & & \downarrow{\scriptstyle y_{\mathcal{C}}} \\
\mathcal{P}(\mathcal{D}) & \xrightarrow{\;\ell^*\;} & \mathcal{P}(\mathcal{C})
\end{array}
$$

for which we have to check if there is a natural equivalence between the functors

$$
d \mapsto
\begin{cases}
c \mapsto \mathrm{map}_{\mathcal{C}}(c, r(d)), \\
c \mapsto \mathrm{map}_{\mathcal{D}}(\ell(c), d).
\end{cases}
$$

But this is a consequence of the fact that r is right-adjoint to ℓ, see Proposition 5.1.5. Since ℓ is a localization, the functor ℓ^* is fully faithful. The proposition thus follows from the fact that the Yoneda functors are fully faithful. □

Definition 5.1.20

A Dwyer–Kan localization which admits a right adjoint is called a *Bousfield localization*.

Definition 5.1.21

An ∞-category is called *essentially small* if it is equivalent to a small ∞-category. An ∞-category is called *locally small* if for any two objects x and y of $\mathcal{C}$, the mapping space $\mathrm{map}_{\mathcal{C}}(x, y)$ is essentially small.

Corollary 5.1.22

Let $\mathcal{C}$ be a locally small ∞-category, and let $\mathcal{C} \to \mathcal{D}$ be a Bousfield localization. Then $\mathcal{D}$ is locally small.

Proof By Proposition 5.1.19, $\mathcal{D}$ can be identified with a full subcategory of $\mathcal{C}$, therefore it is also locally small.   □

Proposition 5.1.23

Let $f \colon \mathcal{C} \to \mathcal{D}$ be any functor which has a fully faithful right adjoint r. Then f is a Bousfield localization.

Proof By Proposition 5.1.19, it suffices to show that f is a Dwyer–Kan localization. Let $\mathcal{E}$ be an auxiliary ∞-category. We need to show that the functor $f^* \colon \mathrm{Fun}(\mathcal{D}, \mathcal{E}) \to \mathrm{Fun}(\mathcal{C}, \mathcal{E})$ is fully faithful, and characterize the essential image. In order to see that f^* is fully faithful, we observe that r^* is a left adjoint to f^* by Proposition 5.1.16. Furthermore, $fr \to \mathrm{id}$ is an equivalence by assumption. From the construction, we find that $r^* f^* \to \mathrm{id}$ is an equivalence as well, so that f^* is fully faithful. It remains to show that f is a Dwyer–Kan localization. If this is the case, then it must be a Dwyer–Kan localization along the set of f-equivalences, i.e., those morphisms which become equivalences after applying f. We thus need to consider a functor $a \colon \mathcal{C} \to \mathcal{E}$ with the property that it sends f-equivalences to equivalences and show that this is equivalent to a composite $\mathcal{C} \to \mathcal{D} \to \mathcal{E}$ for some functor $b \colon \mathcal{D} \to \mathcal{E}$. We claim that $b = ar$ works. In order to confirm this claim, we have to show that there is an equivalence between a and $bf = arf$. The unit of the adjunction is a map $\mathrm{id} \to rf$, which we claim to consist of f-equivalences: Applying f to the map $x \to rf(x)$ yields a map $fx \to frf(x)$, which we may post-compose with the counit to obtain the composite $fx \to frf(x) \to f(x)$. Now according to the triangle identity, the composite is an equivalence, and the fact that the right adjoint r is fully faithful implies that the counit is an equivalence, see Proposition 5.1.8. Hence, the unit is an f-equivalence. Then the fact that a sends f-equivalences to equivalences implies that the canonical map $a \to arf$ is an equivalence as needed.   □

Example The functor $\mathrm{Cat}_\infty \to \mathrm{Spc}$ given by inverting all morphisms has a fully faithful right adjoint, given by the inclusion $\mathrm{Spc} \to \mathrm{Cat}_\infty$. Hence, it is a Bousfield localization.

Proposition 5.1.24

Let $\mathcal{C}$ be an ∞-category and K a simplicial set. If $\mathcal{C}$ admits K-indexed colimits, then their formation assembles into a functor $\operatorname{colim}_K \colon \operatorname{Fun}(K, \mathcal{C}) \to \mathcal{C}$ which is left-adjoint to the constant functor. Conversely, if the constant functor $\operatorname{const} \colon \mathcal{C} \to \operatorname{Fun}(K, \mathcal{C})$ admits a left adjoint F, then $F(p)$ is a colimit of p for any diagram $p \colon K \to \mathcal{C}$.

Proof For the proof of the existence of a left adjoint, we employ Remark 5.1.11. Thus, we first have to specify, for each object p of $\operatorname{Fun}(K, \mathcal{C})$, an object t of $\mathcal{C}$ and a map $p \to \operatorname{const}(t)$. As object, we choose a colimit $\operatorname{colim}_K p$. Therefore, we need to construct a morphism $p \to \operatorname{const}(\operatorname{colim}_K p)$ in $\operatorname{Fun}(K, \mathcal{C})$, where $\operatorname{const}(\operatorname{colim}_K p)$ is the functor which is constant with value $\operatorname{colim}_K p$. By adjunction, such a morphism is a map $K \times \Delta^1 \to \mathcal{C}$. Choosing a colimit cone $\bar{p} \colon K \diamond \Delta^0 \to \mathcal{C}$, we can restrict it along the canonical map $K \times \Delta^1 \to K \diamond \Delta^0$. By construction, the restriction to $K \times \{0\}$ yields p, and the restriction to $K \times \{1\}$ yields the constant functor with value $\bar{p}(\infty)$ as needed. Then we have to show that the composite map

$$\operatorname{map}_{\mathcal{C}}(\operatorname{colim}_K p, x) \to \operatorname{map}_{\operatorname{Fun}(K,\mathcal{C})}(\operatorname{const}(\operatorname{colim}_K p), \operatorname{const}(x)) \to \operatorname{map}_{\operatorname{Fun}(K,\mathcal{C})}(p, \operatorname{const}(x))$$

is an equivalence. Under the equivalences of Proposition 4.3.14, this map corresponds to the canonical map

$$\operatorname{map}_{\mathcal{C}}(\operatorname{colim}_K p, x) \xleftarrow{\simeq} \operatorname{Map}(\bar{p}, x) \longrightarrow \operatorname{Map}(p, x)$$

which is an equivalence by the assumption that $\bar{p}$ is a colimit cone.

Conversely, assume that the constant functor admits a left adjoint $F \colon \operatorname{Fun}(K, \mathcal{C}) \to \mathcal{C}$ and consider a functor $p \colon K \to \mathcal{C}$. We wish to show that $F(p)$ is a colimit of p. For this, note first that the unit of the adjunction gives a map $p \to \operatorname{const} F(p)$ in $\operatorname{Fun}(K, \mathcal{C})$. This map is adjoint to a map $K \times \Delta^1 \to \mathcal{C}$, and as before, one readily checks that this map factors through the projection $K \times \Delta^1 \to K \diamond \Delta^0$. The resulting map $\bar{p} \colon K \diamond \Delta^0 \to \mathcal{C}$ is a cone over p, and it remains to show that it is an initial cone. This is the case if and only if the canonical map

$$\operatorname{Map}_{\mathcal{C}}(\bar{p}, x) \to \operatorname{Map}_{\mathcal{C}}(p, x) \simeq \operatorname{map}_{\mathcal{C}^K}(p, \operatorname{const}_x) \simeq \operatorname{map}_{\mathcal{C}}(F(p), x)$$

is an equivalence. But here again, the composite is equivalent to the canonical map given by restriction along the inclusion $\{\infty\} \to K \diamond \Delta^0$, which is an equivalence. $\square$

Remark 5.1.25

Dually, the same statement holds for limits: If possible, the formation of limits is a functor which is right-adjoint to the constant functor.

Definition 5.1.26

Let $\mathcal{K}$ be a set of simplicial sets. We say that an ∞-category is $\mathcal{K}$-(co)complete, if it admits all $\mathcal{K}$-indexed (co)limits, i.e., (co)limits indexed over simplicial sets K which are contained in $\mathcal{K}$.

Proposition 5.1.27

Let $\mathcal{C}$ be a $\mathcal{K}$-cocomplete, respectively a $\mathcal{K}$-complete, ∞-category and L a simplicial set. Then $\mathrm{Fun}(L, \mathcal{C})$ is again $\mathcal{K}$-cocomplete, respectively $\mathcal{K}$-complete.

Proof Let K be an element of $\mathcal{K}$. We claim that the composite

$$\mathrm{Fun}(K, \mathrm{Fun}(L, \mathcal{C})) \simeq \mathrm{Fun}(L, \mathrm{Fun}(K, \mathcal{C})) \to \mathrm{Fun}(L, \mathcal{C}),$$

where the latter functor is a post-composition with the colimit functor $\mathrm{colim}_K : \mathrm{Fun}(K, \mathcal{C}) \to \mathcal{C}$, is left-adjoint to the constant functor. This follows immediately from the fact that $(\mathrm{colim}_K)_*$ is right-adjoint to const_* by Proposition 5.1.16. □

Corollary 5.1.28

Let $\mathcal{C}$ be $\mathcal{K}$-(co)complete and let L be a simplicial set. Then the constant functor $\mathcal{C} \to \mathrm{Fun}(L, \mathcal{C})$ preserves $\mathcal{K}$-indexed (co)limits.

Proof We wish to show that if K is in $\mathcal{K}$ and $p : K \to \mathcal{C}$ is a diagram with colimit x, then the constant functor with value x is a colimit of the diagram $K \to \mathcal{C} \to \mathrm{Fun}(L, \mathcal{C})$. For this, we first observe that for any functor $\varphi : \mathcal{D} \to \mathcal{E}$ and any simplicial set S, the diagram

$$
\begin{array}{ccc}
\mathrm{Fun}(S, \mathcal{D}) & \xrightarrow{\varphi_*} & \mathrm{Fun}(S, \mathcal{E}) \\
\mathrm{const} \uparrow & & \mathrm{const} \uparrow \\
\mathcal{D} & \xrightarrow{\varphi} & \mathcal{E}
\end{array}
$$

commutes. Applying this observation to the functor $\operatorname{colim}_K \colon \operatorname{Fun}(K, \mathcal{C}) \to \mathcal{C}$ and the simplicial set L, we obtain the commutative diagram

$$
\begin{array}{ccc}
\operatorname{Fun}(K, \operatorname{Fun}(L, \mathcal{C})) & & \\
\Big\uparrow{\simeq} & \searrow^{\operatorname{colim}_K} & \\
\operatorname{Fun}(L, \operatorname{Fun}(K, \mathcal{C})) & \xrightarrow{(\operatorname{colim}_K)_*} & \operatorname{Fun}(L, \mathcal{C}) \\
\Big\uparrow{\text{const}} & & \Big\uparrow{\text{const}} \\
\operatorname{Fun}(K, \mathcal{C}) & \xrightarrow{\operatorname{colim}_K} & \mathcal{C}
\end{array}
$$

where the left vertical composite is given by post-composition with the constant functor $\mathcal{C} \to \operatorname{Fun}(L, \mathcal{C})$. Notice that the upper triangle commutes by the proof of Proposition 5.1.27. The commutativity of the diagram then implies the statement of the corollary. □

Proposition 5.1.29

Let $f \colon \mathcal{C} \to \mathcal{D}$ be a left adjoint. Then f preserves colimits. Likewise, right adjoints preserve limits.

Proof We only prove that left adjoints preserve colimits; the other case follows by passing to opposite categories. So let $F \colon K \to \mathcal{C}$ be a diagram and $\bar{F}$ a colimit cone. We wish to show that $f\bar{F}$ is a colimit cone of fF, which amounts to showing, for an object z of $\mathcal{D}$, that the canonical map

$$
\operatorname{Map}_{\mathcal{D}}(f\bar{F}, z) \to \operatorname{Map}_{\mathcal{D}}(fF, z)
$$

is an equivalence. For this, we claim that for any functor $G \colon L \to \mathcal{C}$, there is a canonical equivalence

$$
\operatorname{Map}_{\mathcal{D}}(fG, z) \simeq \operatorname{Map}_{\mathcal{C}}(G, gz),
$$

where g is the right adjoint of f. Taking this claim for granted for the moment, we then consider the commutative diagram

$$
\begin{array}{ccc}
\operatorname{Map}_{\mathcal{D}}(f\bar{F}, z) & \longrightarrow & \operatorname{Map}_{\mathcal{D}}(fF, z) \\
\Big\downarrow{\simeq} & & \Big\downarrow{\simeq} \\
\operatorname{Map}_{\mathcal{C}}(\bar{F}, gz) & \xrightarrow{\simeq} & \operatorname{Map}_{\mathcal{C}}(F, gz)
\end{array}
$$

where the vertical maps are equivalences by the claim, and where the lower horizontal map is an equivalence by the assumption that $\bar{F}$ is a colimit cone. Therefore, the upper map is also an equivalence as needed.

It remains to prove the claim. For this, we consider the following chain of equivalences:

$$\text{Map}_{\mathcal{D}}(fG, z) \simeq \text{map}_{\mathcal{D}^K}(f_*(G), \text{const}_z) \qquad \text{by Proposition 4.3.14}$$

$$\simeq \text{map}_{\mathcal{C}^K}(G, g_*(\text{const}_z)) \qquad \text{by Proposition 5.1.16}$$

$$\simeq \text{map}_{\mathcal{C}^K}(G, \text{const}_{gz}) \qquad \text{as } g_*(\text{const}_z) = \text{const}_{gz}$$

$$\simeq \text{Map}_{\mathcal{C}}(G, gz) \qquad \text{by Proposition 4.3.14}$$

□

Finally, we need the following proposition in order to establish the adjoint functor theorems in the following section.

Proposition 5.1.30

Consider a pullback diagram of ∞-categories

$$\begin{array}{ccc} \mathcal{C} & \xrightarrow{f} & \mathcal{C}' \\ \downarrow{q} & & \downarrow{p} \\ \mathcal{D} & \xrightarrow{g} & \mathcal{D}' \end{array}$$

where the map p is an isofibration. Suppose that p preserves colimits, and let $F \colon K \to \mathcal{C}$ be a diagram. Then:

(1) A cone $\bar{F} \colon K \diamond \Delta^0 \to \mathcal{C}$ is a colimit cone if its image under f and q is a colimit cone.

(2) If $\mathcal{C}'$ and $\mathcal{D}$ are cocomplete and g preserves colimits, then $\mathcal{C}$ is also cocomplete. Furthermore, f and q preserve colimits.

Proof The first thing to observe is that for any object x of $\mathcal{C}$, there is a homotopy-cartesian diagram

$$\begin{array}{ccc} \text{Map}_{\mathcal{C}}(F, x) & \longrightarrow & \text{Map}_{\mathcal{C}'}(fF, fx) \\ \downarrow & & \downarrow \\ \text{Map}_{\mathcal{D}}(qF, qx) & \longrightarrow & \text{Map}_{\mathcal{D}'}(pfF, pfx) \end{array}$$

This follows from Proposition 4.3.14 and the fact that the mapping spaces in a pullback are given by the pullback of the mapping spaces. (Note that we make use of the fact that applying the functor $\text{Fun}(K, -)$ to the above diagram gives a pullback diagram again, where one leg is an isofibration.)

Next, we wish to analyze whether $\bar{F}$ is a colimit cone. For this, we consider the above squares for $\bar{F}$ and F, and obtain a canonical commutative cube. The assumption that $f\bar{F}$ and

$q\bar{F}$ are colimit cones implies that the comparison maps are equivalences on the left lower corner and on the right upper corner. Using the fact that p preserves colimits, we find that $pf\bar{F}$ is also a colimit cone, so that the comparison map is also an equivalence. Therefore, the comparison map is an equivalence on the upper left corner as well, so that $\bar{F}$ is a colimit cone. This proves (1).

In order to prove (2), it suffices to show that any diagram $F\colon K \to \mathcal{C}$ admits a cone $\bar{F}$ whose image in $\mathcal{D}$ and $\mathcal{C}'$ is a colimit cone: Once this is shown, we can apply (1) to see that $\bar{F}$ is a colimit cone, and by construction q and f send $\bar{F}$ to a colimit cone again.

For the proof the above statement, we consider the composite $F_1\colon K \to \mathcal{C} \to \mathcal{C}'$ and choose a colimit cone $\bar{F}_1\colon K \diamond \Delta^0 \to \mathcal{C}'$. Likewise, we consider the composite $F_2\colon K \to \mathcal{C} \to \mathcal{D}$ and choose a colimit cone $\bar{F}_2\colon K \diamond \Delta^0 \to \mathcal{D}$. Then the images $g\bar{F}_2$ and $p\bar{F}_1$ are also colimit cones by the assumption that both p and g preserve colimits. Hence, there is an equivalence τ between these two cones. Let us say that τ is an equivalence from $p\bar{F}_1$ to $g\bar{F}_2$. Then we obtain a lifting problem

$$
\begin{array}{ccc}
\Delta^0 & \xrightarrow{\bar{F}_1} & \mathcal{C}'_{/fF/} \\
\downarrow & \nearrow^{\hat{\tau}} & \downarrow \\
\Delta^1 & \xrightarrow{\tau} & \mathcal{D}'_{/pfF/}
\end{array}
$$

which admits a solution, since the right vertical map is an isofibration (since p is an isofibration). Furthermore, the dashed arrow is again an equivalence in $\mathcal{C}'$, and hence $\hat{\tau}(1)$ is another cocone of fF. Unravelling the definitions, we obtain a commutative diagram

$$
\begin{array}{ccc}
K \diamond \Delta^0 & \xrightarrow{\hat{\tau}(1)} & \mathcal{C}' \\
\downarrow{\bar{F}_2} & & \downarrow{p} \\
\mathcal{D} & \xrightarrow{g} & \mathcal{D}'
\end{array}
$$

This gives a unique map $K \diamond \Delta^0 \to \mathcal{C}$ which is a cone over F as claimed. $\qquad\square$

5.2 Adjoint Functor Theorems

The goal of this section is to prove two adjoint functor theorems, following the arguments given in [NRS19]. These theorems then imply the adjoint functor theorems of Lurie [Lur09] in the context of presentable ∞-categories.

Definition 5.2.1

A full subcategory $\mathcal{C}_0 \subseteq \mathcal{C}$ of an ∞-category $\mathcal{C}$ is called *colimit-dense*, if every object of $\mathcal{C}$ can be written as a colimit of a diagram $p\colon K \to \mathcal{C}_0 \subseteq \mathcal{C}$.

Theorem 5.2.2
Let $\mathcal{C}$ be a locally small ∞-category which is cocomplete and contains an essentially small, colimit-dense and full subcategory $\mathcal{C}_0 \subseteq \mathcal{C}$. Let $\mathcal{D}$ be a locally small ∞-category and let $F : \mathcal{C} \to \mathcal{D}$ be a functor. Then F admits a right adjoint if and only if F preserves colimits.

Remark 5.2.3
Examples of locally small ∞-categories which admit a small, colimit-dense and full subcategory are *accessible* ∞-categories. An ∞-category is called accessible if it is κ-accessible for some regular cardinal κ. A κ-accessible ∞-category is a locally small ∞-category $\mathcal{C}$ which admits κ-filtered colimits and contains an essentially small subcategory $\mathcal{C}_0$ such that every object of $\mathcal{C}_0$ is κ-compact and every object of $\mathcal{C}$ is a κ-filtered colimit of objects in $\mathcal{C}_0$.
Recall that a κ-filtered colimit refers to a colimit over a functor $f : I \to \mathcal{C}$, where the ∞-category I is κ-filtered. This in turn means that any diagram $J \to I$ admits a cone $J \star \Delta^0 \to I$, provided that J is a κ-small simplicial set (i.e., the cardinality of its non-degenerate simplices is less than κ).

Remark 5.2.4
An accessible ∞-category which is in addition cocomplete is called *presentable*. The above Theorem 5.2.2 can hence be applied to functors between presentable ∞-categories, so that any colimit-preserving functor between presentable categories admits a right adjoint.

Before we dive into the proof of Theorem 5.2.2, let us state an important consequence thereof.

Corollary 5.2.5
Let $\mathcal{C}$ be a locally small ∞-category which is cocomplete and contains an essentially small, colimit-dense and full subcategory $\mathcal{C}_0 \subseteq \mathcal{C}$. Then $\mathcal{C}$ is complete.

Proof Let K be a small simplicial set. Consider the functor const: $\mathcal{C} \to \operatorname{Fun}(K, \mathcal{C})$, which preserves colimits, see Corollary 5.1.28. Moreover, $\operatorname{Fun}(K, \mathcal{C})$ is again locally small (for a justification, see, e.g., [Lur09, Example 5.4.1.8]). In fact, one can write the mapping spaces in the functor category from F to G as the limit of the composite

$$\operatorname{Tw}(K) \longrightarrow K^{\mathrm{op}} \times K \xrightarrow{(F,G)} \mathcal{C}^{\mathrm{op}} \times \mathcal{C} \xrightarrow{\operatorname{map}_{\mathcal{C}}} \operatorname{Spc},$$

and then deduce from the fact that $\mathrm{Tw}(K)$ is small and that each value of this functor is small, that the limit over the functor is a small space as well. Hence, by Theorem 5.2.2, the constant functor admits a right adjoint, which takes a diagram $p\colon K \to \mathcal{C}$ to a limit of p by Proposition 5.1.24.

$\square$

In order to prove Theorem 5.2.2, we need some preliminaries. We will employ the following criterion for obtaining the right adjoint.

Proposition 5.2.6

Let $F\colon \mathcal{C} \to \mathcal{D}$ be a functor. Then F admits a right adjoint if and only if, for all objects d of $\mathcal{D}$, the ∞-category $\mathcal{C}_{/d}$ admits a terminal object.

Proof Let d be an object of $\mathcal{D}$ and consider a terminal object of $\mathcal{C}_{/d}$, given by a pair (Gd, f) where f is a morphism $FGd \to d$ in $\mathcal{D}$. Since we would like to use Proposition 5.1.10 for showing that F admits a right adjoint, we must consider the lower composite in the diagram

$$
\begin{array}{ccccc}
\mathrm{map}_{\mathcal{C}_{/d}}((c,\alpha),(Gd,f)) & \longrightarrow & \mathrm{map}_{\mathcal{D}_{/d}}(\alpha,f) & \longrightarrow & \Delta^0 \\
\downarrow & & \downarrow & & \downarrow{\scriptstyle\alpha} \\
\mathrm{map}_{\mathcal{C}}(c,Gd) & \longrightarrow & \mathrm{map}_{\mathcal{D}}(Fc,FGd) & \xrightarrow{f_*} & \mathrm{map}_{\mathcal{D}}(Fc,d)
\end{array}
$$

Note that the left square is the pullback of mapping spaces induced from the following pullback of ∞-categories:

$$
\begin{array}{ccc}
\mathcal{C}_{/d} & \longrightarrow & \mathcal{D}_{/d} \\
\downarrow & & \downarrow \\
\mathcal{C} & \longrightarrow & \mathcal{D}
\end{array}
$$

Furthermore, the right square is a homotopy pullback by Proposition 3.3.18. Thus, the big square is a homotopy pullback as well, and the upper composite is an equivalence by the assumption that (Gd, f) is a terminal object of $\mathcal{C}_{/d}$. Hence, the lower composite is an equivalence on all components of $\mathrm{map}_{\mathcal{C}}(c, Gd)$ which lie over the component of α in $\mathrm{map}_{\mathcal{D}}(Fc, d)$. Since this holds for all $\alpha\colon Fc \to d$, the claim follows from Proposition 5.1.10.

$\square$

Remark 5.2.7

Likewise, a functor admits a left adjoint if and only if, for all objects d of $\mathcal{D}$, the category $\mathcal{C}_{d/}$ admits an initial object.

Remark 5.2.8

Since ∞-categories with terminal or initial objects are weakly contractible, we can deduce from Theorem 4.4.20 (and its dual version) that a functor which admits a left adjoint is cofinal (and a functor which admits a right adjoint is coinitial).

Therefore, we need to find criteria to ensure that specific categories admit terminal objects. For this, we will make use of the notion of weakly terminal sets.

Definition 5.2.9

Let $\mathcal{C}$ be an ∞-category and S a (small) set of objects. S is said to be *weakly terminal*, if for every object x of $\mathcal{C}$, there exists an object s in S such that the spaces $\mathrm{map}_\mathcal{C}(x, s)$ is not empty. An object t is called weakly terminal if the set $\{t\}$ is a weakly terminal set.

Lemma 5.2.10

Let $\mathcal{C}_0 \subseteq \mathcal{C}$ be an essentially small, full subcategory of a cocomplete category which is colimit-dense. Then $\mathcal{C}$ has a weakly terminal object.

Proof Consider the functor $\mathcal{C}_0 \to \mathcal{C}$ and pick a Joyal equivalence $\mathcal{C}' \simeq \mathcal{C}_0$ with $\mathcal{C}'$ a small simplicial set. Since Joyal equivalences are cofinal by Corollary 4.4.9 and $\mathcal{C}$ is cocomplete, we find that the functor $\mathcal{C}_0 \to \mathcal{C}$ admits a colimit t. We claim that t is a weakly terminal object. In order to see this, let x be another object of $\mathcal{C}$. By assumption, there is a functor $K \to \mathcal{C}_0$ such that the colimit over the composite $K \to \mathcal{C}_0 \to \mathcal{C}$ is given by x. We obtain a canonical map $x \to t$ on colimits. In particular, the space of maps $x \to t$ is not empty. $\square$

Proposition 5.2.11

Let $\mathcal{C}$ be a locally small and cocomplete ∞-category, let S be a weakly terminal set, and let $\mathcal{C}_0$ be the full subcategory spanned by S. Then $\mathcal{C}_0 \to \mathcal{C}$ is cofinal.

Proof By Theorem 4.4.20, it suffices to show that for any object x of $\mathcal{C}$, the slice $(\mathcal{C}_0)_{x/}$ is weakly contractible. If we can show that for any small simplicial set K, any functor $K \to (\mathcal{C}_0)_{x/}$ factors through the inclusion $K \star \Delta^0$ which is contractible, it follows that $(\mathcal{C}_0)_{x/}$ is weakly contractible as needed. In order to prove the above statement, consider a functor $K \to (\mathcal{C}_0)_{x/}$ and the composite

$$K \to (\mathcal{C}_0)_{x/} \to \mathcal{C}_{x/}.$$

Since $\mathcal{C}$ is cocomplete, the same applies to $\mathcal{C}_{x/}$ by Proposition 4.3.35. Therefore, we may choose a colimit cone of the above functor

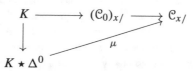

and consider $\mu(\infty)\colon x \to t$. Pick an object s in S for which there exists a map $t \to s$, and pick such a map. Choosing a composite of $x \to t$ and $t \to s$ provides a 2-simplex $\sigma\colon \Delta^2 \to \mathcal{C}$, which is adjoint to a map $\Delta^1 \to \mathcal{C}^{x/}$. Then we consider the lifting problem

$$
\begin{array}{ccc}
K \star \Delta^0 \amalg_{\Delta^0} \Delta^1 & \xrightarrow{(\mu,\sigma)} & \mathcal{C}_{x/} \\
\downarrow & \overset{\mu'}{\nearrow} & \\
K \star \Delta^1 & &
\end{array}
$$

which can be solved, since the vertical map is inner-anodyne by Lemma 1.4.22 and $\mathcal{C}_{x/}$ is an ∞-category. Restricting μ' along the inclusion $K \star \Delta^{\{1\}} \to K \star \Delta^1$ leads to a functor $K \star \Delta^0 \to \mathcal{C}_{x/}$ which factors through $(\mathcal{C}_0)_{x/}$: Since $(\mathcal{C}_0)_{x/} \subseteq \mathcal{C}_{x/}$ is a full subcategory, it suffices to see that all objects of $K \star \Delta^0$ go to $(\mathcal{C}_0)_{x/}$; on K it is true by assumption and on the cone point $\{\infty\}$, by construction, one obtains the map $x \to s$ which is in $(\mathcal{C}_0)_{x/}$, again by construction. Hence, the proposition is proven. □

Corollary 5.2.12

Let $\mathcal{C}$ be a locally small ∞-category which is cocomplete. Assume that a weakly terminal set exists. Then $\mathcal{C}$ admits a terminal object.

Proof Let S be a weakly terminal set and consider the full subcategory $\mathcal{C}_0$ spanned by S. By Proposition 5.2.11, the inclusion $\mathcal{C}_0 \to \mathcal{C}$ is cofinal. Since $\mathcal{C}_0$ is small, the functor $\mathcal{C}_0 \to \mathcal{C}$ admits a colimit. From Corollary 4.4.10, we can thus deduce that the identity functor $\mathcal{C} \to \mathcal{C}$ admits a colimit as well. Such a colimit is a terminal object by Lemma 4.3.15. □

Remark 5.2.13

By passing to opposites, we find that if $\mathcal{C}$ is a locally small and complete category which admits a weakly initial set, then it admits an initial object.

Proof of Theorem 5.2.2 The fact that left adjoints preserve colimits was dealt with in Proposition 5.1.29. Let us therefore prove that F admits a right adjoint if it preserves colimits.

By Proposition 5.2.6, it suffices to show that for every object d of $\mathcal{D}$, the slice $\mathcal{C}_{/d}$ has a terminal object.

For this, we first observe that $\mathcal{C}_{/d}$ is again locally small and cocomplete. The cocompleteness follows from Proposition 5.1.30, because the functor $\mathcal{C} \to \mathcal{D}$ preserves colimits by assumption and the functor $\mathcal{D}_{/d} \to \mathcal{D}$ preserves colimits by Proposition 4.3.34. In order to see that $\mathcal{C}_{/d}$ is again locally small, we calculate the mapping spaces in terms of those in $\mathcal{C}$, $\mathcal{D}$ and $\mathcal{D}_{/d}$: Those in $\mathcal{C}$ and $\mathcal{D}$ are essentially small by assumption, and those in $\mathcal{D}_{/d}$ are also essentially small by Proposition 3.3.18. Therefore, the pullback is also essentially small.

Hence, by Corollary 5.2.12, it suffices to establish the existence of a weakly terminal object, which we will deduce by means of Lemma 5.2.10. In other words, we have to show that $\mathcal{C}_{/d}$ contains an essentially small, full subcategory which is colimit-dense. We claim that $(\mathcal{C}_0)_{/d}$ is such a subcategory. As shown above, $(\mathcal{C}_0)_{/d}$ is locally small, hence it suffices to show that the set of equivalence classes of objects is small. This follows easily from the fact that the objects of $(\mathcal{C}_0)_{/d}$ are given by pairs $(x, \alpha \colon F(x) \to d)$ with $x \in \mathcal{C}_0$. Since $\mathcal{C}_0$ is essentially small, the equivalence classes of such x are a small set. Furthermore, for each such x, the space of maps $\operatorname{map}_{\mathcal{D}}(F(x), d)$ is also essentially small due to the assumption that $\mathcal{D}$ is locally small. Therefore, it remains to show that $(\mathcal{C}_0)_{/d} \to \mathcal{C}_{/d}$ is a colimit-dense subcategory. For this, let $(y, \alpha \colon F(y) \to d)$ be an object of $\mathcal{C}_{/d}$ and write $y = \operatorname{colim}_K p$ for some diagram $p \colon K \to \mathcal{C}_0 \subseteq \mathcal{C}$. We now show that this diagram can be lifted to a diagram $p' \colon K \to (\mathcal{C}_0)_{/d}$. In order to do so, we first note that a colimit cone of the map $K \to \mathcal{C}_0 \xrightarrow{i} \mathcal{C}$ gives rise to a map $K \to \mathcal{C}_{/y}$. Composing this map with F gives us a map $K \to \mathcal{D}_{/F(y)}$, and the map α gives rise to a further map $\mathcal{D}_{/F(y)} \to \mathcal{D}_{/d}$. We thus find a diagram

$$
\begin{array}{ccc}
K & \xrightarrow{\ g\ } & \mathcal{D}_{/d} \\
{\scriptstyle p}\downarrow & & \downarrow \\
\mathcal{C}_0 & \xrightarrow{\ Fi\ } & \mathcal{D}
\end{array}
$$

which commutes by construction. We thus obtain an induced map $p' \colon K \to (\mathcal{C}_0)_{/d}$ whose composite with $(\mathcal{C}_0)_{/d} \to \mathcal{C}_{/d}$ is the diagram p. In order to finish the proof of the theorem, it now suffices to note that the colimit of p' can be calculated by Proposition 4.3.34 to be (y, α). $\qquad\square$

We finish this book with the other adjoint functor theorem that is available for presentable ∞-categories. For this, recall that an ∞-category is called accessible, if it is κ-accessible for some cardinal κ, and that a functor is called accessible if it is κ-accessible for some cardinal κ, i.e., if it preserves κ-filtered colimits for some sufficiently large cardinal κ.

Theorem 5.2.14
Suppose that $\mathcal{C}$ is a locally small, accessible and complete ∞-category and that $\mathcal{D}$ is a locally small and complete ∞-category in which every object is κ-compact for some κ. Then a functor $F \colon \mathcal{C} \to \mathcal{D}$ admits a left adjoint if it preserves limits and is accessible.

Proof We wish to show that for all objects d of $\mathcal{D}$, the slice $\mathcal{C}_{d/}$ has an initial object. By the dual version of Corollary 5.2.12, it suffices to show that $\mathcal{C}_{d/}$ admits a weakly initial set. (Note that $\mathcal{C}_{d/}$ is locally small and complete, since $\mathcal{C}$ and $\mathcal{D}$ are complete by assumption and both $F: \mathcal{C} \to \mathcal{D}$ and $\mathcal{D}_{d/} \to \mathcal{D}$ preserve limits.) We fix a regular cardinal κ such that the following conditions hold:

(1) $\mathcal{C}$ is κ-accessible,
(2) F is κ-accessible, and
(3) d is κ-compact.

For each item, such a κ exists by the assumptions, and by passing to a suitably large cardinal, we also find one that satisfies all three conditions, see Remark 5.2.17 below.

Next, we consider the set S of objects of $\mathcal{C}_{d/}$, consisting of all pairs $(x, \alpha: d \to Fx)$ where x is a κ-compact object of $\mathcal{C}$. Since the subcategory $\mathcal{C}^\kappa$ of κ-compact objects of $\mathcal{C}$ is essentially small, S is in fact a small set. We claim that the set S is weakly initial. In order to see this, we consider an arbitrary object $(z, \beta: d \to Fz)$ of $\mathcal{C}_{d/}$. Since $\mathcal{C}$ is κ-accessible, we can write z as a κ-filtered colimit of a diagram $p: K \to \mathcal{C}^\kappa \subseteq \mathcal{C}$. Since F is κ-accessible, we find that $Fz = \mathrm{colim}_K Fp$. Hence, since d is κ-compact, we find that the map

$$\mathrm{colim}_K \, \mathrm{map}_{\mathcal{D}}(d, Fp(-)) \to \mathrm{map}_{\mathcal{D}}(d, Fz)$$

is an equivalence. Therefore, β must come from some $\mathrm{map}_{\mathcal{D}}(d, Fp(k))$ for k an object of K. Since the image of p lies in the κ-compact objects, this map is present in the set S. This shows that S is weakly initial as claimed, and thus proves the theorem. $\square$

Remark 5.2.15

One can show that under the assumptions of Theorem 5.2.14, a right adjoint is accessible, see, e.g., [NRS19, Theorem 4.1.4 (1)]. Thus, the statement may be promoted to an "if and only if" statement, since we argued in Proposition 5.1.29 that right adjoints preserve limits.

Remark 5.2.16

By Corollary 5.2.5, presentable ∞-categories are complete, so we may apply Theorem 5.2.14 to a presentable ∞-category $\mathcal{C}$. Moreover, an accessible (in particular a presentable) ∞-category $\mathcal{D}$ also has the property that any object is κ-compact for some κ. In conclusion, Theorem 5.2.14 may be applied to an accessible and limit-preserving functor between presentable ∞-categories. The conclusion is that such a functor admits a left adjoint.

Remark 5.2.17

In the proof of Theorem 5.2.14, we freely used some facts about accessible ∞-categories and compact objects, which are subsequently summed up. For this, let $\kappa' > \kappa$ be regular cardinals. Then we have the following properties:

(1) Any κ-compact object is also κ'-compact: This is merely because any κ'-filtered diagram is also κ-filtered.
(2) A κ-accessible functor is also κ'-accessible: Again, this follows from the fact that every κ'-filtered diagram is κ-filtered.
(3) A κ-accessible ∞-category does not need to be κ'-accessible in general, see [Lur09, Remark 5.4.2.12]. However, one can always find many $\kappa'' > \kappa$ such that a κ-accessible ∞-category is also κ''-accessible, see [Lur09, Proposition 5.4.2.11].

Remark 5.2.18

Theorem 5.2.14 also holds without the assumption that $\mathcal{D}$ is complete, see [NRS19, proof of Theorem 4.1.1 (2)].

Corollary 5.2.19

Let $\mathcal{C}$ be a locally small, accessible and complete ∞-category. Then $\mathcal{C}$ is presentable.

Proof We need to show that $\mathcal{C}$ is cocomplete. For this, consider a small simplicial set K and the constant functor $\mathcal{C} \to \mathrm{Fun}(K, \mathcal{C})$. As before, $\mathrm{Fun}(K, \mathcal{C})$ turns out to be locally small and every object is κ-compact for some κ, see [Lur09, Proposition 5.3.4.13]. We may thus apply Theorem 5.2.14 and conclude that the constant functor admits a left adjoint. Therefore, $\mathcal{C}$ admits K-indexed colimits by Proposition 5.1.24.   □

Exercises

Exercise 1 Let $h(\mathrm{CW})$ be the homotopy category of CW-complexes. Show that this category does not have all pushouts. More concretely, show that the diagram

$$* \longleftarrow S^1 \xrightarrow{\cdot 2} S^1$$

does not admit a pushout.

Exercise 2 Work out at least three of the following *simplicial identities*:

(1) $d_i^* d_j^* = d_{j-1}^* d_i^*$ if $i < j$
(2) $d_i^* s_j^* = s_{j-1}^* d_i^*$ if $i < j$
(3) $d_i^* s_j^* = \mathrm{id}$ if $i = j, j+1$
(4) $d_i^* s_j^* = s_j^* d_{i-1}^*$ if $i > i+1$
(5) $s_i^* s_j^* = s_{j+1}^* s_i^*$ if $i \le j$

Here, for any simplicial set $X \colon \Delta^{\mathrm{op}} \to \mathrm{Set}$, we denote the map $X(d_i)$ by d_i^*. Hint: Think about what this means for the maps d_i and s_j in Δ, and prove the corresponding identities there.

Exercise 3 Show that every map in Δ can be uniquely factored as a composition of s_i's followed by a composition of d_j's. Therefore, a simplicial set is equivalently described by a sequence of sets X_n equipped with face and degeneracy maps satisfying the simplicial identities.

Exercise 4 Give examples of simplicial sets where the relation of Definition 1.1.9, leading to $\pi_0^\Delta(X)$, is not symmetric and not transitive.

© The Author(s), under exclusive license to Springer Nature Switzerland AG 2021
M. Land, *Introduction to Infinity-Categories*, Compact Textbooks in Mathematics,
https://doi.org/10.1007/978-3-030-61524-6

Exercise 5 Show that every simplex $x \in X_n$ is of the form $\alpha^*(y)$ for a surjection $\alpha \colon [m] \to [n]$ and a non-degenerate n-simplex y, and show that the pair (α, y) is uniquely determined by x.

Exercise 6 Show that the category Set is bicomplete. Hint: General colimits are constructed as quotients of disjoint unions, and general limits are constructed as subsets of products.

Exercise 7 Let $F \colon I \to \mathcal{C}$ be a functor. Show that a colimit of F can equivalently be described as an initial cocone over F, and that a limit of F can be equivalently described as a terminal cone over F.

Exercise 8 Calculate the limit and colimit of a simplicial set $X \colon \Delta^{\mathrm{op}} \to$ Set.

Exercise 9 Show that the datum of an adjunction in the sense of Definition 1.1.17 is equivalent to the datum of a pair of functors (F, G) together with natural transformations $\varepsilon \colon FG \to$ id and $\eta \colon GF \to$ id satisfying the triangle identities, i.e., the obvious composites

$$F(X) \to F(GF(X)) \cong FG(FX) \to F(X)$$

and

$$G(X) \to GF(G(X)) \cong G(FG(X)) \to G(X)$$

are the identity of $F(X)$ and $G(X)$, respectively.

Exercise 10 Show that a functor $F \colon \mathcal{C} \to \mathcal{D}$ admits a right adjoint if you can specify objects Gy for all $y \in \mathcal{D}$ and maps $\varepsilon_y \colon FGy \to y$, which have the property that the induced map on hom-sets

$$\mathrm{Hom}_{\mathcal{C}}(x, Gy) \xrightarrow{F} \mathrm{Hom}_{\mathcal{D}}(Fx, FGy) \xrightarrow{\varepsilon_y} \mathrm{Hom}_{\mathcal{D}}(Fx, y)$$

is a bijection. Note that there is an obvious dual notion which shows that F admits a left adjoint if one can specify objects Gy for all $y \in \mathcal{D}$ and maps $\eta_y \colon y \to FGy$ which make the induced map on hom-sets

$$\mathrm{Hom}_{\mathcal{D}}(Gy, x) \xrightarrow{F} \mathrm{Hom}_{\mathcal{C}}(FGy, Fx) \xrightarrow{\eta_y} \mathrm{Hom}_{\mathcal{C}}(y, Fx)$$

a bijection.

Exercise 11 Prove that if a simplicial set X has at most n-dimensional non-degenerate simplices, and Y has at most m-dimensional non-degenerate simplices, then their product $X \times Y$ has at most $(n + m)$-dimensional non-degenerate simplices.

Exercise 12 Show that for every simplicial set X, there is a canonical bijection $\pi_0^\Delta(X) \cong \pi_0(|X|)$.

Exercise 13 Show that there are inclusions $I^n \subseteq \Lambda_j^n$ provided $0 < j < n$ or $n \geq 3$, and $\Lambda_j^n \subseteq \partial\Delta^n \subseteq \Delta^n$ for all $n \geq 0$.

Exercise 14 Let I be a category with an initial object i and let J be a category with a terminal object t. Show that a limit of a functor $F \colon I \to \mathcal{C}$ is given by $F(i)$ (together with the canonical maps $F(i) \to F(x)$ for all $x \in I$). Similarly, show that a colimit of a functor $G \colon J \to \mathcal{C}$ is given by $G(t)$ (together with its maps $G(x) \to G(t)$ for all $x \in J$).

Exercise 15 Show that for every $n \geq 0$, there is a pushout

$$
\begin{array}{ccc}
\coprod_{J_n} \partial\Delta^n & \longrightarrow & \mathrm{sk}_{n-1}(X) \\
\downarrow & & \downarrow \\
\coprod_{J_n} \Delta^n & \longrightarrow & \mathrm{sk}_n(X)
\end{array}
$$

where J_n is the set of non-degenerate n-simplices. Furthermore, show that $X \cong \mathrm{colim}_n \, \mathrm{sk}_n(X)$.

Exercise 16 Show that the following simplicial sets are not nerves of categories:

(1) $\partial\Delta^n$ for $n \geq 2$
(2) Λ_j^n for $n \geq 2$ and $0 \leq j \leq n$
(3) I^n for $n \geq 2$

Exercise 17 Suppose that X is a Kan complex. Show that for all $n \geq 0$, the simplicial set $\mathrm{cosk}_n(X)$ is again a Kan complex. Prove that the canonical map $X \to \mathrm{cosk}_n(X)$ induces a bijection

$$
\pi_k^\Delta(X) \to \pi_k^\Delta(\mathrm{cosk}_n(X))
$$

for $k < n$ and that $\pi_k^\Delta(\mathrm{cosk}_n(X)) = 0$ for $k \geq n$.

Exercise 18 Show that a natural transformation between two functors $f, g \colon \mathcal{C} \to \mathcal{D}$ induces a homotopy between $N(f), N(g) \colon N(\mathcal{C}) \to N(\mathcal{D})$. Use this result to show that conjugation with an element determines a self map of BG which is homotopic to the identity. What does conjugation induce on $\pi_1(BG)$? Why does this not show that every group is abelian?

Exercise 19 Show that the nerve of a category $\mathcal{C}$ is 2-coskeletal, i.e., that the canonical map $N(\mathcal{C}) \to \mathrm{cosk}_2(N(\mathcal{C}))$ is an isomorphism of simplicial sets.

Exercise 20 Let X be a simplicial set and let $n \le m$. Show that $\mathrm{sk}_n(\mathrm{sk}_m(X)) = \mathrm{sk}_n(X) = \mathrm{sk}_m(\mathrm{sk}_n(X))$. Deduce that $\mathrm{cosk}_n(\mathrm{cosk}_m(X)) \cong \mathrm{cosk}_m(\mathrm{cosk}_n(X))$. Is it also true that $\mathrm{sk}_n(\mathrm{cosk}_m(X)) \cong \mathrm{cosk}_m(\mathrm{sk}_n(X))$? (If not: Provide a counter example.) Is there a preferred map between these two simplicial sets?

Exercise 21 Let $\mathcal{C}$ be a category and X a simplicial set. Recall that X^{op} is the following simplicial set: $X_n^{\mathrm{op}} = X_n$ and $d_i^{\mathrm{op}} : X_n \to X_{n-1}$ is given by d_{n-i}, and likewise $s_i^{\mathrm{op}} = s_{n-i}$ as a map $X_n \to X_{n+1}$. Prove the following assertions:

(1) $N(\mathcal{C}^{\mathrm{op}}) \cong N(\mathcal{C})^{\mathrm{op}}$
(2) $(\Delta^n)^{\mathrm{op}} \cong \Delta^n$
(3) $(\Lambda_i^n)^{\mathrm{op}} \cong \Lambda_{n-i}^n$
(4) $(\partial\Delta^n)^{\mathrm{op}} \cong \partial\Delta^n$

Exercise 22 Let G be a group and let $\mathbb{B}G$ be the category with one object and G as endomorphisms of that object. Show that $N(\mathbb{B}(G))$ has only one non-trivial homotopy group, namely $\pi_1^{\Delta}(N(\mathbb{B}G))$, and that this group is canonically isomorphic to G.

Exercise 23 Let X be a composer and let $f : x \to y$ be a morphism in X. Show that f is a composition of id_x with f and of f with id_y.

Exercise 24 Consider the map $[0] \to [n]$ in Δ with image $\{0\}$. Show that this determines a map $0 : \Delta^0 \to \partial\Delta^n$. Calculate the simplicial homotopy sets $\pi_i^{\Delta}(\partial\Delta^n, 0)$ for $i \ge 1$ and $n \ge 2$. Deduce that $\partial\Delta^n$ is not a Kan complex.

Exercise 25 Show that the following simplicial sets are not ∞-categories:

(1) $\partial\Delta^n$ for $n \ge 2$
(2) Λ_j^n for $n \ge 3$ and $0 \le j \le n$
(3) I^n for $n \ge 2$

Exercise 26 Determine the homotopy category of the following simplicial sets:

(1) $\partial\Delta^n$ for $n \ge 1$
(2) Λ_j^n for $n \ge 2$ and $0 \le j \le n$
(3) I^n for $n \ge 0$

Exercise 27 Let $f : X \to Y$ be a map of simplicial sets. Prove or give a counter example to the following statements:

(1) If f is a monomorphism, then $hX \to hY$ is fully faithful.
(2) If f is a degree-wise surjection, then $hX \to hY$ is *surjective and full*, i.e., it induces a surjection on objects and on hom-sets.
(3) If f induces a surjection on 0- and 1-simplices, then $hX \to hY$ is surjective and full.

Exercise 28 Prove or disprove the following statement: For any two simplicial sets X and Y, the canonical map $h(X \times Y) \to hX \times hY$ is an isomorphism of categories.

Exercise 29 A category $\mathcal{C}$ is called connected if $\pi_0^\Delta(\mathrm{N}(\mathcal{C}))$ consists of only one element. Show that a groupoid $\mathcal{G}$ is connected if and only if for every two objects $x, y \in \mathcal{G}$, the set $\mathrm{Hom}_{\mathcal{G}}(x, y)$ is non-empty. Show that a connected groupoid is equivalent to $\mathbb{B}G$ for a group G. Show, however, that the category of connected groupoids is *not* equivalent to the category of groups.

Exercise 30 Let X be a topological space. Describe the category $h(\mathcal{S}(X))$. Show that the endomorphisms of each object form a group. Which group is it?

Exercise 31 Suppose that X is a composer with the inner 3-horn extension property. Let $\sigma \colon \Delta^1 \times \Delta^1 \to X$ be a map such that

(1) $\sigma_{|\Delta^1 \times \{0\}} = f$,
(2) $\sigma_{|\Delta^1 \times \{1\}} = g$,
(3) $\sigma_{|\{0\} \times \Delta^1} = \mathrm{id}_x$, and
(4) $\sigma_{|\{1\} \times \Delta^1} = \mathrm{id}_y$,

for morphisms $f, g \colon x \to y$. Show that $f \sim g$ in the sense of Definition 1.2.3.

Exercise 32 Let X be a composer with the extension property for inner 3-horns. Show that for any two composable morphisms $f \colon x \to y$ and $g \colon y \to z$, the simplicial set $\mathrm{Comp}_X(f, g)$ is connected, i.e., that $\pi_0^\Delta(\mathrm{Comp}_X(f, g))$ consists of only one element.

Exercise 33 Let X be a simplicial set and consider the canonical map $X \to \mathrm{N}(hX)$.

(1) Show that this map factors through the canonical map $X \to \mathrm{cosk}_2(X)$.
(2) Show that the induced map $\mathrm{cosk}_2(X) \to \mathrm{N}(hX)$ is an isomorphism if X is isomorphic to the nerve of a category.
(3) Show that the map $\mathrm{cosk}_2(X) \to \mathrm{N}(hX)$ is in general not an isomorphism. Hint: Find an X which is 2-coskeletal, but not the nerve of a category.
(4) Prove or disprove the following statement: The map $\mathrm{cosk}_2(X) \to \mathrm{N}(hX)$ is an isomorphism if and only if X is isomorphic to the nerve of a category.

Exercise 34 Let $(V, \otimes, \mathbb{1})$ be a monoidal category. Then the functor $\mathrm{Hom}_V(\mathbb{1}, -) \colon V \to$ Set is lax monoidal. Is it monoidal? If not: Can you find a condition on $(V, \otimes, \mathbb{1})$ which ensures that it is?

Exercise 35 Let $\mathcal{C}$ be a category with finite products and finite coproducts. We say that $\mathcal{C}$ is *pointed* if the canonical map $\emptyset \to *$ from the initial to the terminal object is an isomorphism. Show that the identity canonically refines to a lax monoidal functor $(\mathcal{C}, \times, *) \to (\mathcal{C}, \amalg, \emptyset)$. When is this functor monoidal? Furthermore, show that any functor $F \colon \mathcal{C} \to \mathcal{D}$ refines canonically to a lax symmetric monoidal functor $(\mathcal{C}, \amalg, \emptyset) \to (\mathcal{D}, \amalg, \emptyset)$. When is it monoidal?

Exercise 36 The goal of this exercise is to show that any essentially surjective and fully faithful functor $F: \mathcal{C} \to \mathcal{D}$ between ordinary categories is an equivalence.

(1) Show that F admits an adjoint G. Hint: Use Exercise 10.
(2) Show that G is itself fully faithful.
(3) Show that an adjoint pair (F, G) of fully faithful functors makes F an equivalence with G an inverse.

Exercise 37 Let $F: \mathcal{C} \to \mathcal{D}$ be a functor with the right adjoint $G: \mathcal{D} \to \mathcal{C}$. Show that they are mutually inverse if F is fully faithful and G is conservative. Here, conservativity means that if $f: x \to y$ is a morphism in $\mathcal{D}$ and $G(f)$ is an isomorphism, then f is an isomorphism.

Exercise 38 Let $F: \mathcal{C} \to \mathcal{D}$ be left-adjoint to $G: \mathcal{D} \to \mathcal{C}$. Show that if G is lax monoidal, then F canonically refines to an oplax monoidal functor. Vice versa, show that if F is oplax monoidal, then G canonically refines to a lax monoidal functor.

Exercise 39 Show that the left adjoint of a monoidal adjunction is in fact monoidal. Recall that a monoidal adjunction consists of lax monoidal functors F and G, which are witnessed to be adjoint by a unit η and a counit ε where both η and ε are monoidal transformations.

Exercise 40 Suppose that F is left-adjoint to G, witnessed by a unit and counit (η, ε). Show that if F is monoidal, then the induced lax monoidal structure on G of Exercise 38 makes $(F, G, \eta, \varepsilon)$ a monoidal adjunction.

Exercise 41 Show that the coherent nerve functor $N: \mathrm{Cat}_\Delta \to \mathrm{sSet}$ commutes with coproducts. Show that $\mathfrak{C}$ is not right-adjoint to N. Does N have a right adjoint at all? Hint: Does the ordinary nerve functor $N: \mathrm{Cat} \to \mathrm{sSet}$ have a right adjoint? And how are these two questions related?

Exercise 42 Show that if the coherent nerve $N(\mathcal{C})$ of a simplicial category is isomorphic to the nerve of an ordinary category, then the underlying category $u\mathcal{C}$ is isomorphic to the homotopy category $\pi(\mathcal{C})$. Make explicit the coherent nerve of the following simplicial category $\mathbb{B}^{\mathrm{simp}}(G)$: There is only one object, and the simplicial set of endomorphisms of this object is given by $N(G)$, where G is a group (of a monoid, if you wish). Deduce from the explicit analysis that $N(\mathbb{B}^{\mathrm{simp}}(G))$ is not isomorphic to the nerve of a category although $u N(\mathbb{B}^{\mathrm{simp}}(G)) \cong \pi(N(\mathbb{B}^{\mathrm{simp}}(G)))$.

Exercise 43 Prove or disprove the following statements:

(1) There exists a simplicial category $\mathcal{C}$ whose coherent nerve $N(\mathcal{C})$ is not an ∞-category.
(2) There exists a simplicial category $\mathcal{C}$ whose coherent nerve $N(\mathcal{C})$ is an ∞-category, but not a Kan complex.

Exercise 44 Suppose that $\mathcal{C}$ is a simplicial category whose hom-simplicial sets are all Kan complexes. Work out what it means concretely that a morphism $f: x \to y$ in a simplicial category is an equivalence. Rephrase the condition of weakly fully faithful functors with the help of this result.

Exercise 45 Show that there exist monomorphisms $X \to Y$ where both X and Y are ∞-categories, but X is not a sub-∞-category in the sense of Definition 1.2.76.

Exercise 46 Let $Y \to X$ be an inclusion of topological spaces. When is the induced map $\mathcal{S}(Y) \to \mathcal{S}(X)$ a subcategory? When is it a full subcategory?

Exercise 47 Let $\mathcal{C}_0 \subseteq \mathcal{C}$ be a full sub-∞-category and let $\mathcal{D}$ be an ∞-category. Show that the functor category $\mathrm{Fun}(\mathcal{D}, \mathcal{C}_0)$ is the full sub-∞-category of $\mathrm{Fun}(\mathcal{D}, \mathcal{C})$ on those functors $f : \mathcal{D} \to \mathcal{C}$ which factor through the inclusion $\mathcal{C}_0 \subseteq \mathcal{C}$.

Exercise 48 Show that a simplicial set X is an ∞-category if and only if X^{op} is an ∞-category, and likewise that X is a Kan complex if and only if X^{op} is a Kan complex. Show that if X is an ∞-category then X is an ∞-groupoid if and only if X^{op} is an ∞-groupoid.

Exercise 49 Let $\mathcal{C}$ be a simplicial category. Show that the coherent nerve $N(\mathcal{C})$ is isomorphic to the nerve of an ordinary category if and only if $\mathcal{C}$ is in the image of the functor $c : \mathrm{Cat} \to \mathrm{Cat}_{\Delta}$.

Exercise 50 Show that the composite

$$\mathrm{Cat}_{\Delta} \xrightarrow{\mathrm{N}} \mathrm{sSet} \xrightarrow{h} \mathrm{Cat}$$

is isomorphic to the functor $\pi : \mathrm{Cat}_{\Delta} \to \mathrm{Cat}$.

Exercise 51 Show the following assertions:

(1) Let $\mathcal{C}$ be an ordinary category and X a simplicial set. Then X is an ∞-category if and only if every map $X \to N(\mathcal{C})$ is an inner fibration.
(2) A map $f : X \to Y$ is an inner fibration if and only if $f^{\mathrm{op}} : X^{\mathrm{op}} \to Y^{\mathrm{op}}$ is an inner fibration.
(3) A map $f : X \to Y$ is a left fibration if and only if the map $f^{\mathrm{op}} : X^{\mathrm{op}} \to Y^{\mathrm{op}}$ is a right fibration.

Exercise 52 Let $S \subseteq S'$ be sets of morphisms. Show that, in the notation of Definition 1.3.3,

(1) $\chi_R(S') \subseteq \chi_R(S)$,
(2) $S \subseteq \chi(S)$, and
(3) $\chi_R(S) = \chi_R(\chi(S))$.

Exercise 53 A category I is called *filtered* if every functor $K \to I$ from a finite category K extends over the inclusion $K \to K^{\triangleright}$. Show that a poset (viewed as a category) is filtered if and only if

(1) for every finite collection of objects $X_1, \ldots, X_n$ of I, there exists an object X of I equipped with maps $X_k \to X$ for all $k = 1, \ldots, n$;
(2) any two morphisms $f, g : X \to Y$ can be equalized, i.e., there exists a morphism $h : Y \to Z$ such that $hf = hg$.

Exercise 54 Let I be a finite category.

(1) Show that I is filtered if it has a terminal object.
(2) Show that there are examples where I is filtered but does not have a terminal object.
(3) Show that I is a poset and filtered if and only if it has a terminal object.

In particular, notice that this shows that there are many filtered categories which are not posets.

Exercise 55 Show that every simplicial set A with only finitely many non-degenerate simplices is *compact*, i.e., the canonical map

$$\operatorname*{colim}_{i \in I} \operatorname{Hom}_{sSet}(A, X_i) \to \operatorname{Hom}_{sSet}(A, \operatorname*{colim}_{i \in I} X_i)$$

is an isomorphism, provided I is a filtered category.

Exercise 56 We call a set S *semi-saturated* if it is closed under pushouts, retracts and countable compositions. Show that a semi-saturated set

(1) contains isomorphisms and is closed under finite coproducts, if it contains the identity of an initial object $\emptyset$;
(2) is closed under composition, i.e., if $f : A \to B$ and $g : B \to C$ are elements of S, then so is $gf : A \to C$;
(3) is closed under countable coproducts if it is closed under finite coproducts, i.e., if $\{f_i : A_i \to B_i\}_{i \in I}$ is a countable family of elements of S, then the map $\coprod_{i \in I} : A_i \to B_i$ is an element of S as well.

Exercise 57 Show that a saturated set S in a category $\mathcal{C}$ contains all isomorphisms. Find an example of a category $\mathcal{C}$ and a semi-saturated set S of morphisms in $\mathcal{C}$ which is non-empty and does not contain all isomorphisms.

Exercise 58 Show the following assertions:

(1) The map $\emptyset \to \{*\}$ in Set generates the set of injections. What is $\chi_R(\emptyset \to *)$? Spell out the factorization obtained by the small object argument for a general map $f : M \to N$ of sets.
(2) The map $\{*, *\} \to \{*\}$ generates the class of surjections. What is $\chi_R(\{*, *\} \to *)$? Spell out the factorization obtained by the small object argument for a general map $f : M \to N$ of sets.

Exercise 59 Consider the set $S = \{\partial\Delta^n \to \Delta^n\}_{n \geq 0}$ given by the boundary inclusions. Show that $\chi(S)$ is given by all monomorphisms of simplicial sets.

Exercise 60 Show that J is not a compact simplicial set, i.e., that there are infinitely many non-degenerate simplices in J.

Exercise 61 Show that if a morphism $f \colon \Delta^1 \to \mathcal{C}$ in an ∞-category extends over the inclusion $\Delta^1 \to J$, then f is an equivalence.

Exercise 62 Fill in the missing steps of Lemma 1.3.35. More precisely, show the following assertions:

(1) The horn inclusion $\Lambda_j^n \to \Delta^n$ for $0 \le j < n$ is a retract of the pushout product map

$$\Delta^n \times \{0\} \amalg_{\Lambda_j^n \times \{0\}} \Lambda_j^n \times \Delta^1 \to \Delta^n \times \Delta^1.$$

(2) The pushout product map

$$\partial\Delta^n \times \Delta^1 \cup \Delta^n \times \{0\} \to \Delta^n \times \Delta^1$$

is left-anodyne.

Exercise 63 Show that a trivial fibration $f \colon X \to Y$ between Kan complexes induces an isomorphism in the category $\pi(\mathrm{Kan})$. Hint: Show that a trivial fibration between Kan complexes is a homotopy equivalence.

Exercise 64 Show that if $f \colon y \to z$ is an equivalence, then

$$\mathrm{map}_\mathcal{C}(x, y) \simeq \mathrm{map}_\mathcal{C}(x, y) \times \Delta^0 \xrightarrow{f} \mathrm{map}_\mathcal{C}(x, y) \times \mathrm{map}_\mathcal{C}(y, z) \to \mathrm{map}_\mathcal{C}(x, z)$$

is a homotopy equivalence.

Exercise 65 Show that that composition as defined right before Definition 1.3.47 is associative up to homotopy, i.e., that composition in an ∞-category determines a category enriched in $h(\mathrm{Kan})$, the homotopy category of Kan complexes. Hint: Consider the diagram

$$
\begin{array}{ccccc}
\Delta^{\{0,1\}} \cup \Delta^{\{1,2\}} \cup \Delta^{\{2,3\}} & \xrightarrow{\ \star\ } & \Delta^{\{0,1,2\}} \cup \Delta^{\{2,3\}} & \longleftarrow & \Delta^{\{0,2\}} \cup \Delta^{\{2,3\}} \\
\downarrow{\scriptstyle\star} & & \downarrow{\scriptstyle\star} & & \downarrow{\scriptstyle\star} \\
\Delta^{\{0,1\}} \cup \Delta^{\{1,2,3\}} & \xrightarrow{\ \star\ } & \Delta^3 & \longleftarrow & \Delta^{\{0,2,3\}} \\
\uparrow & & \uparrow & & \uparrow \\
\Delta^{\{0,1\}} \cup \Delta^{\{1,3\}} & \xrightarrow{\ \star\ } & \Delta^{\{0,1,3\}} & \longleftarrow & \Delta^{\{0,3\}}
\end{array}
$$

and show that all maps labelled with a $\star$ are inner-anodyne.

Exercise 66 Let $f : \mathcal{C} \to \mathcal{D}$ be a functor between ∞-categories. Let $a : x \to x'$ and $b : y \to y'$ be morphisms in $\mathcal{C}$. Show that there is a homotopy commutative diagram of Kan complexes

$$\begin{array}{ccc}
\mathrm{map}_{\mathcal{C}}(x', y) & \longrightarrow & \mathrm{map}_{\mathcal{D}}(fx', fy) \\
\downarrow & & \downarrow \\
\mathrm{map}_{\mathcal{C}}(x', y') & \longrightarrow & \mathrm{map}_{\mathcal{D}}(fx', fy') \\
\downarrow & & \downarrow \\
\mathrm{map}_{\mathcal{C}}(x, y') & \longrightarrow & \mathrm{map}_{\mathcal{D}}(fx, fy')
\end{array}$$

induced by precomposition with a, respectively fa, and post-composition with b, respectively fb.

Deduce that there is a canonical functor F from Cat^1_∞ (the 1-category of ∞-categories) to $\mathrm{Cat}_{\pi(\mathrm{Kan})}$, the category of categories enriched in the homotopy category of Kan complexes, where $F(\mathcal{C})$ has the same objects as $\mathcal{C}$ and the hom-object from x to y is given by the image of $\mathrm{map}_{\mathcal{C}}(x, y)$ in $\pi(\mathrm{Kan})$.

Exercise 67 Show that $[n] \star [m] = [n + m + 1]$. Furthermore, show that $\mathcal{C} \star [0] = \mathcal{C}^{\rhd}$ and $[0] \star \mathcal{C} = \mathcal{C}^{\lhd}$.

Exercise 68 Show that for categories $\mathcal{C}$ and $\mathcal{D}$ we have $N(\mathcal{C}) \star N(\mathcal{D}) \cong N(\mathcal{C} \star \mathcal{D})$. In particular, show that there is a canonical isomorphism $\Delta^i \star \Delta^j \cong \Delta^{i+1+j}$.

Exercise 69 The functors $X \star -$ and $- \star X$ as functors $\mathrm{sSet} \to \mathrm{sSet}$ preserve pushouts. Find an example of a colimit that is not preserved by $X \star -$.

Exercise 70 Show that the slice/join adjunction induces a bijection of lifting problems between diagrams of the kind

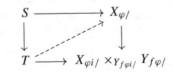

and diagrams of the kind

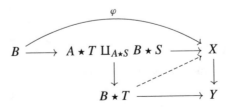

Exercise 71 Show that

$$\Lambda_j^n \star \Delta^m \cup \Delta^n \star \partial\Delta^m = \Lambda_j^{n+1+m}$$

and

$$\partial\Delta^m \star \Delta^n \cup \Delta^m \star \Lambda_j^n = \Lambda_{m+1+j}^{n+1+m}.$$

In order to do so, determine explicitly the following sub-simplicial sets of Δ^{n+1+m}:

(1) $\partial\Delta^n \star \Delta^m$
(2) $\Lambda_j^n \star \Delta^m$
(3) $\Delta^m \star \partial\Delta^n$
(4) $\Delta^m \star \Lambda_j^n$

Exercise 72 For an object x in an ∞-category $\mathcal{C}$, show that the canonical map $\mathcal{C}_{x/} \to \mathcal{C}$ is a left fibration and that $\mathcal{C}_{/x} \to \mathcal{C}$ is a right fibration.

Exercise 73 Show that for an object x of a general simplicial set X, the canonical map $X_{x/} \to X$ is generally not a left fibration.

Exercise 74 Show that an inner fibration $f: X \to Y$ is inner-anodyne if and only if it is an isomorphism.

Exercise 75 Show that $\Delta^0 \to J$ is not inner-anodyne.

Exercise 76 Show that the intersection of left- and right-anodyne maps strictly contains the inner-anodyne maps.

Exercise 77 Show that a functor $F: \mathcal{C} \to \mathcal{D}$ between ∞-categories is conservative if and only if the induced functor $hF: h\mathcal{C} \to h\mathcal{D}$ between the homotopy categories is conservative. Furthermore, show that the canonical functor $\mathcal{C} \to N(h\mathcal{C})$ is conservative.

Exercise 78 Show that a functor $p: \mathcal{C} \to \mathcal{D}$ is conservative if and only if the following diagram is a pullback:

$$\begin{array}{ccc} \mathcal{C}^\simeq & \longrightarrow & \mathcal{C} \\ \downarrow & & \downarrow \\ \mathcal{D}^\simeq & \longrightarrow & \mathcal{D} \end{array}$$

Exercise 79 Show that an inner fibration $\mathcal{C} \to \mathcal{D}$ between ∞-categories is an isofibration if and only if the induced functor $N(h\mathcal{C}) \to N(h\mathcal{D})$ is an isofibration.

Exercise 80 Show that a functor $\mathcal{C} \to \mathcal{D}$ between ∞-categories is an isofibration if and only if $\mathcal{C}^{\mathrm{op}} \to \mathcal{D}^{\mathrm{op}}$ is an isofibration.

Exercise 81 Let $\mathcal{C}_0 \subseteq \mathcal{C}$ be a full subcategory. Show that the inclusion $\mathcal{C}_0 \to \mathcal{C}$ is an isofibration if $\mathcal{C}_0$ is closed under equivalences in $\mathcal{C}$, i.e., that if $x \in \mathcal{C}_0$ and $y \in \mathcal{C}$ is equivalent to x, then y is also in $\mathcal{C}_0$.

Exercise 82 Show that if $f : x \to y$ is an equivalence, then the maps $\mathcal{C}_{/x} \to \mathcal{C}_{/y}$ and $\mathcal{C}_{y/} \to \mathcal{C}_{x/}$ are Joyal equivalences.

Exercise 83 Show that there exists a functor $f : \mathcal{C} \to \mathcal{D}$ between ∞-categories which is conservative, but does not satisfy the RLP with respect to $\Delta^1 \to J$. Hint: Consider the map $J \to \mathcal{S}(|J|)$.

Exercise 84 Show that a left fibration $p : \mathcal{C} \to \mathcal{D}$ is a Kan fibration, provided that $\mathcal{D}$ is an ∞-groupoid.

Exercise 85 Let $p : \mathcal{C} \to \mathcal{D}$ be an isofibration. Show that the set of all monomorphisms $K \to L$ such that the induced map

$$\mathcal{C}^L \to \mathcal{C}^K \times_{\mathcal{D}^K} \mathcal{D}^L$$

is again an isofibration is a saturated set.

Exercise 86 Show that the class of essentially surjective and fully faithful functors satisfies the 3-for-2 property.

Exercise 87 Suppose that $f : \mathcal{C} \to \mathcal{D}$ is a Joyal equivalence. Then show that the restricted map $\mathcal{C}^{\simeq} \to \mathcal{D}^{\simeq}$ is also a Joyal equivalence.

Exercise 88 Show that two functors $f, g : \mathcal{C} \to \mathcal{D}$ are naturally equivalent if and only if f and g represent the same element of $\pi_0(\mathrm{Fun}(\mathcal{C}, \mathcal{D})^{\simeq})$.

Exercise 89 Show that a Joyal equivalence $f : \mathcal{C} \to \mathcal{D}$ induces an equivalence of ordinary categories $h\mathcal{C} \to h\mathcal{D}$.

Exercise 90 Show that a functor $f : \mathcal{C} \to \mathcal{D}$ between ∞-categories is a Joyal equivalence if and only if for every simplicial set K, the induced map

$$f_* : \mathrm{Fun}(K, \mathcal{C}) \to \mathrm{Fun}(K, \mathcal{D})$$

is a Joyal equivalence.

Exercise 91 Let $F : \mathcal{C} \to \mathcal{D}$ be a functor between ordinary categories. Show that the induced map on nerves is inner-anodyne if and only if F is an isomorphism.

Exercise 92 Let $\mathcal{C}$ be an ∞-category and S a set of morphisms of $\mathcal{C}$. Then S is called *saturated* if it coincides with the set $\bar{S}$ of all morphisms that are sent to equivalences under the functor $\mathcal{C} \to \mathcal{C}[S^{-1}]$. Show that:

(1) If $S \subseteq T$ and T is saturated, then $\bar{S} \subseteq T$ as well.
(2) Two sets of morphisms S and T of $\mathcal{C}$ give rise to Joyal-equivalent localizations (compatible with the map from $\mathcal{C}$) if and only if $\bar{S} = \bar{T}$.

Exercise 93 Prove or disprove the following statements:

(1) For every ∞-category $\mathcal{C}$, a set S of morphisms of $\mathcal{C}$ and a set T of morphisms of $\mathcal{C}[S^{-1}]$, the functor $\mathcal{C} \to \mathcal{C}[S^{-1}][T^{-1}]$ is a localization of $\mathcal{C}$.
(2) For every ∞-category $\mathcal{C}$, a set S of morphisms of $\mathcal{C}$ and a set T of morphisms of $\mathcal{C}$, the functor $\mathcal{C} \to \mathcal{C}[S^{-1}][T^{-1}]$ is a localization of $\mathcal{C}$. Here, we view morphisms of $\mathcal{C}$ as morphisms of $\mathcal{C}[S^{-1}]$ via the canonical functor $\mathcal{C} \to \mathcal{C}[S^{-1}]$.

Exercise 94 Let $\mathcal{C}$ be an ordinary category and S a set of morphisms. Show that $\mathcal{C} \to h\mathcal{C}[S^{-1}]$ is the initial functor $\mathcal{C} \to \mathcal{D}$ (up to natural isomorphism) between ordinary categories sending S to isomorphisms.

Exercise 95 Let $\mathcal{C}$ be an ordinary category and S a set of morphisms. Show that every morphism in $h\mathcal{C}[S^{-1}]$ can be represented by a zig zag of morphisms in $\mathcal{C}$, such that the maps pointing in the wrong direction are contained in S.

Exercise 96 Let $\mathcal{C}$ be an ordinary category. Consider a pushout of categories

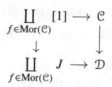

where $\mathrm{Mor}(\mathcal{C})$ denotes the set of all morphisms of $\mathcal{C}$. Show that $\mathcal{D}$ is a groupoid.

Exercise 97 In this exercise, you may use the fact that the unit map $K \to \mathcal{S}(|K|)$ is a homotopy equivalence for any Kan complex K. Recall that a map of simplicial sets is a weak equivalence if its geometric realization is a homotopy equivalence, and let X be a simplicial set. Prove or disprove the following statements:

(1) The unit map $X \to \mathcal{S}(|X|)$ is a monomorphism.
(2) The unit map $X \to \mathcal{S}(|X|)$ is a weak equivalence.
(3) The unit map $X \to \mathcal{S}(|X|)$ is anodyne.

Exercise 98 Let $\mathcal{C}$ be an ∞-category and let $\mathcal{C} \to \mathcal{S}(|\mathcal{C}|)$ be the unit map of the adjunction $(\mathcal{S}, |-|)$. Show that this is a localization of $\mathcal{C}$ along all morphisms.

Exercise 99 Show that the factorization $\mathcal{C} \to P(f) \to \mathcal{D}$ for a functor $f: \mathcal{C} \to \mathcal{D}$ between ∞-categories is functorial, i.e., that for every solid commutative diagram

$$
\begin{array}{ccccc}
\mathcal{C} & \longrightarrow & P(f) & \longrightarrow & \mathcal{D} \\
\downarrow & & \downarrow & & \downarrow \\
\mathcal{C}' & \longrightarrow & P(f') & \longrightarrow & \mathcal{D}'
\end{array}
$$

a dashed arrow exists which makes both small squares commute.

Exercise 100 Consider a pushout diagram of simplicial sets

$$
\begin{array}{ccc}
X & \longrightarrow & Y \\
\downarrow & & \downarrow \\
X' & \longrightarrow & Y'
\end{array}
$$

in which $X \to X'$ is a monomorphism and $X \to Y$ is a Joyal equivalence. Show that the map $X' \to Y'$ is also a Joyal equivalence.

Exercise 101 Let $X \to X'$ and $Y \to Y'$ be Joyal equivalences between simplicial sets. Show that both maps $X \sqcup Y \to X' \sqcup Y'$ and $X \times Y \to X' \times Y'$ are Joyal equivalences.

Exercise 102 Show that:

(1) A retract of a Joyal equivalence is a Joyal equivalence.
(2) The set of monomorphisms which are also Joyal equivalences is saturated.

Prove or disprove that the set of Joyal equivalences is saturated.

Exercise 103 Recall that a map $f: X \to Y$ is said to admit a pre-inverse if there exist maps $g: Y \to X, \tau: \Delta^1 \to \mathrm{Hom}(X, X)$ and $\tau': \Delta^1 \to \mathrm{Hom}(Y, Y)$ such that

(1) $\tau_\varepsilon = \mathrm{id}_X$ and $\tau_{1+\varepsilon} = gf$, where $\varepsilon \in \{0, 1\} \cong \mathbb{Z}/2$,
(2) $\tau'_\varepsilon = \mathrm{id}_Y$ and $\tau'_{1+\varepsilon} = fg$, where again $\varepsilon \in \{0, 1\} \cong \mathbb{Z}/2$,
(3) for all objects x of X, the morphism $\tau(x): \Delta^1 \to X$ represents a degenerate edge of X, and for all objects y of Y, $\tau'(y): \Delta^1 \to Y$ represents a degenerate edge of Y.

Show that a map $f: X \to Y$ which admits a pre-inverse is a Joyal equivalence.

Exercise 104 Let $p: \mathcal{C} \to \mathcal{D}$ be an inner fibration between ∞-categories which induces a surjection on 0-simplices and is a Joyal equivalence. Show that p is a trivial fibration.

Exercise 105 Show that for any two simplicial sets, $X \star Y$ is a retract of $X \diamond Y$.

Exercise 106 Show that the canonical map

$$|\underline{\mathrm{Hom}}(A, B)| \longrightarrow \mathrm{map}(|A|, |B|)$$

is a homotopy equivalence. You may use the fact that both the unit map $A \to S(|A|)$ and the counit map $|S(X)| \to X$ are weak equivalences.

The goal of the following exercises is to (almost) give a proof of the fact that anodyne maps are precisely those monomorphisms which are weak equivalences. More precisely, we will show that it is implied by the following statement: Let $p: X \to Y$ be a Kan fibration. Then there exists a factorization of p as

$$X \xrightarrow{\alpha} Z \xrightarrow{\beta} Y$$

where β is a trivial fibration (i.e., it has the RLP with respect to monomorphisms) and α is a minimal fibration. In this context, we need to know the following facts about minimal fibrations:

- A minimal fibration is a Kan fibration.
- A minimal fibration $\alpha: X \to Z$ is locally trivial, i.e., for every simplex $\Delta^n \to Z$, the pulled-back fibration is isomorphic (over Δ^n) to a projection $\Delta^n \times B \to \Delta^n$.

Exercise 107 Show that a Kan fibration $p: X \to Y$ is a trivial fibration if and only if its fibres are contractible. Hint: For the interesting direction, consider a lifting problem

$$
\begin{CD}
\partial\Delta^n @>a>> X \\
@VVV @VVV \\
\Delta^n @>b>> Y
\end{CD}
$$

For future reference, we let $y = b(n)$. Consider the map $g: \Delta^n \times \Delta^1 \to \Delta^n$ determined by $g(k, 0) = k$ and $g(k, 1) = n$. Show that the map $\partial\Delta^n \times \{0\} \to \partial\Delta^n \times \Delta^1$ is anodyne, and use g to obtain a map

$$\Delta^1 \to X^{\partial\Delta^n} \times_{Y^{\partial\Delta^n}} Y^{\Delta^n}$$

sending 0 to the original square and 1 to the square

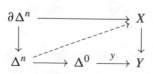

where the map $\Delta^0 \to Y$ is the object y. Show that this lifting problem can be solved. Finally, show that this implies that the original lifting problem can be solved as well.

Exercise 108 In this exercise, we will use minimal fibrations to show that the geometric realization of a Kan fibration $p: X \to Y$ is a Serre fibration, i.e., it has the RLP with respect to the inclusions $D^n \times \{0\} \to D^n \times D^1$. Here, D^n is the n-dimensional topological cube.

(1) Show that the geometric realization of a minimal fibration is a Serre fibration whose fibre is given by the geometric realization of the fibre of the Kan fibration.
(2) Show that the geometric realization of a trivial fibration is a Serre fibration.

Hints: For (1), show that the realization of a locally trivial map is also locally trivial. Then show that a locally trivial map of spaces is a Serre fibration. For (2), show that there exists a monomorphism $X \to W$ with W a contractible Kan complex. Consider the maps $X \to W \times Y \to Y$. Show that the latter is a trivial fibration and deduce that p is a retract of $W \times Y \to Y$.

Exercise 109 Show that a Kan fibration $p: X \to Y$ which is additionally a weak equivalence has contractible fibres.

Exercise 110 Show that a monomorphism $i: A \to B$ is a weak equivalence if and only if it is anodyne.

Exercise 111 Show that a cocartesian fibration $p: X \to Y$ whose fibres are Joyal-equivalent to Δ^0 is a trivial fibration.

Exercise 112 Let P be a poset and $\mathcal{C}$ an ∞-category. Suppose that you are given a function $f: P \to \mathrm{ob}(\mathcal{C})$ with the following property: Whenever $x \leq y$ are elements of P, then the spaces of maps $\mathrm{map}_{\mathcal{C}}(fx, fy)$ are contractible.

(1) Show that there exists a functor $F: P \to \mathcal{C}$ extending the given function f on objects.
(2) Show that any two such extensions are equivalent.

Of course, the "correct" version of (2) is the following: Show that there is a contractible space parametrizing all possible choices of such extensions, i.e., that in the pullback diagram

$$
\begin{array}{ccc}
\mathrm{Ext}(f) & \longrightarrow & \mathrm{Fun}(P, \mathcal{C}) \\
\downarrow & & \downarrow \\
\Delta^0 & \longrightarrow & \mathrm{Fun}(P_0, \mathcal{C})
\end{array}
$$

the simplicial set $\mathrm{Ext}(f)$ is a contractible Kan complex. Can one replace P by an arbitrary 1-category and obtain the same results?

Exercise 113 Let $\mathcal{C}$ be an ∞-category and consider for each $i \geq 0$ a morphism $x_i \to x_{i+1}$ between objects in $\mathcal{C}$. Show that these maps assemble into a functor $\mathbb{N} \to \mathcal{C}$.

Exercise 114 Prove or disprove that an isofibration $p\colon \mathcal{C} \to \mathcal{D}$ between ∞-categories is a trivial fibration if and only if its fibres are Joyal-equivalent to Δ^0.

Exercise 115 Let $\mathcal{C}$ be an ∞-category and consider the inner fibration $p\colon \mathcal{C} \to \Delta^0$. Show that a morphism f in $\mathcal{C}$ is p-(co)cartesian if and only if f is an equivalence.

Exercise 116 Show that if a morphism between two squares of ∞-categories is given in which each comparison map is a Joyal equivalence, then the one square is homotopy-cartesian if and only if the other square is.

Exercise 117 Suppose that you are given a pullback diagram

$$
\begin{array}{ccc}
X' & \longrightarrow & X \\
\downarrow{\scriptstyle p'} & & \downarrow{\scriptstyle p} \\
Y' & \longrightarrow & Y
\end{array}
$$

in which the map p is a (co)cartesian fibration. Show that the map p' is a (co)cartesian fibration as well.

Exercise 118 Let $p\colon X \to S$ be an inner fibration of simplicial sets and $f\colon x \to y$ an edge in X. Suppose that f is p'-cocartesian, where p' is given by the following pullback:

$$
\begin{array}{ccc}
X' & \dashrightarrow & X \\
\downarrow{\scriptstyle p'} & & \downarrow{\scriptstyle p} \\
S' & \longrightarrow & S
\end{array}
$$

Show that f is locally cocartesian.

Exercise 119 Prove Lemma 3.2.7

Exercise 120 Consider the simplicial set $S = \Delta^2 / \Delta^{\{0,1\}}$ and the canonical map $p\colon \Delta^{\{0,2\}} \to S$. Show that the diagram

$$
\begin{array}{ccc}
\Lambda_0^2 & \xrightarrow{\ f\ } & \Delta^{\{0,2\}} \\
\downarrow & & \downarrow \\
\Delta^2 & \xrightarrow{\ c\ } & S
\end{array}
$$

is a pullback, where c is the canonical projection. Then proceed to show:

(1) p is an inner fibration.
(2) The edge $f_{|\Delta^{\{0,1\}}}$ is degenerate, but not p-cocartesian.

•

Exercise 121 Show that the functors $(-)^\flat, (-)^\sharp \colon \mathrm{sSet} \rightarrow \mathrm{sSet}_+$ and the functors $u, m \colon \mathrm{sSet}_+ \rightarrow \mathrm{sSet}$ are involved in various adjunctions. Here, u is the forgetful functor and m is the functor which sends a marked simplicial set (X, S) to the smallest sub-simplicial set $X_0 \subseteq X$ containing S.

Exercise 122 Let $p \colon \mathcal{E} \rightarrow \mathcal{C}$ be a cocartesian fibration and K a marked simplicial set equipped with a map $f \colon K \rightarrow \mathcal{C}^\sharp$. Consider the sub-$\infty$-category of $\mathrm{Fun}_f^{\mathrm{mcc}}(K, \mathcal{E})$ on those 1-simplices whose corresponding map $K \times \Delta^1 \rightarrow \mathcal{E}$ is a map of marked simplicial sets $K \times (\Delta^1)^\sharp \rightarrow \mathcal{E}^\flat$, i.e., we consider only those transformations of functors whose components are pointwise p-cocartesian. Show that this sub-∞-category is given by $\mathrm{Fun}_f^{\mathrm{mcc}}(K, \mathcal{E})^\simeq$.

Exercise 123 Show that if $K = \Delta^0$ and $f \colon \Delta^0 \rightarrow \mathcal{C}$ picks out an object z of $\mathcal{C}$, then $\mathrm{Fun}^{\mathrm{cc}}(K, \mathcal{E}) \simeq \mathcal{E}$ and $\mathrm{Fun}_f^{\mathrm{cc}}(K, \mathcal{E}) \cong \mathcal{E}_z$.

Exercise 124 Let $X \rightarrow Y$ be a cocartesian fibration and $f \colon K \rightarrow Y^\sharp$ a map of marked simplicial sets. Show that $\mathrm{Map}_f^\flat(K, X^\natural)$ is an ∞-category and that $\mathrm{Map}_f^\sharp(K, X^\natural)$ is the largest sub-∞-groupoid inside $\mathrm{Map}_f^\flat(K, X^\natural)$.

Exercise 125 Show that the functor $(-)^\flat \colon \mathrm{sSet} \rightarrow \mathrm{sSet}^+$ sends Joyal equivalences to marked equivalences.

Exercise 126 Show that the category of marked simplicial sets is canonically enriched in simplicial sets.

Exercise 127 Show that the functor $LF \colon \mathrm{sSet} \rightarrow \mathrm{sSet}$ of Theorem 3.3.8 preserves colimits and monomorphisms.

Exercise 128 Show that the initial vertex maps assemble to a natural transformation $LF \Rightarrow \mathrm{id}$.

Exercise 129 Suppose to be given a commutative diagram

$$
\begin{array}{ccccccc}
A_0 & \longrightarrow & A_1 & \longrightarrow & A_2 & \longrightarrow & \cdots \\
\downarrow{\scriptstyle f_0} & & \downarrow{\scriptstyle f_1} & & \downarrow{\scriptstyle f_2} & & \\
B_0 & \longrightarrow & B_1 & \longrightarrow & B_2 & \longrightarrow & \cdots
\end{array}
$$

in which the maps f_i are Joyal equivalences and all horizontal maps are monomorphisms. Show that the induced map $f \colon A = \mathrm{colim}\, A_i \rightarrow \mathrm{colim}\, B_i = B$ is a Joyal equivalence.

Exercise 130 Show that for an ∞-groupoid X, there is a canonical Joyal equivalence $X^{\mathrm{op}} \simeq X$.

Exercise 131 Show that the functor $N(\Delta^{op}_{/\mathcal{C}}) \to \mathcal{C}$ is full in the sense that every morphism of $\mathcal{C}$ is the image of a morphism under this functor. Use this result to show that the composition $\mathcal{C} \to \mathrm{Cat}_\infty$ induces on objects and morphisms the constructions we have done earlier.

Exercise 132 Suppose that you are given a cocartesian fibration $p \colon \mathcal{E} \to \mathcal{C}$. Recall that we have constructed for every edge $f \colon \Delta^1 \to \mathcal{C}$ a functor $\mathcal{E}_x \times \Delta^1 \to \mathcal{E}$, whose restriction to $\mathcal{E}_x \times \{1\}$ is given by $f_!$. Show that this functor may equivalently be constructed by considering the diagram

$$
\begin{array}{ccc}
\mathcal{E}_x \times \Delta^{\{0\}} & \longrightarrow & \mathcal{E} \\
\downarrow & \nearrow & \downarrow \\
\mathcal{E}_x \times \{1\} \longrightarrow \mathcal{E}_x \times \Delta^1 & \xrightarrow{f \circ \mathrm{pr}} & \mathcal{C}
\end{array}
$$

and show the existence of a dashed arrow having all the above properties.

Exercise 133 Given a cocartesian fibration $\mathcal{E} \to \mathcal{C} \times \Delta^1$, construct a functor $\mathcal{E}_0 \to \mathcal{E}_1$ which commutes with the projections to $\mathcal{C}$. Here, $\mathcal{E}_i$ is the pulled-back cocartesian fibration along the inclusion $\mathcal{C} \times \{i\} \to \mathcal{C} \times \Delta^1$. Show that this functor is a morphism of cocartesian fibrations. Likewise, construct for a cocartesian fibration $\mathcal{E} \to \mathcal{C} \times \Delta^2$ a 2-simplex in the ∞-category $(\mathrm{Cat}_\infty)_{/\mathcal{C}}$. If you are eager, do this for a general n instead of 2.

Exercise 134 Let $p \colon \mathcal{C} \to \mathcal{D}$ be an inner fibration between ∞-categories. Suppose that for every map $f \colon \Delta^1 \to \mathcal{D}$, the induced map $\Delta^1 \times_{\mathcal{D}} \mathcal{C} \to \Delta^1$ is a Joyal equivalence. Show that p is a Joyal equivalence.

Exercise 135 Let $\mathcal{C}$ be an ∞-category and x and object of $\mathcal{C}$. Show that the object represented by $\Delta^0 \to \mathcal{C}_{/x}$ which is adjoint to the map $\Delta^0 \star \Delta^0 \to \mathcal{C}$ given by id_x is a terminal object.

Exercise 136 Let $\mathcal{C}$ be an ∞-category and x an initial object of $\mathcal{C}$. Show that if y is equivalent to x, then y is also initial in $\mathcal{C}$.

Exercise 137 Suppose that $\mathcal{C}_{\mathrm{term}}$ is not empty. Show that any terminal object of $\mathrm{Fun}(K, \mathcal{C})$ takes values in $\mathcal{C}_{\mathrm{term}}$.

Exercise 138 Show that there exists $\mathcal{C}$ and K such that $\mathrm{Fun}(K, \mathcal{C})$ has a terminal object but $\mathcal{C}$ does not.

Exercise 139 Show that there exists a simplicial set X such that map $\mathrm{Tw}(X) \to X \times X^{op}$ of Definition 4.2.3 is not a right fibration.

Exercise 140 Prove or disprove that the diagram

$$\begin{array}{ccc} \mathrm{Fun}(\Delta^1, \mathcal{C}) & \longrightarrow & \mathrm{Fun}(\Delta^1, \mathcal{D}) \\ \downarrow & & \downarrow \\ \mathcal{C} \times \mathcal{C} & \longrightarrow & \mathcal{D} \times \mathcal{D} \end{array}$$

is homotopy-cartesian if and only if f is fully faithful.

Exercise 141 Let $p\colon K \to \mathcal{C}$ be a diagram. Suppose that there is a colimit cone $\tilde{p}$ of p with colimit x in $\mathcal{C}$. Let y be an object of $\mathcal{C}$ which is equivalent to x. Show that y is also the colimit of a colimit cone of p.

Exercise 142 Let $p\colon K \to \mathcal{C}$ be a diagram and let $q\colon I \to \mathcal{C}_{p/}$ be a further diagram. Let $\bar{q}\colon I \star K \to \mathcal{C}$ be the associated map. Show that there is an isomorphism $(\mathcal{C}_{p/})_{q/} \cong \mathcal{C}_{\bar{q}/}$. Likewise, show that there is an isomorphism $(\mathcal{C}_{/p})_{q/} \cong (\mathcal{C}_{q'/})_{/p'}$ where q' is the restriction of $q\colon I \star K \to \mathcal{C}$ to I and p' is adjoint to $\bar{q}$.

Exercise 143 Let K be the coequalizer of two monomorphisms $f, g\colon A \to B$ and let $F\colon K \to \mathcal{C}$ be a functor. Suppose that the restrictions of F to B and A admit colimits and that $\mathcal{C}$ admits coequalizers. Show that F admits a colimit in this case.

Exercise 144 Let $F\colon \mathcal{C} \to \mathcal{D}$ be a functor between ∞-categories. Show that F preserves small colimits if and only if it preserves small coproducts and pushouts (or coequalizers).

Exercise 145 Show that a map $f\colon K \to L$ is cofinal if and only if $f^{\mathrm{op}}\colon K^{\mathrm{op}} \to L^{\mathrm{op}}$ is coinitial.

Exercise 146 Show that left-anodyne maps do not satisfy the left cancellation property among monomorphisms.

Exercise 147 Let $f\colon \mathcal{C} \to \mathcal{D}$ and $g\colon \mathcal{D} \to \mathcal{C}$ be functors. Suppose that $\varepsilon\colon fg \to \mathrm{id}$ is a natural transformation. Show that the map

$$\mathrm{map}(x, g(y)) \to \mathrm{map}(f(x), f(g(y))) \to \mathrm{map}(f(x), y)$$

is natural in x.

Exercise 148 Show that the set of right-anodyne maps $i\colon A \to B$ whose pullback along any left fibration is again right-anodyne is a saturated set and satisfies the right-cancellation property.

Exercise 149 A right deformation retract is a monomorphism $i: A \rightarrow B$ such that there exists a retraction $p: B \rightarrow A$ and a simplicial homotopy $H: \Delta^1 \times B \rightarrow B$ with $H(0) = \mathrm{id}_B$, $H(1) = ip$ and whose restriction to $\Delta^1 \times A$ is constant at the identity of A.

(1) Show that for every simplicial set, the map $\{1\} \times K \rightarrow \Delta^1 \times K$ is a right deformation retract.
(2) Show that a right deformation retract is a right-anodyne map.
(3) Show that the pullback of a right deformation retract along a left fibration is again a right deformation retract.

Exercise 150 Give an example of a proper (or smooth) map which is not universally proper (or smooth).

Exercise 151 Let $\mathcal{C}$ be a cocomplete ∞-category and let $F: \Delta^n \rightarrow \mathcal{C}$ be a functor. Calculate the colimit of the restriction $\partial \Delta^n \rightarrow \Delta^n \rightarrow \mathcal{C}$.

Exercise 152 Let $p: \mathcal{E} \rightarrow \mathcal{D}$ be a cartesian fibration. Show that the canonical map $\mathcal{E}_d \rightarrow \mathcal{E}_{d/}$ admits a right adjoint. Deduce that for a cartesian fibration, the canonical map $\mathcal{E}_d \rightarrow \mathcal{E}_{d/}$ is a weak equivalence.

Exercise 153 Show that a fully faithful and essentially surjective functor is invertible.

Exercise 154 Let $F: K \rightarrow \mathrm{Cat}_\infty$ be a functor and $p: \mathcal{E} \rightarrow K$ the associated cocartesian fibration. Show that the colimit of F is given by $\mathcal{E}[\mathrm{cc}^{-1}]$, where the set cc is the set of p-cocartesian edges. Likewise, show that the limit $\mathrm{Fun}^{\mathrm{cc}}_K(K, \mathcal{E})$ is given by the category of cocartesian sections of p. Deduce the analogs for functors with values in Spc.

Exercise 155 Consider, for a diagram $F: K \rightarrow \mathcal{C}$ and an object x of $\mathcal{C}$, the functor

$$\mathrm{map}_{\mathcal{C}}(F(-), x): K \rightarrow \mathrm{Spc}^{\mathrm{op}}.$$

Show that its limit is given by $\mathrm{Map}_{\mathcal{C}}(F, x)$. Deduce that for a functor $f: \mathcal{C} \rightarrow \mathcal{D}$ with the right adjoint g, there exists a canonical equivalence $\mathrm{Map}_{\mathcal{D}}(fF, z) \simeq \mathrm{Map}_{\mathcal{C}}(F, gz)$.

Exercise 156 Let $f: \mathcal{C} \rightarrow \mathcal{D}$ and $f': \mathcal{D} \rightarrow \mathcal{E}$ be composable functors which admit right adjoints g and g', respectively. Show that in this case gg' is a right adjoint of $f'f$.

Exercise 157 Show that, for each object x of an ∞-category $\mathcal{C}$, the functor $\mathrm{map}_{\mathcal{C}}(x, -): \mathcal{C} \rightarrow \mathrm{Spc}$ preserves limits.

Bibliography

[Cis19]   D.-C. Cisinski, *Higher categories and homotopical algebra*, Cambridge Studies in Advanced Mathematics, vol. 180, Cambridge University Press, Cambridge, 2019.

[GJ09]   P. G. Goerss and J. F. Jardine, *Simplicial homotopy theory*, Modern Birkhäuser Classics, Birkhäuser Verlag, Basel, 2009, Reprint of the 1999 edition.

[Hau17]   R. Haugseng, *Introduction to ∞-categories*, http://folk.ntnu.no/runegha/notes.html, 2017.

[Heb20]   F. Hebestreit, *Yoneda's lemma, adjunctions, and (co)limits*, http://www.math.uni-bonn.de/people/fhebestr/HigherCats2/, 2020.

[Hir03]   P. S. Hirschhorn, *Model categories and their localizations*, Mathematical Surveys and Monographs, vol. 99, American Mathematical Society, Providence, RI, 2003.

[Hir15]   _____, *The homotopy groups of an inverse limit of a tower of fibrations*, arXiv:1507.01627 (2015).

[HK20]   F. Hebestreit and A. Krause, *Mapping spaces in homotopy coherent nerves*, arXiv:2011.09345 (2020).

[Joy07]   A. Joyal, *Quasi-categories vs simplicial categories*, http://www.math.uchicago.edu/~may/IMA/Incoming/Joyal/QvsDJan9(2007).pdf, 2007.

[Joy08]   _____, *Notes on quasicategories*, https://www.math.uchicago.edu/~may/IMA/Joyal.pdf, 2008.

[Lur09]   J. Lurie, *Higher topos theory*, Annals of Mathematics Studies, vol. 170, Princeton University Press, Princeton, NJ, 2009.

[Lur17]   _____, *Higher algebra*, https://www.math.ias.edu/~lurie/papers/HA.pdf, 2017.

[Mil19]   H. Miller (ed.), *Handbook of homotopy theory*, Chapman and Hall/CRC Press, 2019.

[Ngu18]   H. K. Nguyen, *Theorems in higher categories and applications*, https://epub.uni-regensburg.de/38448/, 2018.

[NRS19]   H. K. Nguyen, G. Raptis, and C. Schrade, *Adjoint functor theorems for ∞-categories*, J. London Math. Soc. (2019).

[Rez19]   C. Rezk, *Degenerate edges of cartesian fibrations are cartesian edges*, https://faculty.math.illinois.edu/~rezk/degenerate-cartesian.pdf, 2019.

[Rez20]   _____, *Stuff on quasicategories*, https://faculty.math.illinois.edu/~rezk/quasicats.pdf, 2020.

[RV18]   E. Riehl and D. Verity, *The comprehension construction*, High. Struct. **2** (2018), no. 1, 116–190.

[Ste17]   D. Stevenson, *Covariant model structures and simplicial localization*, North-West. Eur. J. Math. **3** (2017), 141–203.

© The Author(s), under exclusive license to Springer Nature Switzerland AG 2021
M. Land, *Introduction to Infinity-Categories*, Compact Textbooks in Mathematics,
https://doi.org/10.1007/978-3-030-61524-6

[Ste18a]   _____ , *Model structures for correspondences and bifibrations*, arXiv:180708226
             (2018).
[Ste18b]   _____ , *Stability for inner fibrations revisited*, Theory Appl. Categ. **33** (2018),
             Paper No. 19, 523–536.